Optical Vortices

Fundamentals and applications

Online at: https://doi.org/10.1088/978-0-7503-5844-6

IOP Series in Advances in Optics, Photonics and Optoelectronics

SERIES EDITOR

Professor Rajpal S Sirohi Consultant Scientist

About the Editor

Rajpal S Sirohi is currently working as a faculty member in the Department of Physics, Alabama A&M University, Huntsville, AL, USA. Prior to this, he was a consultant scientist at the Indian Institute of Science, Bangalore, and before that he was Chair Professor in the Department of Physics, Tezpur University, Assam. During 2000–2011, he was an academic administrator, being vice-chancellor to a couple of universities and the director of the Indian Institute of Technology, Delhi. He is the recipient of many international and national awards and the author of more than 400 papers. Dr Sirohi is involved with research concerning optical metrology, optical instrumentation, holography, and the speckle phenomena.

About the series

Optics, photonics, and optoelectronics are enabling technologies in many branches of science, engineering, medicine, and agriculture. These technologies have reshaped our outlook and our ways of interacting with each other, and have brought people closer together. They help us to understand many phenomena better and provide deeper insight into the functioning of nature. Further, these technologies themselves are evolving at a rapid rate. Their applications encompass very large spatial scales, from nanometers to the astronomical scale, and a very large temporal range, from picoseconds to billions of years. This series on advances in optics, photonics, and optoelectronics aims to cover topics that are of interest to both academia and industry. Some of the topics to be covered by the books in this series include biophotonics and medical imaging, devices, electromagnetics, fiber optics, information storage, instrumentation, light sources, charge-coupled devices (CCDs) and complementary metal oxide semiconductor (CMOS) imagers, metamaterials, optical metrology, optical networks, photovoltaics, free-form optics and its evaluation, singular optics, cryptography, and sensors.

About IOP ebooks

The authors are encouraged to take advantage of the features made possible by electronic publication to enhance the reader experience through the use of color, animation, and video and by incorporating supplementary files in their work.

A full list of titles published in this series can be found here: https://iopscience.iop.org/bookListInfo/series-on-advances-in-optics-photonics-and-optoelectronics.

Optical Vortices

Fundamentals and applications

Yuanjie Yang
University of Electronic Science and Technology of China, Chengdu, China

Yu-Xuan Ren
Institute for Translational Brain Research, Fudan University, Shanghai, China

Carmelo Rosales-Guzmán
Centro de Investigaciones en Optica A.C., Leon Guanajuato, Mexico

IOP Publishing, Bristol, UK

ISBN   978-0-7503-5844-6 (ebook)
ISBN   978-0-7503-5842-2 (print)
ISBN   978-0-7503-5845-3 (myPrint)
ISBN   978-0-7503-5843-9 (mobi)

DOI   10.1088/978-0-7503-5844-6

Version: 20241101

IOP ebooks

British Library Cataloguing-in-Publication Data: A catalogue record for this book is available from the British Library.

Published by IOP Publishing, wholly owned by The Institute of Physics, London

IOP Publishing, No.2 The Distillery, Glassfields, Avon Street, Bristol, BS2 0GR, UK

US Office: IOP Publishing, Inc., 190 North Independence Mall West, Suite 601, Philadelphia, PA 19106, USA

To those who love optical vortices.

Contents

Preface   xii

Acknowledgments   xiii

Author biographies   xiv

Symbols   xv

1   Introduction   1-1
1.1   Vortices in nature   1-1
1.2   Historical review of optical vortices   1-2
1.3   Orbital angular momentum of vortex beams   1-3
      References   1-3

2   Generation of optical vortex beams   2-1
2.1   Typical vortex beams   2-2
      2.1.1   Laguerre–Gaussian beams   2-2
      2.1.2   Bessel vortex beams   2-2
      2.1.3   Perfect vortex beams   2-4
      2.1.4   Anomalous vortex beams   2-6
      2.1.5   Fractional vortex beams   2-7
      2.1.6   Open vortex beams   2-10
2.2   Methods for vortex generation   2-11
      2.2.1   Mode conversion   2-11
      2.2.2   Spiral phase plate   2-12
      2.2.3   Computer-generated holograms   2-13
      2.2.4   Digital devices   2-15
      2.2.5   Photon sieves   2-17
      2.2.6   Metasurfaces   2-18
      2.2.7   Other techniques   2-20
      References   2-23

3   Measuring OAM of vortex beams   3-1
3.1   Interference or diffraction method   3-1
3.2   Geometric coordinate transformation method   3-5
3.3   Deep learning method   3-6
3.4   Surface plasmon polaritons method   3-8
      References   3-9

4   Partially coherent vortex beams   **4-1**

4.1   Orbital angular momentum of partially coherent beams   4-1

4.2   Singularities in partially coherent vortex beams   4-2

4.3   Measuring the topological charge of partially coherent vortex beams   4-3

    4.3.1 Partially coherent Laguerre–Gaussian beams   4-3

    4.3.2 Partially coherent elegant Laguerre–Gaussian beams   4-6

4.4   Applications of partially coherent vortex beams   4-8

    4.4.1 Optical communication   4-8

    4.4.2 Optical manipulation   4-8

    4.4.3 Imaging   4-9

    4.4.4 Other promising applications   4-9

    References   4-9

5   Vector vortex beams   **5-1**

5.1   Basic concepts   5-1

    5.1.1 The vectorial nature of light   5-2

5.2   Definition of vector vortex beams   5-6

    5.2.1 The higher-order Poincaré sphere   5-9

5.3   Generation of vector vortex beams   5-10

    5.3.1 Dynamic phase control   5-11

    5.3.2 Geometric phase control   5-19

5.4   Characterization of vector beams   5-21

    5.4.1 Stokes polarimetry   5-21

    5.4.2 Quantum-like nonseparability of vector beams   5-22

    References   5-28

6   Spatio-temporal vortex and applications   **6-1**

6.1   Concept of the spatiotemporally structured light   6-1

6.2   Temporal analog of lens   6-2

6.3   Spatiotemporal optical vortex   6-4

6.4   Generation of spatiotemporal optical vortex   6-6

6.5   Spatiotemporal solitons and soliton molecules   6-9

6.6   Spin–orbit interaction of spatiotemporal optical vortex   6-13

6.7   Summary   6-16

    References   6-16

7	**Plasmonic vortices**	**7-1**
7.1	Vortices in surface plasmon polaritons: an introduction	7-1
7.2	Generation of plasmonic vortices	7-2
	7.2.1 Generation of SPP vortices using PVLs	7-2
	7.2.2 Spin-to-orbit coupling of SPP vortices	7-5
	7.2.3 Deuterogenic plasmonic vortices	7-6
	7.2.4 Generation of SPP vortices using plasmonic metasurfaces	7-8
7.3	Applications of SPP vortices and beyond	7-10
	7.3.1 Applications of SPP vortices	7-10
	7.3.2 Surface plasmonic vortices beyonds optical frequency	7-11
7.4	Conclusion	7-14
	References	7-15

8	**Tailoring structured light with spatial light modulators**	**8-1**
8.1	The liquid crystal spatial light modulators	8-1
8.2	Tailoring light with spatial light modulators	8-4
	8.2.1 Phase modulation	8-5
	8.2.2 Complex amplitude modulation	8-7
	References	8-9

9	**Tailoring optical vortex with digital micromirror device**	**9-1**
9.1	Introduction of amplitude DMD	9-1
9.2	Algorithms to generate amplitude hologram for DMD	9-2
9.3	Tailoring super-Gaussian beam with DMD	9-5
9.4	Tailoring perfect vortex beam with DMD	9-8
9.5	Tailoring the vortex Hermite–Gaussian beam	9-11
9.6	Tailoring vortex Ince–Gaussian beam with DMD	9-15
9.7	Tailoring the vortex Airy beam with DMD	9-17
9.8	Summary	9-22
	References	9-22

10	**Structured light for optical communication**	**10-1**
10.1	Classical communications with structured light	10-1
	10.1.1 Optical communications in optical fibers	10-2
	10.1.2 Optical communications in free space	10-4
	10.1.3 Underwater optical communications	10-7

10.2   Secure quantum communications with structured light   10-8

  10.2.1 Quantum key distribution with polarization   10-10

  10.2.2 Quantum key distribution in higher dimensions   10-13

  10.2.3 Quantum communications with vector beams   10-13

  References   10-15

11   Optical trapping with optical vortices   11-1

11.1   Principle of optical tweezers   11-1

11.2   Optical trapping with nondiffracting beams   11-2

11.3   Optical trapping with vortex beams   11-3

11.4   Holographic optical tweezers   11-6

11.5   Optical manipulation with metallic nanoparticles   11-9

11.6   Optical trapping with structured light   11-10

11.7   Mind-controlled optical rotation   11-12

11.8   Conclusion   11-13

  References   11-14

12   Biomedical imaging with optical vortex   12-1

12.1   Grant challenges in biomedical imaging   12-1

12.2   Volumetric microscopy with structured light   12-2

12.3   Reolustion improvement in nonlinear microscope   12-5

12.4   Axial resolution in volumetric microscopy   12-8

12.5   Structured light to resolve the axial information   12-10

12.6   Improve the signal-to-background ratio   12-13

12.7   Conclusion   12-14

  References   12-14

13   Metrology with structured light   13-1

13.1   Laser remote sensing   13-1

  13.1.1 The longitudinal Doppler shift   13-1

  13.1.2 The transverse Doppler shift   13-2

13.2   Other techniques   13-6

  13.2.1 Determining all velocity components in a   13-6
          full 3D helical motion

  13.2.2 Single-shot determining all velocity components in a   13-7
          full 3D helical motion

13.2.3 Determination of the vorticity of fluids   13-7

References   13-9

14   Past, present, and future   14-1

14.1 Overview of the past researches   14-1

14.2 Present researches on the optical vortex   14-2

14.3 Future perspectives   14-3

Preface

Vortices are inherent to any wave phenomena, and optical vortex beams have become one of the most important light modes today, due to each photon of vortex beams being able to carry $\ell\hbar$ orbital angular momentum (OAM), where ℓ is known as the topological charge and $\hbar$ denotes the reduced Plank constant. Thus far, vortex beams have found a wide variety of applications, for example, classical optical communications and quantum communications, optical manipulation, optical metrology and imaging, to mention but a few. Besides optical vortex beams, some other vortices, such as electron vortex beams, neutron atom vortex beams, plasmonic vortices and radio vortices can carry OAM as well, and may lead to a new wide range of applications in many fields.

Over the past three decades, significant fundamental studies on vortex beams have been carried out, from the paraxial model to the non-paraxial model, from scalar beams to vectorial beams. Dynamical characteristics of polarization singularities, coherence singularities, and speckle fields have been extensively studied as well. Meanwhile, different approaches and technologies have been proposed to generate vortex beams and measure their OAM. Consequently, with the advance of optical vortices, the rapid development of optical technologies, such as optical manipulation, optically driven motors, optical communications and remote sensing, imaging, optical delivering, etc., has been achieved, although many applications are still in their nascent stage.

This book provides a broad review of this remarkable field, aiming to introduce the physical principles in the optical vortices field, and to summarize the most important applications as well. We freely acknowledge that we can only overview a fraction of the field as a whole. The goal of this book is to present the topic to a broad audience, both specialist and non-specialist, in related areas.

In writing this book, we have been guided by the pioneering works of many great scientists, and we warmly thank our co-workers Prof. Kishan Dholakia, Prof. Andrew Forbes, Prof. Juan P Torres, Prof. Lluis Torner, Prof. Yangjian Cai and Prof. Chengliang Zhao for their help. Further thanks goes to our graduate students, Miao Dong, Yihua Bai, Jun Yao, Haoran Lv, and Jiandian Yan for the help and feedback during the preparation of the manuscript.

Yuanjie Yang
Yu-Xuan Ren
Carmelo Rosales-Guzmán
April 2024

Acknowledgments

To all of those who have significantly contributed to the development of this fascinating field, from the fundamental concepts to the applications.

Author biographies

Yuanjie Yang

Yuanjie Yang is currently a professor of Physics and vice dean in the School of Physics, University of Electronic Science and Technology of China. He got his PhD degree from Sichuan University (China) in 2008, and after that he carried out postdoctoral research at the University of St Andrews (UK), University of York (UK) and National University of Singapore. His research interest mainly focuses on optical vortex beams, orbital angular momentum, plasmonic vortices and optical trapping, etc. He has published 80 papers including in *Science*, *Physical Review Letters*, *Nature Communications*, *Nano Letters*, etc. He currently serves as editorial board member of *Frontiers in Physics*, *Chinese Optics Letters*, etc.

Yu-Xuan Ren

Yu-Xuan Ren is currently a research associate professor in the Institute for Translational Brain Research at Fudan University. He obtained his PhD at the University of Science and Technology of China in 2012, and has visited Yale University as visiting student. He has worked in the National Center for Protein Sciences Shanghai, San Francisco State University, Hong Kong University as a research assistant and a post-doc. He currently serves as associate editor of '*Frontiers in Physics*', and is the young editorial member of '*Chinese Journal of Lasers*', '*Journal of Infrared and Millimeter Waves*', and '*Journal of Fujian Normal University*'. He has authored over 70 research papers published in scientific journals including, *Light: Science & Applications*, *Nanophotonics*, *Photonics Research*, *ACS Photonics*, *Advanced Photonics*, *Optica*, *Applied Surface Science*, and served as external reviewers for many scientific journals.

Carmelo Rosales-Guzmán

Carmelo Rosales-Guzmán is a principal investigator at Centro de Investigaciones en Optica (CIO, Mexico) since 2020. From 2018 to 2020 he was a professor at the University of Science and Technology of Harbin (China). From 2015 to 2018 he was a postdoctoral research fellow at the structured light laboratory of the University of Witwatersrand (South Africa). Dr Rosales obtained his Ph.D. degree in 2015 from ICFO-The institute of photonics sciences (Barcelona, Spain). He is currently on the editorial panels of *Journal of Optics Frontiers in Physics* and *APL Photonics*. His main research interest involves the generation, characterization and applications of structured light, with over 90 research articles. His significant contributions to the field of structured light, granted him the 2023 SPIE Early Career Achievement Award.

Symbols

c	Speed of light
$\mathbf{p}$	Linear momentum density
$\mathbf{j}$	Angular momentum density
$\mathbf{S}$	Time-dependent Poynting vector
l	Topological charge of vortex beam
p	Radial mode index
ε	Dielectric permittivity
μ	Magnetic permeability
λ	Wavelength of light
ρ	Radial coordinate of cylindrical coordinates
φ	Azimuthal coordinate of cylindrical coordinates
z_R	Rayleigh range
f	Frequency
$\mathbf{K}$	Wave-vector
W	Beam power
w_0	Beam waist

IOP Publishing

Optical Vortices
Fundamentals and applications
Yuanjie Yang, Yu-Xuan Ren and Carmelo Rosales-Guzmán

Chapter 1

Introduction

1.1 Vortices in nature

Vortices are very common phenomena and can be seen everywhere in the natural world, such as water whirlpools, smoke rings, tornadoes, mesocyclones, the Great Red Spot of Jupiter and spiral galaxies. Generally speaking, vortices are regions within a fluid (liquid, gas or plasma) where the flow spins around an axis line. When vortices are formed, they carry angular momentum. In fluids, vortices are startlingly common, for instance, a water whirlpool can be formed in a cup simply by a quick stir, and can be found when water is being sucked down a bath or sink drain. Whirlpools can also be found in a river or sea at the meeting of opposing currents, which can be powerful and dangerous, and can suck down boats. Tornadoes are another kind of famous vortices with destructive forces, which form when large currents of warm air collide with currents of cold air. Small tornadoes may snap flowers or trees, and some bigger tornadoes can wipe out cars, houses or even whole towns. Besides the macroscopical vortices aforementioned in nature, they can be seen on extremely small scales such as quantum mechanical Bose–Einstein condensates. From the 1990s, optical vortices have been readily found as well.

In recent years, intensive studies of vortices were carried out in optics, since the discovery that optical vortex topological charge ℓ can carry an orbital angular momentum (OAM) $\ell\hbar$ per photon [1]. An optical vortex has a donut profile with a phase singularity, namely, a nodal point with a zero of the complex field, and one round-trip around the singularity by any closed contour will change the phase by $2\ell\pi$. The topological charge ℓ, sometimes called winding number, denotes the net change of phase in a circuit C enclosing the phase singularity, quantized in units of 2π:

$$l = \frac{1}{2\pi} \times \oint_C \nabla\phi(\vec{r})\mathrm{d}r. \tag{1.1}$$

The unique nature of optical vortices has led to applications in many areas such as optical manipulation, super-resolution microscopy, optical data storage, free-space

doi:10.1088/978-0-7503-5844-6ch1

communication and quantum information [2]. Therefore, the study of optical vortices is of importance in both the fundamental aspects and the applications. Interestingly, the study of vortex has brought together researchers from disparate fields with different backgrounds. So far, besides optical vortex, we have generated electron vortex [3], neutrons vortex [4], atom vortex [5], plasmon vortex [6], etc. It has drawn broad interest from the scientific community, not only for their common properties and similar methods, but also for their attractive application prospects in different fields.

Thus far, the vortices have been studied extensively, from optical vortex beams, namely, electromagnetic waves, to electron vortex beams, viz., matter waves. However, the study of vortices is still in its early days. For example, we now know that spiral galaxies have a vortex structure, but the formation of spiral galaxies and why supermassive black holes can always be seen in the nuclei of disk spiral galaxies are still open questions. The study of vortex beams may provide some clues to understand some unknown natural phenomena.

1.2 Historical review of optical vortices

Since the vortex phenomenon is ubiquitous, the study of vortex or phase singularities has a long history [7]. In 1833, Whewell observed phase singularities in the tide waves [8], and Hamilton discovered polarization singularities of light beams as the effect of conical refraction in 1838 [9]. In 1919, Ignatowskii noted the feasibility of a local backward light flow, which was analyzed clearly by Richards and Wolf [10] in 1959, using the Poynting vector distribution. Shortly after that, Boivin, Dow and Wolf found that the energy flow has vortices in the focal plane, and the intensity of light vanishes in the centre of a vortex [11].

The universally recognized seminal paper in this field is the paper 'Dislocations in wavetrains' by Nye and Berry [12] in 1974, which revealed that phase singularity is a general feature of wave interference. It is said that Arnold Sommerfeld noted that the phase has an unusual properties near the points of zeros of intensity, but he ignored this phenomenon since there is very little light in such regions [13]. Therefore, Sommerfeld missed the opportunity to uncover the very nature of optical vortices. In 1979, Vaughan and Willetts studied the interference properties of light beam with a helical phase front [14], which is actually an optical vortex beam. However, the phrase 'optical vortex' had not been seen in literature until 1989. In this paper, based on the Maxwell–Bloch equation, Coullet *et al* [15] showed the existence of a phase singularity in laser beams and that an optical vortex rotates around the zero of the intensity. A Laguerre–Gauss (LG) beam is one of the most common vortex beams and can be expressed as:

$$\mathrm{LG}_p^l(\rho,\,\varphi,\,z) = \frac{\omega_0}{\omega(z)}\sqrt{\frac{2p!}{\pi(|l|+p)!}}\left(\frac{\sqrt{2}\rho}{\omega(z)}\right)^{|l|} L_p^{|l|}\left[2\left(\frac{\rho}{\omega(z)}\right)^2\right]\exp\left[i(2p+|l|+1)\xi(z)\right]$$
$$\times \exp\left[-\left(\frac{\rho}{\omega(z)}\right)^2\right]\exp\left[\frac{-\mathrm{i}k\rho^2}{2R(z)}\right]\exp\left(\mathrm{i}l\varphi\right),$$

(1.2)

where $L_p^{|l|}(x)$ is the associated Laguerre polynomial, l and p are azimuthal and radial mode indices, respectively $(l \in \mathbb{Z}, \quad p \in \mathbb{N}), \quad \xi(z) = arctan(z/z_R)$, $\omega(z) = \omega_0 \sqrt{1 + (z/z_R)^2}$, $R(z) = z\sqrt{1 + (z/z_R)^2}$, z_R is the Rayleigh range of the beam, ω_0 is the waist of Gaussian beam.

1.3 Orbital angular momentum of vortex beams

In 1992, Allen *et al* [1] showed that an optical vortex beam with an azimuthal phase structure $exp(il\phi)$ can carry an orbital angular momentum (OAM) of $l\hbar$ per photon, where denotes the topological charge of the vortex beam. After that, the study of optical vortices has grown in an incredible rate, and thus far, optical vortices have found numerous applications in optical micro-manipulation [16], free-space communication [17], and quantum information [18], etc.

Assuming the complex amplitude of a vortex beams can be expressed as:

$$u(\vec{r}) = u_0(\vec{r}) \exp(il\phi). \tag{1.3}$$

The expression of OAM distribution for a beam is [1]

$$M = \frac{i\omega\varepsilon_0}{2}\vec{r} \times [u \, \nabla \, u^* - u^* \, \nabla \, u]. \tag{1.4}$$

For a vortex beam, as shown in equation (1.3), the expression of OAM distribution can be written as

$$M = \varepsilon_0 \omega l \, |u|^2. \tag{1.5}$$

The total OAM value L is the integral of M over the beam cross-section:

$$L = \varepsilon_0 \omega l \iint |u|^2 \rho d\rho d\phi, \tag{1.6}$$

which results in the expression

$$L = \frac{lW}{\omega} = lN\hbar, \tag{1.7}$$

where W is the beam energy, N is the number of photons and $W = N\hbar\omega$.

From equation (1.7) we can see that each photon in a vortex beam carries an OAM $l\hbar$.

References

[1] Allen L, Beijersbergen M W, Spreeuw R J C and Woerdman J P 1992 Orbital angular momentum of light and the transformation of Laguerre–Gaussian laser modes *Phys. Rev.* A **45** 8185–9

[2] Yao A M and Padgett M J 2011 Orbital angular momentum: origins, behavior and applications *Adv. Opt. Photon* **3** 161

[3] Verbeeck J, Tian H and Schattschneider P 2010 Production and application of electron vortex beams *Nature* **467** 301

[4] Clark C, Barankov R, Huber M, Arif M, Cory D and Pushin D 2015 Controlling neutron orbital angular momentum *Nature* **525** 504

[5] Lembessis V E, Ellinas D, Babiker M and Al-Dossary O 2014 Atom vortex beams *Phys. Rev. A* **89** 053616

[6] Kim H, Park J, Cho S, Lee S, Kang M and Lee B 2010 Synthesis and dynamic switching of surface plasmon vortices with plasmonic vortex lens *Nano Lett.* **10** 529

[7] Soskin M S and Vasnetsov M V 2001 Singular optics *Progress in Optics* ed E Wolf (Amsterdam: Elsevier) 219–76

[8] Berry M 1981 Singularities in waves and rays *Physics of Defects* ed R Balian, M Kleman and J-P Poirier (Amsterdam: North-Holland) 453

[9] Born M and Wolf E 1999 *Principles of Optics* 7th edn (New York: Pergamon)

[10] Richards B and Wolf E 1959 Electromagnetic diffraction in optical systems, II. Structure of the image field in an aplanatic system *Proc. R. Soc. Lond.* A **253** 358

[11] Boivin A, Dow J and Wolf E 1967 *J. Opt. Soc. Am.* **57** 1171

[12] Nye J F and Berry M V 1974 Dislocations in wave trains *Proc. Roy. Soc.* A **336** 165–90

[13] Gbur G J 2016 *Singular Optics* (Boca Raton, FL: CRC Press)

[14] Vaughan J M and Willetts D V 1979 Interference properties of a light beam having a helical wave surface *Opt. Commun.* **30** 263–7

[15] Coullet P, Gil I and Rocca F 1989 Optical vortices *Opt. Commun.* **73** 403–8

[16] Dholakia K and Cizmar T 2011 Shaping the future of manipulation *Nat. Photon.* **5** 335

[17] Wang J *et al* 2012 Terabit free-space data transmission employing orbital angular momentum multiplexing *Nat. Photon.* **6** 488

[18] Molina-Terriza G, Torres J P and Torner L 2007 Twisted photons *Nat. Phys.* **3** 305

IOP Publishing

Optical Vortices
Fundamentals and applications
Yuanjie Yang, Yu-Xuan Ren and Carmelo Rosales-Guzmán

Chapter 2

Generation of optical vortex beams

Light beams, as electromagnetic waves, can carry both energy and momentum. We know that momentum can be classified into linear and angular momentum, and there are two particular kinds of angular momenta: spin angular momentum (SAM) and orbital angular momentum (OAM). More than a century ago, Poynting proposed that the SAM is related to the photon spin, namely, the circularly polarized light carries a SAM of $\pm\hbar$ per photon [1], whereas linearly polarized light carries no SAM. In 1932, Darwin realized photon can carry both SAM and OAM [2]. But it was not until 1992 that Allen *et al* revealed the OAM of a light beam clearly [3]. It was shown that vortex beams with an azimuthal phase factor exp ($il\theta$) can carry OAM of $l\hbar$ per photon, where l can be any integer number, named topological charge. The discovery of the OAM of light has changed the way we understand and employ light.

The electric expression of an optical vortex beam along the z axis can be expressed as $E(r, \theta, z) = u_0(r, \theta, z)\exp(il\theta)\exp(-ikz)$, where $u_0(r, \theta)$ is the field amplitude distribution at the z plane, z is the propagation distance, k is the wavenumber, and l is the topological charge [4]. Different from traditional Gaussian beams, vortex beams exhibit a phase singularity in the center, which leads to a doughnut-shaped intensity profile and helical wavefronts, in which the number of intertwined helices and the handedness are dependent on the topological charge l. Moreover, there exists a phase singularity in the center, where the phase value is undefined.

In this chapter, a brief introduction about various kinds of vortex beams is presented in section 2.1. In section 2.2, generation methods including classical optical elements, digital devices, photon sieve, metasurfaces and other techniques are summarized subsequently.

doi:10.1088/978-0-7503-5844-6ch2

2.1 Typical vortex beams

2.1.1 Laguerre–Gaussian beams

Laguerre–Gaussian (LG) mode, containing the phase factor exp($il\theta$), is a complete and orthogonal set of solutions for the paraxial wave equation, which is inherently a kind of vortex beams [5].

$$LG_{pl}(r, \phi, z) = \sqrt{\frac{2p!}{\pi(p + |l|)!}} \frac{1}{\omega(z)} \left(\frac{r\sqrt{2}}{\omega(z)}\right)^{|l|} \exp\left[\frac{-r^2}{\omega^2(z)}\right]$$

$$L_p^{|l|}\left(\frac{2r^2}{\omega(z)^2}\right) \exp\left(-\frac{r^2}{\omega(z)^2}\right) \exp(il\phi) \tag{2.1}$$

$$\exp\left[\frac{ik_0 r^2 z}{2\left(z^2 + z_R^2\right)}\right] \exp\left[-i\eta \tan^{-1}\left(\frac{z}{z_R}\right)\right],$$

where $\eta = 2p + |l| + 1$, z_R is the Rayleigh range, $L_p^{|l|}(x)$ is the associated Laguerre polynomial with p being the number of radial nodes, $\omega(z) = \omega(0)[(z^2 + z_R^2)/z_R^2]^{1/2}$ denotes the $1/e$ radius of the beam with $\omega(0)$ being the beam waist. Besides, there are other types of vortex beams such as non-zeroth order Bessel beams [6] and Mathieu beams [7]. Recent decades have witnessed several ground-breaking demonstrations of vortex beams in many aspects, ranging from optical communications [8], optical micromanipulation [9] to quantum information [10], etc.

The intensity and phase distribution of LG beams for different p and l are shown in figure 2.1. When $p = 0$, $l = 1$, one can see a donut-shaped intensity pattern in figure 2.1(a) and an increase of 2π around the center of the phase pattern in figure 2.1 (d). When p is nonzero, one can see from figures 2.1(b) and (c) that there exists $p + l$ concentric rings in intensity patterns. Figures 2.1(e) and (f) show that the phase changes $2\pi l$ around the singularities. Until now, optical vortex beams have been artificially produced by many approaches. The use of diffractive optical elements, including the spiral phase plate and computer-generated holograms, are most widely used.

2.1.2 Bessel vortex beams

Perhaps one the most famous solutions to the exact Helmholtz equation in cylindrical coordinates is the higher-order Bessel modes. Two of the most prominent properties of these modes when propagated in free space are their tendency to maintain a non-spreading intensity profile and the recovery of the same, when an opaque obstruction is placed in its path. Mathematically, high-order Bessel modes are given by

$$B(\rho, \phi, z) = E_0 J_l(k_t \rho)\exp(ik_z z)\exp(il\phi) \tag{2.2}$$

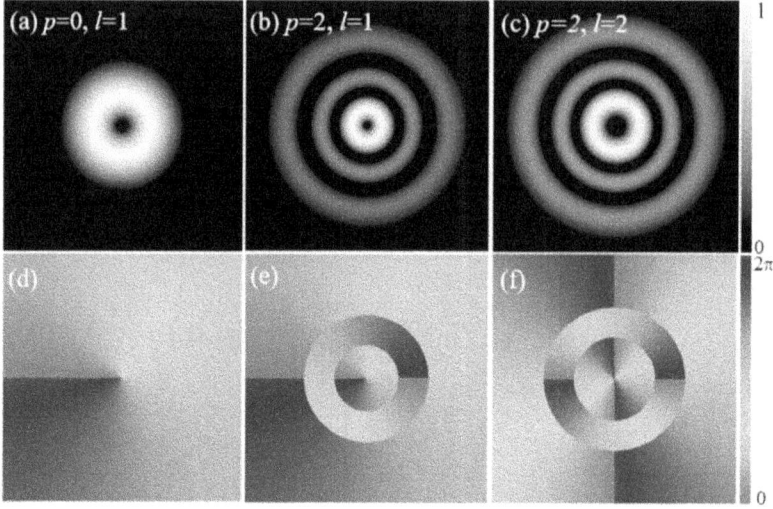

Figure 2.1. The intensity (a–c) and the corresponding phase distribution (d–f) of an LG beam for different indices.

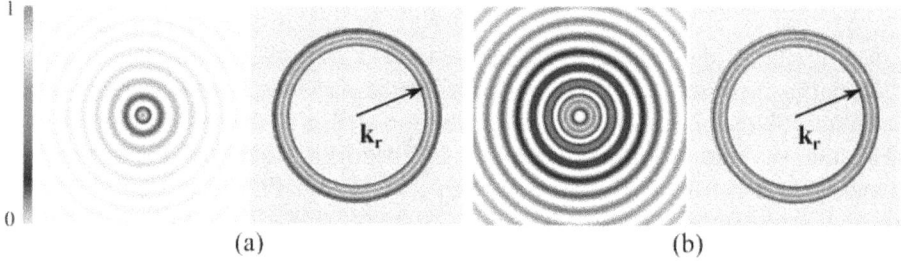

Figure 2.2. Transverse intensity profiles of Bessel beams and its Fourier transform with (a) $l = 0$ and (b) $l = 1$. The color denotes the normalized intensity distribution. Reproduced from [11]. CC BY 4.0.

where $J_l(x)$ represents the lth order Bessel polynomial. Moreover, k_t, k_z are the transverse and longitudinal components of the wave vector, respectively, obeying the relations $k = 2\pi / \lambda$ and $k^2 = k_t^2 + k_z^2$.

A more intuitive way of describing a Bessel beam is by considering these as the result of a superposition of plane waves propagating on a cone, where each of them undergoes identical phase shifts, $k_z \Delta z$ over a distance Δz. This interpretation can be observed in the angular spectrum of the Bessel beam, which takes the form of a ring or radius k_r in the k-space. Therefore, the optical Fourier transform of the Bessel beams is a ring and vice versa, the optical Fourier transform of a ring will result in a Bessel beam. Therefore, Bessel beams can be generated using a ring-slit aperture. Using this method, Durnin *et al* produced a Bessel beam experimentally for the first time. Figure 2.2 illustrates this concept schematically; we show the transverse intensity profiles of a Bessel beam with topological charge $l = 0$ and its Fourier transform (figure 2.2(a)), while in figure 2.2(b) we show that of a Bessel beam of topological charge $l = 1$ along with its Fourier transform.

Two of the most prominent properties of Bessel modes are, on one hand, their tendency to maintain an invariant intensity profile, namely, $I_B(\rho, \phi, z \geqslant 0) = I_B(\rho, \phi, 0)$, and on the other, their tendency to recover its original form when an opaque obstruction is placed in its path. Such properties can be explained by invoking again the plane waves propagating on a cone approach, as detailed next. When the opaque object or radius a is placed in the center of the Bessel beam, some of the waves that create the beam are blocked by this object, creating a shadow region. Nonetheless, some other plane waves can pass the object unaffected, which ultimately are the ones that reconstruct the intensity profile of the beam after a certain propagation distance. As mentioned earlier, Bessel beams can be generated by placing a slit aperture in front of a ring-slit aperture, none the less, this is a very inefficient way since most of incident beam's intensity is blocked. A far more efficient way to produce a Bessel beam is using an axicon. The on-axis intensity is formed by conical wavevectors that propagate on the surface of a cone.

These modes are known as Bessel–Gauss modes and are given, at $z = 0$, by the expression,

$$B(\rho, \phi, 0) = J_l(k_t\rho)\exp\left[-\left(\frac{\rho}{\omega_0}\right)^2\right]\exp(il\phi) \qquad (2.3)$$

Experimentally they can be synthetized using a near field or a far-field approach. In the first approach, a Gaussian beam is passed through a conical lens known as an axicon, while in the second it is passed through an opaque ring aperture, which renders a less efficient method. Finally, as will be discussed later, the 0th order Bessel–Gauss has been demonstrated to have great relevance in optical trapping to stack multiple particles along the beam's central core.

2.1.3 Perfect vortex beams

This section was reproduced with permission from [50].

Over the past two decades, there has been a remarkable surge in the development of versatile and diverse optical vortex beams, alongside the aforementioned optical vortex beams [12, 13]. In certain applications, such as micromanipulation, vortex beams are usually required as small rings with large topological charge. Conventional vortex beams do not meet this criterion since the ring size increases with the topological charge. Perfect vortex beams, however, maintain constant ring radius regardless of the topological charge. Figure 2.3 exhibits the transverse intensity profiles of perfect vortex beams with topological charges of $l = 1, 4, 10$ and 15.

The notion of a 'perfect vortex beam' was initially proposed by Ostrovsky et al in 2013. These beams possess intensity profiles that are independent of their topological charge [14, 15], which can be mathematically expressed as follows [14]:

$$g_v(\rho, \theta) \propto \text{circ}\left(\frac{\rho}{a}\right)\exp(iv\theta)\sum_{n=1}^{\infty}\frac{J_v(\alpha_{v,n}\rho_o/a)}{J_{v+l}^2(\alpha_{v,n})}J_v\left(\alpha_{v,n}\frac{\rho}{a}\right) \qquad (2.4)$$

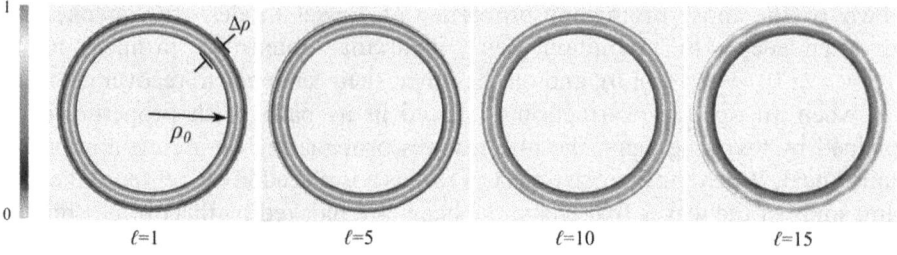

Figure 2.3. Transverse intensity distribution of perfect vortex beams with topological charge $l = 1$, 4, 10 and 15, where the intensity profile remains constant as the topological charge increases. Reproduced from [11]. CC BY 4.0.

where v is the topological charge, J_v is the vth-order Bessel function of the first kind, $\alpha_{v,n}$ is the nth zero of function J_v, and a is the upper limit of the radial coordinate ρ.

It is known that a twisted nematic liquid spatial light modulator (LC-SLM) can be utilized for phase distribution modulation [14]. In the experiment, the LC-SLM's transmittance is represented through a configuration comprising distinct, concentric ring-shaped slits with varying radii and widths, forming an annular topology. Additionally, periodic azimuthal phase modulation is incorporated within these slits. Consequently, the transmittance of the LC-SLM can be expressed as [14]

$$t_v(r, \phi) \propto \sum_{n=1}^{N} \alpha_{v,n} \beta_{v,n} \exp(iv(\phi - \phi_{v,n})) \delta(r - r_{v,n}) \tag{2.5}$$

The finite-sum approximation of perfect vortex beam can be further expressed to replace equation (2.4), i.e.,

$$U_v(\rho, \theta) \propto \exp(iv\theta) \sum_{n=1}^{N} \frac{J_v(\alpha_{v,n}\rho_0/a)}{J_{v+l}^2(\alpha_{v,n})} J_v\left(\alpha_{v,n}\frac{\rho}{a}\right) \tag{2.6}$$

The finite-sum approximation of the vortex beam (equation (2.6)) exhibits a central ring with a higher intensity and additional weak bright rings [14]. Upon careful observation, it becomes evident that only the central ring corresponds to a perfectly formed vortex, while the presence of unwanted additional rings hinders its potential applications. Moreover, the generated perfect vortex beam demonstrates poor quality due to the non-ideal modulation characteristics of the LC-SLM and the actual width of the ring slits.

Subsequently, the purity of the generated perfect vortex beam has been enhanced through the utilization of the width pulse approximation of Bessel function in the LC-SLM [16]. Additionally, another technique involving the utilization of the Fourier transform property of the Bessel beam has been proposed to generate a perfect vortex beam [17]. In this method, the SLM is designed to produce Bessel–Gauss beams, and the perfect vortex beam is experimentally generated by optical Fourier transform of a fourth-order Bessel–Gauss beam. The ring radius of the perfect vortex beam decreases as the axicon parameter decreases, denoted as a.

Recently, an amplitude digital micromirror device (DMD) [18, 19] has been introduced to generate high-quality perfect vortex beams with large topological charges, reaching up to 90. It has been demonstrated that the generated perfect vortex beams maintain a high fidelity and do not exhibit undesired additional rings until the topological charge exceeds 90. This improvement can be attributed to the DMD's higher refreshing frequency and higher fill factor rate of approximately 90%, which enables more accurate amplitude modulation compared to the SLM.

Furthermore, Vaity *et al* proposed that a perfect vortex beam can be regarded as the Fourier transform of a Bessel beam, and it can be generated by employing a phase hologram where the transmittance is determined by the phase of a Bessel beam [17]. It is worth noting that although the concept of a 'perfect vortex beam' was not introduced until 2013, such beams had been previously utilized by Roichman and Grier in 2007 for 3D optical trapping and particle transport. Roichman *et al* highlighted that the radius of the 3D ring trap is independent of the topological charge, which is the first experimental demonstration of optical trapping using a 'perfect vortex beam' under the name of a holographic ring trap [20]. The perfect vortex beam has been utilized by the optical trapping as a specific case of a structured light beam to study the dynamics of driven particles in the form of optical matter.

2.1.4 Anomalous vortex beams

This section was reproduced with permission from [50].

Traditional vortex beams such as high-order LG beams and Bessel beams remain consistent profiles during their propagation. In 2013, a unique vortex beam called the anomalous vortex (AV) beam was introduced [21]. The AV beam exhibits an intriguing characteristic, it will evolve into an elegant Laguerre–Gaussian (ELG) beam in the far field when propagating in free space.

At the initial plane ($z = 0$), the electric field of the AV beam can be described as [21]

$$E_{n,m}(\rho_0, \theta_0, 0) = E_0 \frac{\rho_0}{\omega_0^{2n+|m|}} \exp \frac{-\rho_0^2}{\omega_0^2} \exp(-im\theta_0) \tag{2.7}$$

where E_0 is a constant, m is the topological charge, n is the beam order of the AVB, and w_0 is the beam waist size of the fundamental Gaussian beam, ρ_0 and θ_0 are radial and azimuthal coordinates, respectively. Equation (2.7) will reduce to the electric field for a hollow Gaussian beams if $m = 0$ and $n \neq 0$. If letting $n = 0$, it involves to an ordinary Gaussian vortex beams.

The electric field of the AV beam through the ABCD optical system can be simplified using far-field approximation, which can be expressed as [21]

$$E_{n,m}(\rho, \theta, z) = \frac{i^{m+1}\pi n!}{\lambda z} \exp(-ikz)\omega_0^{2n+|m|+2}$$
$$\exp(im\theta)\frac{\rho_0^{2|m|}}{\omega} \exp \frac{-\rho_0^2}{\omega_0^2} L_n^{|m|}\left(\frac{\rho^2}{\omega^2}\right) \tag{2.8}$$

In equation (2.8), the complex amplitude of an AV beam in the far field is consistent with that of an ELG beam. At the source plane, the AV beam exhibits a distinct single ring profile with a helical phase. However, in the far field, the amplitude distribution of the AV beam manifests multiple rings, which significantly diverges from the characteristics observed in LG vortex beams.

2.1.5 Fractional vortex beams

This section was reproduced with permission from [50].

Vortex beams with non-integer topological charges, also known as fractional vortex beams, has aroused significant interest in recent years [22–24]. Several studies have demonstrated that the intensity profiles of fractional vortex beams exhibit a mixed screw-edge dislocation and possess a radial opening within the annular ring. These unique structures and characteristics, distinct from integer topological charge, have been extensively explored and contributed to a deeper understanding of angular momentum. Exploiting the distinctive features of such beams has provided practical applications in diverse fields, including optical trapping [24], quantum information [25], and optical communication [26].

Various methods have been proposed to generate fractional vortex beams. Conceptually, one of the simplest approaches is using the spiral phase plate with a fractional step height [22]. When a plane wave with unit amplitude passes through a spiral phase plate, a wave possessing a phase singularity of strength $2\pi\alpha$ at the plane z = 0 can be expressed as

$$\Phi_\alpha(R, 0) = \exp(i\alpha\phi) \tag{2.9}$$

where $R = x$, $y = R\cos\phi$, $\sin\phi$, $\rho = R/z^{1/2} = \varepsilon$, η. When $\alpha = n$ is an integer, there exists a screw dislocation along the z axis R = 0. When α is non-integer, there are mixed screw-edge dislocations. The corresponding paraxial propagating wave is expressed as [22]

$$\Phi_\alpha(r) = \frac{\exp(i(z + \pi\alpha)\sin(\pi\alpha))}{\pi}\sum_{-\infty}^{\infty}\frac{\exp(in\phi)P_n(\rho)}{\alpha - n} \tag{2.10}$$

where $P_n(\rho)$ is given in [22], z is the propagation distance, and n is integer topological charge.

Fractional vortex beams of different types, such as fractional Bessel beams [27, 28], can also be generated using optical elements. Furthermore, the use of two such devices enabled the observation of high-dimensional spatial entanglement of twin photons possessing fractional OAM [29]. Another widely employed method for generating fractional vortex beams is the holography method. By employing a computer-generated hologram with an l-pronged fork dislocation, a vortex beam with a topological charge of l can be produced. Such a hologram can generate vortex beams with fractional topological charges as well, where an additional radial discontinuity appears along the $\phi = 0$ due to the phase step of the fractional part [30].

In addition, a technique based on the production of sub-harmonic diffraction orders has been proposed for generating fractional vortex beams [31]. Typically, the generated fractional vortex beams are unstable. However, a novel method utilizing the synthesis of LG modes has been introduced to generate stable fractional vortex beams, ensuring their structural stability during propagation [32]. Aforementioned beam generation techniques are primarily applicable in the visible and near-infrared spectra. For shorter wavelengths, such as extreme-ultraviolet (EUV) and soft x-rays, alternative generation mechanisms for fractional OAM beams have been proposed. Hernández García *et al* has investigated the generation of EUV attosecond pulse beams carrying fractional orbital momentum through high-order harmonic generation [33, 34]. Furthermore, it is widely known that the OAM of vortex beams typically remains constant during propagation. However, recent research has revealed novel characteristics of vortex beams. It has been discovered that the OAM of these beams can not only vary in spatial regions [35, 36] but also in time [37]. In 2018, Yang *et al* introduced the concept of an anomalous Bessel vortex beam, which exhibits the novel feature that topological charge varies as it propagates in free space [35]. Figure 2.4 illustrates the generation of such a vortex beam using a Fermat's spiral slit. The structure of the Fermat's spiral slit can be described as [35]

$$r_a = \left(r_0^2 + \frac{lz\lambda\alpha}{\pi} \right)^{1/2} \tag{2.11}$$

where r_0 is the initial radius of the spiral slit, r_a and α are the radial and the angular coordinates, respectively, z is the propagation distance, and l is the topological charge.

Assuming the spiral slit is placed at $z = 0$, when a plane wave passes through the slit, the transmitted wave originating from different parts of the slit experience different optical paths and converge at the center of the observation plane. The cumulative phase difference at the coordinates $(0, 0, z)$ can be expressed as

$$\int_0^{2\pi} d\theta = \frac{2\pi}{\lambda} \int_0^{2\pi} \frac{l\lambda}{2\pi} d\alpha = 2\pi l \tag{2.12}$$

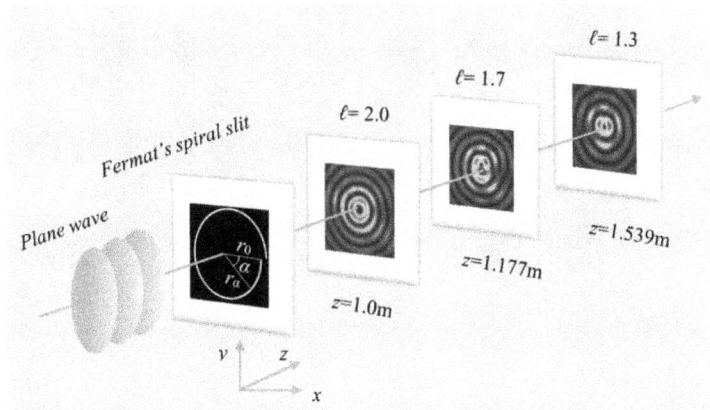

Figure 2.4. The notation of the spiral slit for achieving an anomalous Bessel vortex beam. Reproduced with permission from [35].

There will be an optical vortex in the center of the observation plane. For a given spiral slit, the value of $lz\lambda$ can be regarded as a constant. Therefore, we can see with increasing propagation distance z or λ, topological charge l will decrease. Figure 2.5 shows theoretical and experimental results of the intensity and phase pattern of the anomalous Bessel vortex beam during propagation from $z = 1$ m to $z = 2$ m. From figure 2.5, we can see clearly that the OAM of such vortex beams varies continuously during propagation.

Moreover, all aforementioned vortex beams are static, i.e., the OAM does not change with time. Rego $et\ al$ introduced a new property of vortex beams, manifested as a time-varying OAM EUV laser pulsed beam [37]. Two linearly polarized infrared pulse vortex beams exhibiting the same frequency content and pulse width, but with different topological charge $l_1 = 1$ and $l_2 = 2$, are considered. They are separated by a time delay of a pulse width. Then a temporal envelope of the superposition of the delayed beams is focused into an argon gas jet, where high-order harmonics are emitted. The EUV harmonic beams with a 17th harmonic are generated in the far field. According to the OAM conservation law that $l_q = ql$, where l_q and l are the topological charge of the qth order harmonic and the driving field, the OAM of superposition of the enveloped is varying and can be expressed as $n_1 l_1 + n_2 l_2$, and $n_1 + n_2 = q$. Thus, the OAM of EUV harmonic beams with a 17th harmonic can vary with time, i.e., from $17l_1 + 0l_2 = 17$ to $0l_2 + 17l_2 = 34$. Such a

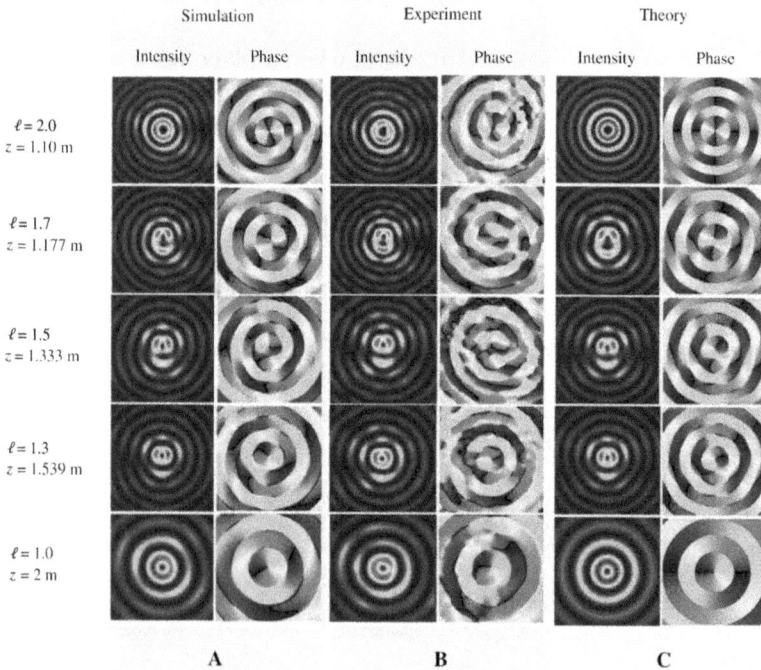

Figure 2.5. Intensity and phase patterns of vortex beams with different topological charge at different distance z. (a) and (b) Simulations and experimental results of factional Bessel vortex beams using spiral slit. (c) The corresponding theoretical results of an ideal factional Bessel vortex beam. Reproduced with permission from [35].

self-torqued high harmonic generation beam shows a brand new property of vortex beams, and it can be used for laser manipulation on attosecond time and nanometer spatial scales.

2.1.6 Open vortex beams

Generally, the intensity distribution of a vortex beam is a closed ring, such as a Laguerre–Gaussian beam. It shows that open vortex beams (OVBs) have non-trivial OAM spectra and have potential in many applications, such as metrology. In 2021, Zeng et al studied the OVB's structural stability, including the intensity invariance and the topological charge conservation [38]. By using coherent wave-packets and geometrical ray-like trajectories, a generalized model of OVBs is theoretically proposed to validate their structural stability. Moreover, it was experimentally showed that, using a partial fork grating (PFG), OVBs can be generated and demonstrated structural characteristics.

For a regular vortex beam, a centered circle $\rho = \rho_0$ in the waist plane can be used as an orbit of ray r, thus the shape of beams is determined through the envelope of rays. As shown in figure 2.6(a), two OVBs are generated in the ±1 orders of the diffraction with beams illuminating the PFG, where different orders show opposite opening orientations. The intensity and the phase distribution of a vortex array are shown in figures 2.6(b) and (c), respectively. Moreover, figures 2.6(g)–(i) show the OVBs' far-field distributions by using different gratings (figures 2.6(d)–(f)), resulting in OVBs with a quarter ring, a half ring, and a three-quarter ring, respectively.

It is noted that the topological charge of OVB is conserved with a fixed PFG. Figures 2.6(j)–(l) show the corresponding phase distributions, where the local phase in the open-ring area is stable. Different from conventional vortex beams, the structure of OVB is nonuniform and asymmetric, and the ultimate topological

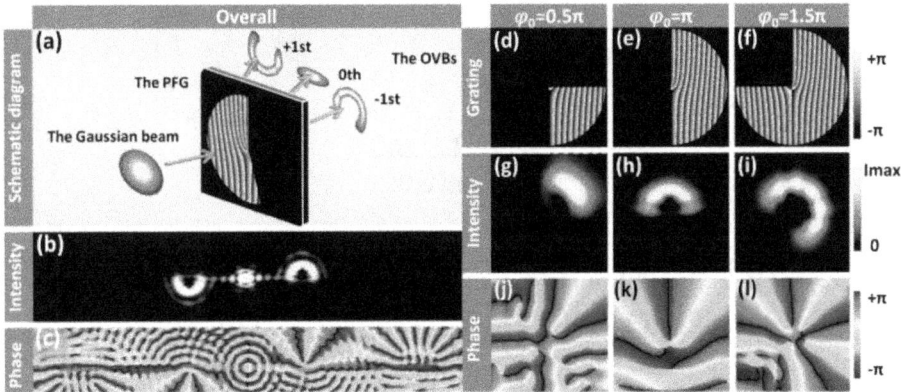

Figure 2.6. (a) Schematic diagram of the OVB generated using a Gaussian beam diffracted by the partial fork grating. The output (b) intensity and (c) phase pattern at the Fourier plane. partial fork gratings with the parameters: $l_0 = 4$; (d) $\phi_0 = \pi/2$, (e) $\phi_0 = \pi$, and (f) $\phi_0 = 3\pi/2$. (g)–(i) Far-field intensity distributions corresponding to (d)–(f). (j)–(l) Phase profiles corresponding to (g)–(i). Reprinted from [38], with the permission of AIP Publishing.

charge of OVB remains the same as the initial one, which means the topological charge is conserved during the propagation process.

2.2 Methods for vortex generation

2.2.1 Mode conversion

In the early years, laser beams carrying OAM was quickly developed both theoretically and experimentally. Particularly, Beijersbergen *et al* first generated such beams experimentally by the use of two cylindrical lenses called mode converters [39], where LG modes of arbitrary order can be transformed directly from the Hermit–Gaussian (HG) modes by introducing Gouy phase shift between the horizontal and vertical directions, and vice versa. For this reason, we will introduce this first-discovered method briefly in the following part.

This kind of optical element is called a mode converter. When the cylindrical lens is placed at a specific distance, as shown in figure 2.7(a), it can realize a $\pi/2$ mode converter. When the HG mode with indices m and n passes through this mode converter, an LG mode with the radial index $p = \min(m,n)$ and azimuthal index $l = m - n$ is obtained. This phenomenon can be explained from the mathematical relationship between the Hermite and Laguerre polynomial, which is expressed as [40]

$$u_{mn}^{LG}(x, y, z) = \sum_{k=0}^{N} i^k b(n, m, k) u_{N-k, k}^{HG}(x, y, z) \qquad (2.13)$$

where $b(n, m, k)$ is a real coefficient. The factor ik shows that there exists $\pi/2$ phase difference between the HG mode and LG mode, realizing the transformation. It is

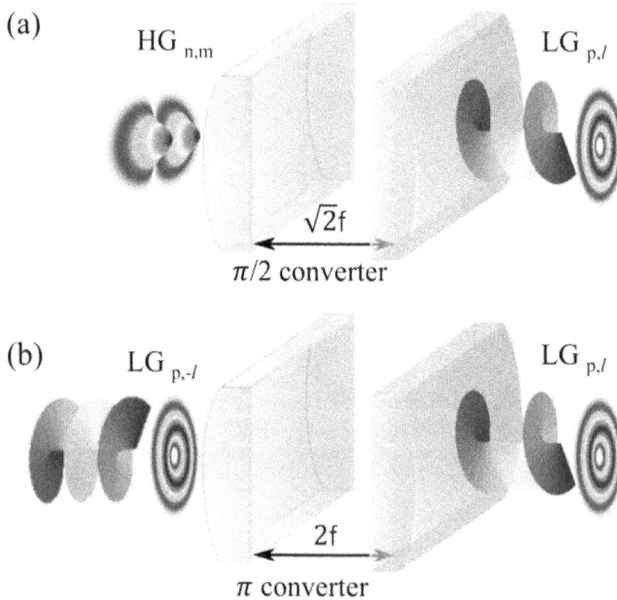

Figure 2.7. Schematic diagram of (a) $\pi/2$ converter, (b) π converter.

noted that although this method can produce high-quality optical vortex beams, complex optical systems and high-order HG beams are still required, which limits its application.

Similarly, as illustrated in figure 2.7(b), the π converter was proposed and the distance between the lenses is $2f$ instead. If a LG mode is incident on the π converter, there will be no OAM exchange on both lenses but the sign of which will be reversed.

2.2.2 Spiral phase plate

This section was reproduced with permission from [50].

As previously mentioned, phase modulation can be realized by introducing spatial variations in thickness or refractive index. In this section, we will focus on the spiral phase plate, which is one of the most commonly used refractive optical elements and is often considered as the foundation for various alternative phase-modulation techniques. The spiral phase plate is a transparent plate with a spiral thickness profile, where the optical height is directly proportional to the azimuthal angle [41, 42]. By imparting a spiral phase retardation, optical beams with helical wavefronts can be effectively transformed using a spiral phase plate.

The spiral phase plate is characterized by a transparent plate with a refractive index of n, where the optical thickness, denoted as h, increases in direct proportion to the azimuthal angle θ, as depicted in figure 2.8(a) [44, 45]. The functional relationship between h and θ can be expressed as [44]

$$h = h_0 + h_s\frac{\theta}{2\pi} \tag{2.14}$$

where h_0 is the base height of optical element and h_s is the step height. The phase delay can be expressed as

$$\Phi(\theta, \lambda) = \frac{2\pi}{\lambda}\left[\frac{(n - n_0)h_s\theta}{2\pi} + nh_0\right] \tag{2.15}$$

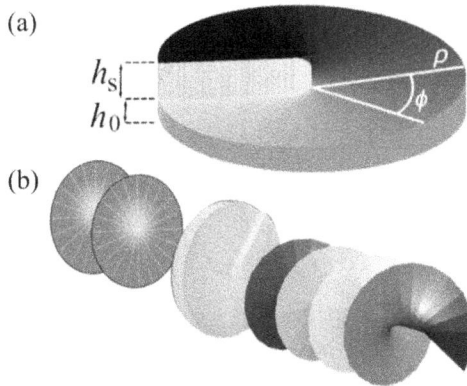

Figure 2.8. (a) The diagram of spiral phase plate and (b) a vortex beam generated by spiral phase plate.

where n_0 is the refraction index of the surrounding medium and λ is the vacuum wavelength. When a plane beam passes through such a spiral phase plate, a vortex with a topological charge $l = h_s(n - n_0)/\lambda$ is obtained, as shown in figure 2.8(b). Thus, the output beam carries OAM $l\hbar$ per photon. The spiral phase plate allows for the direct and effective transformation of a non-helical beam into a helical beam, whereas the drawbacks are also obvious. For example, the quality of the spiral phase plate significantly impacts the generation of vortex beams. In practice, fabricating a smooth helical surface for the spiral phase plate is challenging due to the high precision required in manufacturing techniques. Instead, the helical surface is often approximately fabricated using discrete steps, resulting in a reduction in quality. Additionally, the spiral phase plate is highly sensitive to the optical wavelength, limiting its effectiveness for vortex beams' generation [46].

In recent years, several approaches have been proposed to overcome these limitations [47]. For example, an adjustable spiral phase plate has been developed, allowing to produce vortex beams with various topological charges while accommodating a wide range of wavelengths [47]. Another method involves the use of flat spiral phase plate, the height of which remains constant while the refractive index increases with the azimuthal angle [48]. It is worth noting that the accuracy of the generated optical vortices improves with an increasing number of phase steps. To simplify the fabrication process, recent studies have shown that only three phase steps are sufficient to accurately generate the desired OAM mode, rather than using continuous variation through modal decomposition [49].

2.2.3 Computer-generated holograms

This section was reproduced with permission from [43]. © 2022 Chinese Laser Press.

A more advanced technique for generating structured fields is based on diffractive optical elements, which was pioneered by Soskin *et al* in the early 1990s. Holography is a method for recording and reconstructing the phase and intensity of an optical field. Holographic optical elements, such as fork gratings and spiral zone plates [50], are commonly employed to generate optical vortices. The concept of the fork grating was first introduced by Soskin *et al* [51–53]. The *l*-pronged fork dislocations represent the topological charge *l*, with the reversed direction indicating an opposite sign of topological charge. To reconstruct the desired light pattern, the interference pattern of a tilted plane light with a complex amplitude of $\Phi_1(x, y) = A_1(x, y)\exp[i\phi(x, y)]$ and the vortex light with a complex amplitude of $\Phi_2(x, y) = A_2(x, y)\exp(il\theta)$ should be obtained as the first step. After interference, the intensity is

$$
\begin{aligned}
I = {}& B_1(x, y) + B_2(x, y)\exp\{i[\phi(x, y) - l\theta]\} \\
& + B_2(x, y)\exp\{-i[\phi(x, y) - l\theta]\},
\end{aligned}
\tag{2.16}
$$

where $B_1(x, y) = |A_1(x, y)|^2 + |A_2(x, y)|^2$ and $B_2(x, y) = A_1(x, y)A_2(x, y)$. Then, as shown in figure 2.9, the fork-shaped grating is binarized and recorded on a photographic film using a lithography technique.

Suppose $\Phi_1(x, y)$ is the reference wave; the following beams can be generated:

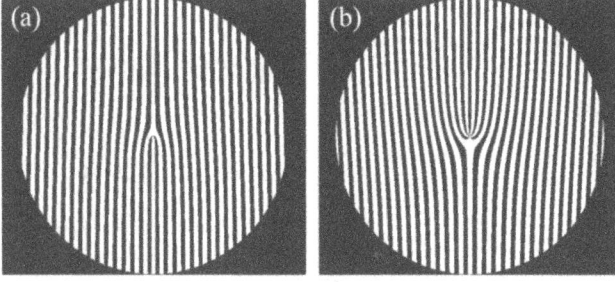

Figure 2.9. A fork-shaped diffraction grating produced by computer simulation. (a) $l = 2$, (b) $l = 5$.

$$B_1(x, y) \exp[i\phi(x, y)] + B_2(x, y) \exp\{i[2\phi(x, y) - l\theta]\} + B_2(x, y) \exp(il\theta) \quad (2.17)$$

where the first term represents the unmodulated plane wave, the second term is the combination of vortex and reference wave and the third term is the desired vortex wave. Optical vortices with different topological charges can be generated by a single fork grating, while the efficiency can be different.

Despite of fork grating, spiral zone plate is another mostly used computer-generated hologram for the generation of vortex beams [54, 55]. In a cylindrical coordinate system (r, ϕ, z), the spherical vortex beam can be simply expressed as the spherical wave multiplied by the azimuthal term [56],

$$\Phi_{SVW} \propto \frac{\exp[2\pi i k \sqrt{r^2 + (z - z_1)^2}]}{\sqrt{r^2 + (z - z_1)^2}} \exp(2\pi i l\phi), \quad (2.18)$$

where the center of the spherical wave is $r = 0$ and $z = z_1$. The hologram of a reference plane wave constructed with this spherical vortex beam can be expressed as

$$I_{holo} = \left| \frac{\exp[2\pi i k \sqrt{r^2 + (z - z_1)^2}]}{r^2 + (z - z_1)^2} \exp(2\pi i l\phi) + 1 \right|^2. \quad (2.19)$$

Particularly, the spiral zone plate can be degraded to a regular Fresnel zone plate with $l = 0$. In practice, these computer-generated holograms are normally binarized for the fabrication of masks that can be given as follows

$$b(r) = \begin{cases} 1, & 0 \leqslant \sqrt{L^2 + r_n^2} - \sqrt{L^2 + r^2} < \dfrac{\lambda}{2}, \\ 0, & \dfrac{\lambda}{2} \leqslant \sqrt{L^2 + r_n^2} - \sqrt{L^2 + r^2} < \lambda, \end{cases} \quad (2.20)$$

where L is the focal length and r_n is the innermost radius of the nth zone. Figure 2.10 illustrates the generation of vortex beams using a spiral zone plate. Furthermore, a single micro-fabricated spiral zone plate can produce a series of vortex beams at different positions along the propagation direction. The topological charge of an nth order vortex beam can be n times larger than that of a first-order vortex beam.

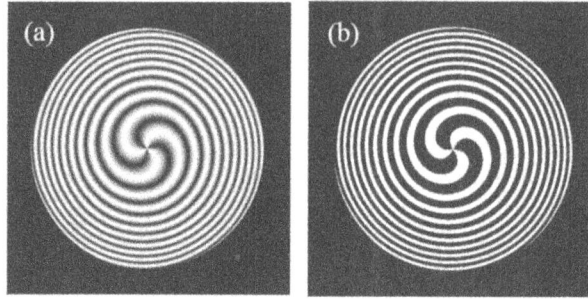

Figure 2.10. (a) Layout structures of the spiral zone plate. (b) Binarized spiral zone plate.

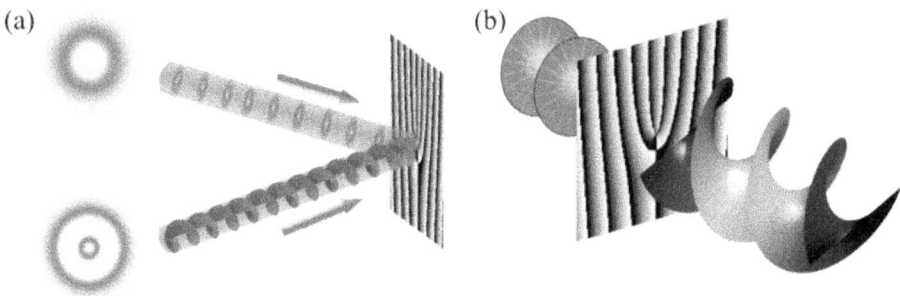

Figure 2.11. (a) Two light waves, a reference and the desired output, are interfered to generate a hologram (b) A Gaussian beam is passed through the hologram to generate the desired beam.

Additionally, an electron vortex beam with a topological charge of 90 was observed in a higher diffracted order.

The key difference between the fork grating and the spiral zone plate lies in the type of beam used to generate the hologram, whether it is a plane wave or a spherical wave. As mentioned earlier, by using a refractive spiral phase plate, the topological charge increases with an increasing thickness, whereas computer-generated holograms only require an increase in the number of lines in the pronged dislocation or spirals. Therefore, particularly for higher topological charges, fabricating computer-generated holograms can be much easier than fabricating spiral phase plates.

In this approach, the interference pattern between a plane wave and the desired beam is generated and recorded as a hologram, typically on photographic film, as depicted schematically in figure 2.11(a). Consequently, when the hologram is illuminated by a plane wave, the desired beam can be produced in the first diffraction order, as shown in figure 2.11(b).

2.2.4 Digital devices

This section was reproduced with permission from [50].

The utilization of diffractive optical elements was fully exploited after the emergence of digital holography, such as computer-controlled devices spatial light modulators (SLMs) and digital micromirror devices (DMDs). SLMs are liquid crystal devices with a pixelated structure that allows for the modulation of light properties, including amplitude and phase. By replacing the experimental calculation of holograms with mathematical derivations, SLMs have facilitated the generation of structured light fields [57]. This approach provides high accuracy in the produced modes and offers tremendous flexibility for rapid reconfiguration of the digital hologram, enabling modifications of the output field's shape and position. This capability enabled various applications, such as optical tweezers, particle tracking, and characterizing complex light fields.

Furthermore, SLMs can generate multiple light beams simultaneously by multiplexing various holograms into a single hologram, opening up possibilities for novel applications. For instance, in optical tweezers, arrays of multiple optical traps can be generated. In optical communications, simultaneous generation and detection of multiple modes can be realized. SLMs have also allowed for on-demand generation of almost any beam shape directly from a laser cavity by replacing one of the mirrors with an SLM, known as the digital laser. Note that SLMs are wavelength-dependent and can only modulate horizontal polarization, a characteristic that has been exploited in the generation of vector beams.

SLMs are electro-optical devices capable of modulating the optical characteristics of light. Liquid crystal (LC) SLMs are one of the most common devices, which comprise liquid crystal molecules whose transmission properties can be independently modulated by the applied electric field [58–60]. The incident light approaches the LC molecules through the glass substrate, transparent electrode and alignment film. Each LC pixel is controlled by the applied voltage from the corresponding electrode. Therefore, one only needs to encode the grayscale image into a computer to manipulate the wavefront of a light beam in real-time.

Considering an incident beam as a Gaussian beam, a vortex beam can be produced by imposing an azimuthal phase with topological charge l on the incident beam by SLM. Additionally, a blazed grating is needed to separate the first order from other undesired diffracted lights. Comparing with static optical elements that only produce specific intensity and phase distribution, the LC-SLM can modulate its phase distribution in real time [61]. Using SLM to implement the fabrication of aforementioned holograms greatly simplifies the experimental process with high efficiency. Meanwhile, the quality of the vortex beam is affected by the pixels of the SLM.

As another typical digital device to generate vortex beams, DMDs can manipulate incident light with high framerate and a great number of spatial degrees of freedom. A DMD consists of numerous high-speed digital switches, such as aluminum reflective mirrors, that can be independently modulated pixel by pixel. The resolution of the DMD depends on the number of pixels, and a larger number of micromirrors is preferred to minimize quantization errors in encoding intensity and phase, thereby ensuring high-quality beam generation. Binary data is transmitted to the static random access memory of the DMD, resulting in an electrostatic charge distribution that deflects each individual micromirror to an 'on' or 'off' state. Rapidly switching the micromirrors on and off leads to the realization of beam shaping.

Compared to liquid crystal displays that absorb up to 90% of the incident light with low transmittance and contrast, DMDs serve as a reflective SLM that improves image quality with higher throughput and superior diffraction efficiency [62]. Liquid crystal-based techniques are typically limited by the switching rate due to the viscosity of the liquid crystal. As an alternative method for beam shaping, DMDs enable binary control (on or off) of the beams at an extremely high switching rate exceeding 20kHz, which is approximately two orders of magnitude faster than typical phase-only SLMs [63].

Likewise, the dynamic generation of optical vortices can be achieved by using a DMD with spatially controllable phase and intensity. One commonly employed technique for phase modulation using the DMD is Lee holography [64]. It has been demonstrated that Lee holography with pixel dithering can generate LG modes with a purity exceeding 94% for low-resolution fields [65]. In 2010, Ren *et al* successfully transformed Gaussian beams into LG beams with different topological charges by projecting fork-shaped gratings onto the DMD [66]. Subsequently, in 2013, Mirhosseini *et al* proposed a technique to dynamically encode both phase and amplitude onto the laser beam using a DMD, enabling the generation of OAM modes with rapid switching capabilities [67]. The fidelity and efficiency of Lee and superpixel methods with topological charges ranging from 1 to 100 have also been investigated [68].

Compared to conventional techniques, generating OAM modes using the DMD allows for high-speed mode switching with fewer optical elements. The superpixel method, which combines a spatial filter and a DMD, enables independent control of the amplitude and phase of light with high resolution [69]. By employing a clock signal for switching and loading computer-generated holograms onto the DMD memory, the quantum number of generated OAM modes can be switched at speeds up to 4 kHz. In fact, commercially available DMDs can achieve switching speeds as high as 32 kHz. Furthermore, the cost of DMDs is generally lower compared to phase-only SLMs. However, it is important to note that the efficiency of OAM mode generation is measured to be only 1.5%, and the maximum theoretical efficiency of binary amplitude gratings is approximately 10%. In contrast, phase-only SLMs typically exhibit efficiency levels above 50% and can even reach 90%. Compared to the Lee holography method, the fidelity and efficiency of superpixel method can be improved.

2.2.5 Photon sieves

This section was reproduced with permission from [43]. © 2022 Chinese Laser Press.

In addition to the methods introduced above, binary amplitude masks such as the photon sieve are commonly employed for shaping the phase and amplitude of light. The photon sieve consists of a large number of transmissive pinholes with specifically designed diameters and distribution. It was first proposed by Kipp *et al* in 2001 as a method to overcome limitations in spatial resolution, reduce unwanted diffraction orders, and eliminate scattered intensity associated with conventional zone plates [70]. An advantageous feature of the photon sieve is its simplicity in manufacturing, attributable to its self-supporting structure.

Originally developed as an optical diffractive element for improving spatial resolution and focusing ability in soft x-rays, the photon sieve adopts a design with annularly distributed pinholes based on the Fresnel zones. Unlike conventional zone plates that employ concentric rings as the basic diffracting elements, the photon sieve utilizes pinholes with varying diameters. This nonabsorbent characteristic enables its use across a broad spectrum, ranging from far infrared to visible light, extreme ultraviolet, and x-rays. The light field passing through the photon sieve is controlled by parameters such as the position, number, and size of the pinholes. The focal length of the photon sieve is inversely proportional to the wavelength, resulting in potential chromatic aberration. Additionally, the diameter of the outermost ring decreases with a larger radius, allowing for higher resolution capability.

Later, Qiu *et al* proposed an ultrahigh-capacity photon sieve consisting of subwavelength holes arranged in two different structural orders: aperiodicity and randomness [71]. The aperiodic photon sieve, composed of 7240 holes with varying sizes, was utilized to achieve sub-diffraction-limit focal spots (with a full width at half maximum of ∼0.32λ). On the other hand, the random photon sieve acted as a uniform optical hologram with high diffraction efficiency. The combination of these two types of photon sieves enabled the manipulation of polarization, amplitude, and phase of visible light, aiming for superfocusing and high-quality holograms.

Furthermore, Liu *et al* demonstrated generalized photon sieves based on the original photon sieve introduced by Kipp *et al* in 2001, to generate arbitrarily structured complex fields using simple pinhole arrays [72]. Compared to the original photon sieve, the designed generalized photon sieve consisted of only a few hundred, rather than thousands, of pinholes to create structured beams at the focal point.

Similarly, electron sieves can also be used to produce vortex beams. In 2017, Yang *et al* proposed electron sieves with rotationally symmetric structures to flexibly and systematically generate electron vortex beams [73]. Unlike spiral computer-generated holograms, where different OAM modes are focused at different on-axis positions while others exist only in the background, these electron sieves can generate compound vortex beams with three bright rings ($l = -11$, 44, and 55). Later in 2019, the same research group demonstrated the production of optical vortex beams with unique OAM modes using a simple pinhole plate [74]. Additionally, pinhole plates were used to generate superpositions of OAM modes with opposite topological charges, known as photonic gears.

2.2.6 Metasurfaces

This section was reproduced with permission from [43]. © 2022 Chinese Laser Press.

In conventional methods, generating optical vortex beams for OAM-based applications at miniature scales is challenging due to the bulkiness of elements required to accumulate the desired phase change through propagation. Metasurfaces are ultrathin devices constructed by artificial arrays of meta-atoms, which can control the phase, amplitude, and polarization state of incident light. By imparting a space-variant abrupt phase change to the incident light through an ultrathin metasurface, propagation-induced phase accumulation is unnecessary. This revolutionary concept of light manipulation has paved the way for the development of ultrathin optical devices.

Primary approaches to design metasurfaces are the Pancharatnam–Berry/geometry phase, dynamic phase and detour phase [75–78]. Metasurfaces utilizing the geometry phase consist of anisotropic subwavelength nanoantennas with spatially varying orientations, enabling precise tailoring of the light field on a pixel-by-pixel basis. For instance, metasurfaces composed of rectangular-hole nanoantennas can generate focused perfect vortex beams with annular ring intensity that is independent of the topological charges. These metasurfaces combine the phase profiles of a spiral phase plate, Fourier transform lens, and axicon based on the geometry phase [79].

One advantage of metasurfaces based on the geometry phase is their independence from inherent dispersion. When the incident light is right-hand circularly polarized with a normalized Jones vector of $1/\sqrt{2}\,[1,\,i]$, the transmitted light passing through the metasurface can be decomposed into two components. These include a polarization-residual term, which maintains the original polarization state, and a polarization-converted term, which has an orthogonal polarization state. The electric field of the polarization-converted term can be expressed as [80]

$$
\begin{aligned}
\overrightarrow{E}_{out}^{RCP} &= \overrightarrow{E}_{con}^{RCP} + \overrightarrow{E}_{res}^{RCP} \\
&= \frac{1}{\sqrt{2}}\left(\frac{A}{A+B}e^{i\left(\frac{\pi}{2}-2\phi\right)}\begin{bmatrix} 1 \\ -i \end{bmatrix} + \frac{B}{A+B}\begin{bmatrix} 1 \\ i \end{bmatrix} \right),
\end{aligned}
\tag{2.21}
$$

where A and B represent amplitudes of the converted and residual parts, respectively, and $\overrightarrow{E}_{con}^{RCP}$ and $\overrightarrow{E}_{res}^{RCP}$ denote the electric field of the converted and residual terms, respectively. One can note from the expression that there is an abrupt phase change in the converted part due to the spin-orbit transformation in anisotropic and inhomogeneous media whereas there is no phase delay in the residual part. Moreover, the polarization-converted term can be imparted with an phase shift variation 2ϕ with a spatial orientation angle ϕ.

Desipte of geometry phase, another prominent approach for designing metasurfaces involves modifying the transmission phase through changes in the meta-atoms, which is known as the dynamic phase. The geometries of meta-atoms can be various shapes, such as cross slits [81, 82], nanofins [83], u-shaped slots [84], and v-shaped antennas [85, 86]. A typical metasurface utilizing the dynamic phase consists of nanoholes of different sizes distributed in eight fan-shaped sections, enabling phase tuning within the range of $(0,2\pi)$ [87]. However, owing to the strong absorption in metallic materials, metasurfaces etched on metallic structures suffer much loss in the visible spectrum, whereas it can be used in higher wavelength spectra such as infrared light.

Moreover, vortex beams can be generated by metasurfaces based on the combination of dynamic and geometry phases. Metasurfaces composed of rectangular-shaped nanoantennas with different sizes can covert right- and left-circular polarizations into states with independent values of OAM [88]. Consider that two orthogonal polarization states can be expressed in the linear polarization basis $|\lambda^+\rangle = \begin{bmatrix} \cos\chi \\ e^{i\sigma}\sin\chi \end{bmatrix}$, $|\lambda^-\rangle = \begin{bmatrix} -\sin\chi \\ e^{i\sigma}\cos\chi \end{bmatrix}$, where σ and χ are related to the polarization states. The Jones matrix for the J plate can be written as

$$
J(\phi) = e^{i\sigma}
\begin{bmatrix}
e^{i\sigma}\mu_1(\phi) & \dfrac{\sin 2\chi}{2}\nu(\phi) \\[2ex]
\dfrac{\sin 2\chi}{2}\nu(\phi) & e^{-i\sigma}\mu_2(\phi)
\end{bmatrix},
\tag{2.22}
$$

where ϕ is the azimuthal angle, n and m are integers that control the output OAM, $\mu_1(\phi) = e^{im\phi}\cos^2\chi + e^{in\phi}\sin^2\chi$, $\nu(\phi) = e^{im\phi} - e^{in\phi}$, and $\mu_2(\phi) = e^{im\phi}\sin^2\chi + e^{in\phi}\cos^2\chi$. Thus it can be found that $J(\phi)|\lambda^+\rangle = e^{im\phi}|(\lambda^+)^*\rangle$ and $J(\phi)|\lambda^-\rangle = e^{in\phi}|(\lambda^-)^*\rangle$, and light with spin state $|\lambda^+\rangle(|\lambda^-\rangle)$ illuminates on the J plate and a vortex beam with topology of $m(n)$ can be generated with flipped handedness.

In addition to the generation of a single vortex beam, metasurfaces composed of three parts can generate multiple optical vortex beams at different focal planes by combining the functionalities of a lens and a spiral phase plate [89]. However, each approach has its limitations in generating vortex beams. For metasurfaces based on the geometric phase, the purity of the generated vortex beam is sensitive to the dichroism of the material, especially for vortex beams with high-order topological charges. Nonetheless, metasurfaces have fundamentally revolutionized the generation and manipulation of optical vortex beams, holding great potential in various fields such as optical communications and integrated optics.

2.2.7 Other techniques

This section was reproduced with permission from [43]. © 2022 Chinese Laser Press.

In practical experiments, traditional devices are susceptible to damage when exposed to high-power laser illumination. As a result, alternative approaches such as foil-made light fans and s-plates have been introduced to generate optical vortex beams [90, 91]. Optical retarder elements with azimuthal rotation of principle axes, such as s-plates and q-plates, have gained significant attention. The s-plate is a super-structured half-wave plate polarization converter that enables the production of optical vortices and beams with azimuthal or radial polarization [92]. Liquid crystal q-plates, as a spatially nonuniform birefringent optical element, have been researched due to their ability to manipulate both the SAM and OAM of photons [93]. The q-plate consists of liquid crystal retarders, where the optical axes of each retarder are rotated with respect to the center of the device, and the retardance of each retarder can be controlled by the applied voltage. Define the direction at the x axis as ϕ_0, the rotation angle of the liquid crystal retarder can be expressed as $\alpha(r, \phi) = q\phi + \phi_0$, where q is the topological charge of the q-plate.

Similar to the half- and quarter-wave plate, the Jones matrix of a q-plate can be written as [94]

$$J_q = \begin{bmatrix} \cos 2\alpha & \sin 2\alpha \\ \sin 2\alpha & -\cos 2\alpha \end{bmatrix}, \tag{2.23}$$

with the incidence of a right-hand circularly polarized light $E_{in} = E_0 \times [1, i]$, the resultant light field can be expressed as

$$E_{out} = J_q E_{in} = E_0 \begin{bmatrix} \cos 2\alpha & \sin 2\alpha \\ \sin 2\alpha & -\cos 2\alpha \end{bmatrix} \begin{bmatrix} 1 \\ i \end{bmatrix},$$

$$= E_0 \exp(i2q\phi) \exp(i2\phi_0) \begin{bmatrix} 1 \\ -i \end{bmatrix} \tag{2.24}$$

Note that a vortex light field with an opposite circular polarization plus an azimuthal phase is transformed, the topological charge of which is $2q$. Analogously, in the case of a left-hand circularly polarized incident light, the sign of topological charge can be reversed, i.e., the output helical wavefront is determined by the incident polarization state. Moreover, the circular aperture is a simple optical element that can be integrated and used to generate an optical vortex. The effect of diffraction from circular aperture is of importance for many optical instrumentations. When a circularly polarized Gaussian beam is diffracted by a circular aperture, optical vortices are generated in the Fresnel region. The generation of an optical vortex can be attributed to spin-to-OAM conversion. Counterintuitively, the OAM of light diffracted by a circular aperture is derived from Belinfante's spin momentum rather than the time-averaged Poynting momentum.

Consider a circularly polarized Gaussian beam ($\sigma = \pm 1$) diffracted by a circular aperture with radius a located at plane $z = 0$ in the Cartesian coordinate system, where z axis is taken to be the propagation axis. According to the vectorial Rayleigh-Sommerfeld diffraction integral, the electric field of the diffraction light can be obtained after integration. There exists a spin-dependent helical phase term $\exp(i\sigma\phi)$ with topological charge $l = \sigma$ in the z component of the electric field, indicating that the diffraction light carries OAM induced by spin-to-orbital angular momentum conversion.

The intensity and phase distributions of the diffraction light are studied theoretically. Figures 2.12(a)–(d) show theoretical results of the Fresnel diffraction light with $\sigma = 1$, both in xz and xy planes. Figures 2.12(a) and (b) show the total intensity distributions, namely, $I = |E_x|^2 + |E_y|^2 + |E_z|^2$. Figures 2.12(b)–(d) show the intensity and phase distributions of the diffraction light in planes $z = z1$, $z2$ and $z3$. It can be seen clearly that the intensity distributions of the longitudinal component in xy planes at $z = z_1$, z_2 and z_3 are rings with a dark center (figure 2.12(c)), which is caused by phase singularity. Figure 2.12(d) shows that the topological charge $l = 1$. Figure 2.12(e) shows the experimental results of total intensity distributions corresponding to figure 2.12(b), from which we can see the dark center clearly as well. Here, the wavelength of incident light $\lambda_0 = 532$ nm, the waist width $\omega_0 = 0.7$ mm and the radius of the circular aperture a $= 1.8 \ \mu$m.

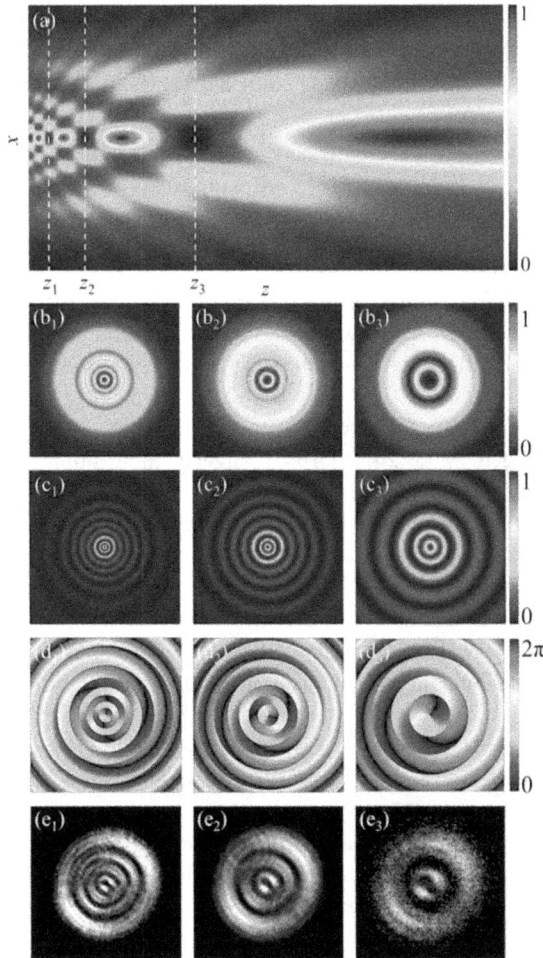

Figure 2.12. Theoretical and experimental results of the light field diffracted by a circular aperture. (a) Theoretical intensity distribution of the diffraction light in xz plane using equations (1) and (2). (b) The total intensity distributions in xy planes at $z = z_1$, (b1) z_2 (b2) and z_3 (b3). (c) The corresponding longitudinal components of (b). (d) The corresponding phase distributions to (c). (e) Experimental results of the total intensity distributions corresponding to (b). Reprinted with permission from [95]. Copyright (2023) American Chemical Society.

Furthermore, as well as optical vortex beams and electron vortex beams, plasmonic waves can also possess OAM, referred to as plasmonic vortices. Surface plasmon polaritons are collective oscillations of incident electromagnetic waves coupled with electrons on a metal surface, localized at the interface and exhibiting evanescence in the direction perpendicular to the interface. The energy of the plasmonic field is confined to a subwavelength scale, allowing it to surpass the conventional diffraction limit and provide significantly enhanced local field intensity. To generate plasmonic vortices with specific topological charges, elements such as plasmonic metasurfaces and plasmonic vortex lenses have been extensively investigated [96]. It has been demonstrated that plasmonic waves carrying well-defined OAM can be generated by illuminating incident light onto metal surfaces with various subwavelength structures, such as nanoslits [97], nanoholes [98–100], and gratings [100].

References

[1] Poynting J H 1909 The wave motion of a revolving shaft, and a suggestion as to the angular momentum of a beam of circularly polarized light *Proc. R. Soc. Lond.* A **82** 560–7

[2] Darwin C G 1932 Notes on the theory of radiation *Proc. R. Soc. Lond.* A **136** 36–52

[3] Spreeuw R J C, Allen L, Beijersbergen M W and Woerdman J P 1992 Orbital angular momentum of light and the transformation of Laguerre-Gaussian laser modes *Phys Rev.* A **45** 8185

[4] Curtis J E and Grier D G 2003 Structure of optical vortices *Phys. Rev. Lett.* **90** 133901

[5] Yao A and Padgett M J 2011 Orbital angular momentum: origins, behavior and applications *Adv. Opt. Photonics* **3** 161–204

[6] Miceli J J Jr, Durnin J and Eberly J 1987 Diffraction-free beams *Phys. Rev. Lett.* **58** 1499–501

[7] Iturbe-Castillo M, Gutierrez-Vega J and Chavez-Cerda S 2000 Alternative formulation for invariant optical fields: Mathieu beams *Opt. Lett.* **25** 1493–5

[8] Ren Y, Kristensen T P, Huang H, Willner A E, Bozinovic N, Yue Y and Ramachandran S 2013 Terabit-scale orbital angular momentum mode division multiplexing in fibers *Science* **340** 1545–8

[9] Heckenberg N R, He H, Friese M E and Rubinsztein-Dunlop H 1995 Direct observation of transfer of angular momentum to absorptive particles from a laser beam with a phase singularity *Phys. Rev. Lett.* **75** 826

[10] Romero J, Jha A K, Yao A, Franke-Arnold S, Ireland D G, Boyd R W, Barnett S M, Leach J, Jack B and Padgett M J 2010 Quantum correlations in optical angle-orbital angular momentum variables *Science* **329** 662–5

[11] Yang Y, Ren Y-X, Chen M, Arita Y and Rosales-Guzmán C 2021 Optical trapping with structured light: a review *Adv. Photon.* **3** 034001

[12] Curtis J E and Grier D G 2003 Structure of optical vortices *Phys. Rev. Lett.* **90** 133901

[13] Shen Y, Wang X, Xie Z, Min C, Fu X, Liu Q, Gong M and Yuan X 2019 Optical vortices 30 years on: OAM manipulation from topological charge to multiple singularities *Light Sci. Appl.* **8** 90

[14] Ostrovsky A S, Rickenstorff-Parrao C and Arrizon V M 2013 Generation of the 'perfect' optical vortex using a liquid-crystal spatial light modulator *Opt. Lett.* **38** 534–6

[15] Chen M, Mazilu M, Arita Y, Wright E M and Dholakia K 2013 Dynamics of micro-particles trapped in a perfect vortex beam *2014 Conf. on Lasers and Electro-Optics (CLEO) —Laser Science to Photonic Applications* pp 1–2

[16] García-García J, Rickenstorff-Parrao C, Ramos-García R, Arrizon V M and Ostrovsky A S 2014 Simple technique for generating the perfect optical vortex *Opt. Lett.* **39** 5305–8

[17] Vaity P and Rusch L A 2015 Perfect vortex beam: Fourier transformation of a bessel beam *Opt. Lett.* **40** 597–600

[18] Chen Y, Fang Z, Ren Y, Gong L and Lu R 2015 Generation and characterization of a perfect vortex beam with a large topological charge through a digital micromirror device *Appl. Opt.* **54** 8030–5

[19] Ren Y, Li M, Huang K, Wu J, Gao H, Wang Z and Li Y 2010 Experimental generation of Laguerre-Gaussian beam using digital micromirror device *Appl. Opt.* **49** 1838–44

[20] Roichman Y, Sun B, Roichman Y, Amato-Grill J and Grier D G 2008 Optical forces arising from phase gradients *Phys. Rev. Lett.* **100** 013602

[21] Yang Y, Dong Y, Zhao C and Cai Y 2013 Generation and propagation of an anomalous vortex beam *Opt. Lett.* **38** 5418–21

[22] Berry M V 2004 Optical vortices evolving from helicoidal integer and fractional phase steps *J. Opt.* **6** 259–68

[23] Martinez-Castellanos I and Gutiérrez-Vega J C 2015 Shaping optical beams with non-integer orbital-angular momentum: a generalized differential operator approach *Opt. Lett.* **40** 1764–7

[24] Tao S, Yuan X, Lin J, Peng X and Niu H 2005 Fractional optical vortex beam induced rotation of particles *Opt. Exp.* **13** 7726–31

[25] Chen L, Lei J and Romero J 2014 Quantum digital spiral imaging *Light Sci. Appl.* **3** e153

[26] Zhang W, Wang L, Wang W and Zhao S 2019 Propagation property of Laguerre–Gaussian beams carrying fractional orbital angular momentum in an underwater channel *OSA Continuum* **2** 3281–7

[27] Wen J, Wang L, Yang X, Zhang J and Zhu S 2018 Vortex strength and beam propagation factor of fractional vortex beams *Opt. Exp.* **27** 5893–904

[28] Yang Z, Zhang X, Bai C and Wang M 2018 Nondiffracting light beams carrying fractional orbital angular momentum *J. Opt. Soc. Am.* A **35** 452–61

[29] Dorrah A H, Zamboni-Rached M and Mojahedi M 2016 Controlling the topological charge of twisted light beams with propagation *Phys. Rev.* A **93** 063864

[30] Leach J, Yao E and Padgett M J 2004 Observation of the vortex structure of a non-integer vortex beam *New J. Phys.* **6** 71

[31] Basistiy I V, Pas 'ko V, Slyusar V V, Soskin M S and Vasnetsov M V 2004 Synthesis and analysis of optical vortices with fractional topological charges *J. Opt. A: Pure Appl. Opt.* **6** S166–9

[32] Götte J B, O'Holleran K, Preece D C, Flossmann F, Franke-Arnold , Barnett S M and Padgett M J 2008 Light beams with fractional orbital angular momentum and their vortex structure *Opt. Exp.* **16** 993–1006

[33] Turpin A, Rego L, Picón A, San Román J and Hernández-García C 2017 Extreme ultraviolet fractional orbital angular momentum beams from high harmonic generation *Sci. Rep.* **7** 43888

[34] Hernández-García C, Picón A, San Román J and Plaja L 2013 Attosecond extreme ultraviolet vortices from high-order harmonic generation *Phys. Rev. Lett.* **111** 083602

[35] Yang Y, Zhu X, Zeng J, Lu X, Zhao C and Cai Y 2018 Anomalous bessel vortex beam: modulating orbital angular momentum with propagation *Nanophotonics* **7** 677–82

[36] Wang H, Liu L, Zhou C, Xu J, Zhang M, Teng S and Cai Y 2019 Vortex beam generation with variable topological charge based on a spiral slit *Nanophotonics* **8** 317–24

[37] Rego L, Dorney K M, Brooks N J *et al* 2019 Generation of extreme-ultraviolet beams with time-varying orbital angular momentum *Science* **364** 6447

[38] Zeng R, Zhao Q, Shen Y, Liu Y and Yang Y 2021 Structural stability of open vortex beams *Appl. Phys. Lett.* **119** 171105

[39] van der Veen H, Beijersbergen M W, Allen L and Woerdman J P 1993 Astigmatic laser mode converters and transfer of orbital angular-momentum *Opt. Commun.* **112** 321–7

[40] Beijersbergen M W, Allen L, van der Veen H E L O and Woerdman J P 1993 Astigmatic laser mode converters and transfer of orbital angular momentum *Opt. Commun.* **96** 123–32

[41] Kristensen M, Beijersbergen J M W, Coerwinkel R P C and Woerdman P 1994 Helical-wavefront laser beams produced with a spiral phaseplate *Opt. Commun.* **112** 321–7

[42] Eliel E R, Woerdman J P, Verstegen E J K, Oemrawsingh S S R, Van Houwelingen J A W and Kloosterboer J G 2004 Production and characterization of spiral phase plates for optical wavelengths *Appl. Opt.* **43** 688–94

[43] Bai Y, Lv H, Fu X and Yang Y 2022 Vortex beam: generation and detection of orbital angular momentum *Chin. Opt. Lett.* **20** 012601

[44] Beijersbergen M W, Coerwinkel R P C, Kristensen M and Woerdman J P 1994 Helical-wavefront laser beams produced with a spiral phaseplate *Opt. Commun.* **112** 321–7

[45] Oemrawsingh S S R, van Houwelingen J A W, Eliel E R, Woerdman J P, Verstegen E J K, Kloosterboer J G and 't Hooft G W 2004 Production and characterization of spiral phase plates for optical wavelengths *Appl. Opt.* **43** 688–94

[46] Peele A G, McMahon P J, Paterson D, Tran C Q, Mancuso A P, Nugent K A, Hayes J P, Harvey E, Lai B and McNulty I 2002 Observation of an x-ray vortex *Opt. Lett.* **27** 1752–4

[47] Rotschild C, Zommer S, Moed S, Hershcovitz O and Lipson S G 2004 Adjustable spiral phase plate *Appl. Opt.* **43** 2397–9

[48] Sheng Z, Wu W and Wu H 2019 Design and application of flat spiral phase plate *Acta Phys. Sin-Ch. Ed.* **68** 054102

[49] Rai M R, Rosen J, Minin O V, Minin I V, Forbes A, Vijayakumar A and Rosales-Guzmán C 2019 Generation of structured light by multilevel orbital angular momentum holograms *Opt. Exp.* **27** 6459–70

[50] Jiang Z and Werner D H 2021 *Electromagnetic Vortices: Wave Phenomena and Engineering Applications* (New York: Wiley)

[51] Vasnetsov M, Bazhenov V and Soskin M 1990 Laser beams with screw dislocations in their wavefronts *Jetp. Lett.* **52** 429–31

[52] Soskin M S, Bazhenov V and Vasnetsov M V 1992 Screw dislocations in light wavefronts *J. Mod. Optic.* **39** 985–90

[53] Stefanov I, Janicijevic L, Stoyanov L, Topuzoski S and Dreischuh A 2015 Far field diffraction of an optical vortex beam by a fork-shaped grating *Opt. Commun.* **350** 301–8

[54] Smith C P, Heckenberg N R, McDuff R and White A G 1992 Generation of optical phase singularities by computer-generated holograms *Opt. Lett.* **17** 221–3

[55] Sakdinawat A and Liu Y 2007 Soft-X-ray microscopy using spiral zone plates *Opt. Lett.* **32** 2635–7

[56] Tanaka N, Saitoh K, Hasegawa Y and Uchida M 2012 Production of electron vortex beams carrying large orbital angular momentum using spiral zone plates *J. Electron. Microsc.* **61** 171–7

[57] Rosales-Guzmán C and Forbes A 2017 How to shape light with spatial light modulators (Bellingham, WA: SPIE) https://doi.org/10.1117/3.2281295

[58] Curtis J E, Koss B A and Grier D G 2002 Dynamic holographic optical tweezers *Opt. Commun.* **207** 169–75

[59] Savage N 2009 Digital spatial light modulators *Nat. Photon.* **3** 170–2

[60] Rosales-Guzmán C, Bhebhe N and Forbes A 2017 Simultaneous generation of multiple vector beams on a single SLM *Opt. Exp.* **25** 25697–706

[61] Rosales-Guzmán C and Forbes A 2017 How to shape light with spatial light modulators

[62] Nesbitt R S, Smith S L, Benton S A and Molnar R A 1999 Holographic recording using a digital micromirror device *Proc. SPIE* **3637** 1–9

[63] Padgett M J, Cizmar T, Mitchell K J, Turtaev S and Phillips D B 2016 High-speed spatial control of the intensity, phase and polarization of vector beams using a digital micro-mirror device *Opt. Exp.* **24** 29269–82

[64] Lee W H 1974 Binary synthetic holograms *Appl. Opt.* **13** 1677–82

[65] Drori D S, Lerner V and Katz N 2012 Shaping Laguerre-Gaussian laser modes with binary gratings using a digital micromirror device *Opt. Lett.* **37** 4826–8

[66] Wang Z, Wu J, Ren Y, Huang M and Li Y 2010 Experimental generation of Laguerre-Gaussian beam using digital micromirror device *Appl. Opt.* **49** 1838–44

[67] Chen C, Rodenburg B, Malik M, Mirhosseini M, Loaiza O S M and Boyd R W 2013 Rapid generation of light beams carrying orbital angular momentum *Opt. Exp.* **21** 030204

[68] Ren Y, Gong L, Chen Y, Fang Z and Lu R 2015 Generation and characterization of a perfect vortex beam with a large topological charge through a digital micromirror device *Appl. Opt.* **54** 8030–5

[69] Bertolotti J, Goorden S A and Mosk A P 2014 Superpixel-based spatial amplitude and phase modulation using a digital micromirror device *Opt. Express* **22** 17999–8009

[70] Johnson R L, Berndt R, Adelung R, Harm S, Kipp L, M S and Seemann R 2001 Sharper images by focusing soft X-rays with photon sieves *Nature* **414** 184–8

[71] Francisco J G V, Hong M, Luk'yanchuk B, Teng J, Qiu C, Huang K and Liu H 2015 *Nat. Commun.* **6** 7059

[72] Padgett M J, Liu R, Li F and Phillips D B 2015 Generalized photon sieves: fine control of complex fields with simple pinhole arrays *Optica* **2** 1028

[73] Babiker M, Yang Y, Thirunavukkarasu G and Yuan J 2017 Orbital-angular-momentum mode selection by rotationally symmetric superposition of chiral states with application to electron vortex beams *Phys. Rev. Lett.* **119** 094802

[74] Liu L, Liu Y, Rosales-Guzmán C, Yang Y, Zhao Q and Qiu C 2019 Manipulation of orbital-angular-momentum spectrum using pinhole plates *Phys. Rev. A.* **12** 064007

[75] Kats M A, Antoniou N, Lin J, Genevet P and Capasso F 2013 Nanostructured holograms for broadband manipulation of vector beams *Nano. Lett.* **13** 4269–74

[76] Lei T, Si G, Xie Z, Lin J, Du L, Min C, Liu J and Yuan X 2016 Plasmonic nano-slits assisted polarization selective detour phase meta-hologram *Laser Photon. Rev.* **10** 978–85

[77] Si G, Wang X, Lin J, Min C, Xie Z, Lei T and Yuan X 2017 Meta-holograms with full parameter control of wavefront over a 1000nm Bandwidth *ACS Photon.* **4** 2158–64

[78] Zhuang X *et al* 2018 Facile metagrating holograms with broadband and extreme angle tolerance *Light-Sci. Appl.* **7** 78

[79] Gao J, Zhang Y, Liu W and Yang X 2018 Generating focused 3D perfect vortex beams by plasmonic metasurfaces *Adv. Opt. Mater.* **6** 1701228

[80] Wen D, Yue F and Xin J 2016 Vector vortex beam generation with a single plasmonic metasurface *ACS Photon.* **3** 1558–63

[81] Katayama I, Ng S H, Vongsvivut J, Tobin M J, Kuchmizhak A, Nishijima Y, Bhattacharya S, Juodkazis S, Dharmavarapu R and Izumi K 2019 Dielectric cross-shaped resonator based metasurface for vortex beam generation in mid-IR and THz wavelengths *Nanophotonics* **8** 1263–70

[82] Lei T, Si G, Xie Z, Lin J, Du L, Min C, Liu J and Yuan X 2016 Plasmonic nano-slits assisted polarization selective detour phase meta-hologram *Laser Photon. Rev.* **10** 978–85

[83] Devlin R C, Oh J, Zhu A, Capasso F, Khorasaninejad M and Chen W 2016 Metalenses at visible wavelengths: diffraction-limited focusing and subwavelength resolution imaging *Science* **352** 1190–4

[84] Jiang L, Chen M L N and Sha W E I 2017 Ultrathin complementary metasurface for orbital angular momentum generation at microwave frequencies *IEEE. Trans. Antenn. Propag.* **65** 396–400

[85] Pham A, Zhao A and Drezet A 2021 Plasmonic fork-shaped hologram for vortex-beam generation and separation *Opt. Lett.* **46** 689–92

[86] Boltasseva A, Kildishev A V and Shalaev V M 2013 Planar photonics with metasurfaces *Science* **339** 1232009

[87] Xu T, Kudyshev Z A, Cartwright A N, Sun J, Wang X and Litchinitser N M 2014 Spinning light on the nanoscale *Nano. Lett.* **14** 2726–9

[88] Rubin N A, Mueller J P B, Capasso F, Devlin R C and Ambrosio A 2017 Arbitrary spin-to-orbital angular momentum conversion of light *Science* **358** 896–901

[89] Hussain S *et al* 2016 Visible-frequency metasurface for structuring and spatially multi-plexing optical vortices *Adv. Mater.* **28** 2533–9

[90] Zhang L, Zhang X, Wang W, Shi Y, Shen B and Xu Z 2014 Light fan driven by a relativistic laser pulse *Phys. Rev. Lett.* **112** 235001

[91] Gecevicius M, Beresna M and Kazansky P G 2011 Polarization sensitive elements fabricated by femtosecond laser nanostructuring of glass *Opt. Mater. Exp.* **1** 783–95

[92] Beresna M, Gecevicius M, Drevinskas R and Kazansky P G 2014 Single beam optical vortex tweezers with tunable orbital angular momentum *Appl. Phys. Lett.* **104** 231110

[93] Hurtado E, Pierce M, Sanchez-Lopez M M, Badham K, Davis J A, Kurihara N H M and Moreno I 2015 Analysis of a segmented q-plate tunable retarder for the generation of first-order vector beams *Appl. Opt.* **54** 9583–90

[94] Manzo C, Marrucci L and Paparo D 2006 Optical spin-to-orbital angular momentum conversion in inhomogeneous anisotropic media *Phys. Rev. Lett.* **96** 163905

[95] Yao J, Qin Y, Bai Y, Gao M, Zhou L, Li J, Jiang Y and Yang Y 2023 Generation of optical vortices by diffraction from circular apertures *ACS Photon.* **10** 4267–72

[96] Liu Y, Xie D, Jin Z, Li J, Hu G, Yang Y, Wu L and Qiu C 2020 Deuterogenic plasmonic vortices *Nano. Lett.* **20** 6774–9

[97] Li X, Wu L and Yang Y 2019 Generation of surface plasmon vortices based on double-layer Archimedes spirals *Acta Phys. Sin-Ch. Ed.* **68** 234201

[98] Devaux E, Linke R A, Martin-Moreno L, Garcia-Vidal F J, Lezec H J, Degiron A and Ebbesen T W 2002 Beaming light from a subwavelength aperture *Science* **297** 820

[99] Yanai A, Lerman G M and Levy U 2009 Demonstration of nanofocusing by the use of plasmonic lens illuminated with radially polarized light *Nano Lett.* **9** 2139

[100] Escalante M, Segerink F B, Korterik J P, Offerhaus H L, van den Bergen B and van Hulst N F 2005 Creating focused plasmons by noncollinear phasematching on functional gratings *Nano Lett.* **5** 2144–8

Chapter 3

Measuring OAM of vortex beams

The measurement of the topological charge of vortex beams holds significant importance, given their wide-ranging applications in fields such as optical communication, optical trapping [1, 2], and quantum information [3, 4]. This chapter presents various approaches for detecting topological charge, such as the interference method, diffraction method, geometric coordinate transformation and other new measurement methods.

3.1 Interference or diffraction method

The interference can give rise to novel waveforms, which can be used to measure the topological charge of a vortex beam. For example, the interference between a tilted plane wave and a vortex beam can generate a distinctive fork-shaped interference pattern [5]. The interference pattern, which resembles a fork-shaped grating as discussed in previous chapters, displays l-dislocation fringes. The sign and number of the topological charge dictate the orientation and shape of the l-dislocation fringe, rendering vortex beams effectively detectable. When the coaxial plane wave interferes with the vortex beam, the resulting interference pattern will exhibit a petal-shaped structure, characterized by $|l|$ petals. When the interfering beam transitions from a plane wave to a spherical wave, the corresponding interference pattern transforms into a spiral stripe configuration, where the number of stripes corresponds to the value of l [6, 7]. Additionally, a vortex beam can also interfere with its own mirror image [8, 9], resulting in an interference pattern consisting of $2l$ petals in the shape of a flower, where l is the topological charge of the vortex beam.

In addition to interfering with other beams, vortex beams can also be detected for their topological charge through double-slit interference [10, 11]. As shown in figure 3.1, the incident of a vortex beam with a topological charge $l = 1$ on the double slit (represented by white solid lines) induces a phase difference associated with the azimuthal angle at the corresponding y position. This leads to twisted interference fringes along the x-direction, where the displacement indicates the

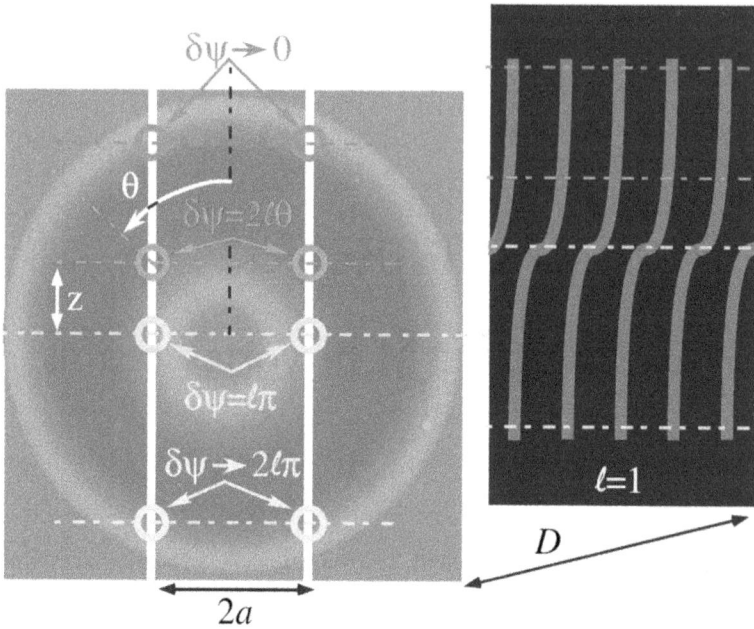

Figure 3.1. The phase distribution of a vortex beam with $l = 1$, the spiral phase passing through the double slits, and interference intensity distribution [11]. Reprinted by permission from Springer Nature Customer Service Centre GmbH: Springer [Applied Physics B], Copyright (2014).

topological charge of the incident light and the twist direction signifies its sign. The measurement of topological charge using double-slit interference is straightforward and convenient, as it only requires a single optical component without the need for an additional beam of light [11]. This method has the potential to be applied to situations involving partially coherent beams [12], single photons [13], and other similar scenarios.

Methods for detecting vortex beams based on interference also include dynamic angle double slits [14–16] and multipoint interferometers [17, 18]. When a vortex beam is incident vertically on an opaque screen with two non-parallel slits, the phase difference between the two slits determines the far-field interference pattern. The variation in the angle between the two slits results in an alternating transition of the central output's intensity between darkness and brightness. Moreover, this approach enables the measurement of the vortex beam's significantly large topological charge due to the continuous scanning. When a vortex beam with different topological charges is incident on a multipoint interferometer composed of uniformly distributed circular holes [17], it produces distinct interference patterns. However, this method can only be applied when the topological charge is low. As the topological charge increases, the interference patterns will periodically repeat themselves, as illustrated in figure 3.2(a). For a vortex beam with topological charges l_1 and l_2, if $l_1 \pm l_2 = Nm$ is obtained, the far-field complex amplitude distribution of the beam through the nth pinhole can be expressed as

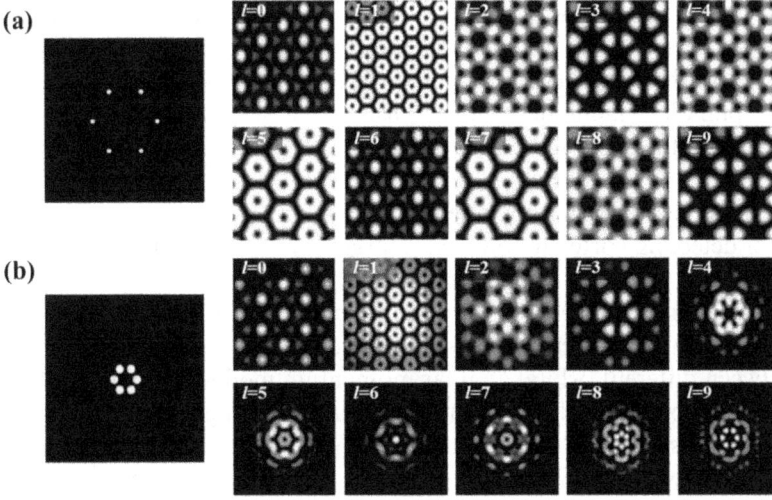

Figure 3.2. Far-field intensity patterns for vortex beams with different topological charges diffracted by (a) multipoint interferometer and (b) improved multipoint interferometer with $N = 6$. Adapted with permissions from [18]. © 2020 Chinese Laser Press.

$$
\begin{aligned}
E_n^{l_1} &= \exp(-il_1 a_n)\exp\left[-i\frac{ka}{f}(x\cos a_n + y\sin a_n)\right] \\
&= \exp[-ia_n(l_2 + Nm)]\exp\left[-i\frac{ka}{f}(x\cos a_n + y\sin a_n)\right] \\
&= \exp(-il_2 a_n)\exp(-i2mn\pi)\exp\left[-i\frac{ka}{f}(x\cos a_n + y\sin a_n)\right] \\
&= E_n^{l_2}
\end{aligned}
\tag{3.1}
$$

$$
\begin{aligned}
E_n^{l_1} &= \exp(-il_1 a_n)\exp\left[-i\frac{ka}{f}(x\cos a_n + y\sin a_n)\right] \\
&= \exp[-ia_n(Nm - l_2)]\exp\left[-i\frac{ka}{f}(x\cos a_n + y\sin a_n)\right] \\
&= \exp(-il_2 a_n)\exp(-i2mn\pi)\exp\left[-i\frac{ka}{f}(x\cos a_n + y\sin a_n)\right] \\
&= E_n^{l_2 *}
\end{aligned}
\tag{3.2}
$$

Hence, with a high topological charge in the incident vortex beam, the interference pattern will exhibit periodic repetition. As a result of these constraints, an improved multipoint interferometer [18] was proposed, incorporating larger pinholes. In this case, the aperture radius (r_0) is not considered negligible, as depicted in figure 3.2(b). Such a small change allows each pinhole to capture more phase information of the incident beams. This is due to the larger size of the aperture and the fact that the apertures are closer to the center of the beam axis, where the phase changes more rapidly than in the outer region. The improved multipoint interferometer, enabled by larger pinholes, facilitates the acquisition of additional information about incident light, allowing for the differentiation of vortex beams with higher OAM [18].

A Mach–Zehnder interferometer equipped with a Dove prism is a highly effective method for detecting vortex topological charges [19, 20]. The phase rotation ψ of the incident beam is twice the angle of rotation of the Dove prism along its long axis. When the topological charge of the incident beam is l, the phase difference between the two arms of the interferometer is $l\psi$. The proposed method effectively discriminates between different orbital angular momentum modes of single photons. When the parity of the photon is different, it can be output through different ports [20]. The capability of this method can be further enhanced by cascading Mach–Zehnder interferometers with varying rotation angles, enabling the measurement of an arbitrary number of OAM states. Apart from these methods, alternative approaches such as the Sagnac interferometer [21, 22] have been proposed to enhance topological charge measurement stability.

The identification of topological charge through diffraction is commonly achieved using techniques like aperture diffraction [23–26], grating diffraction [27–29], cylindrical lenses [30, 31], and other similar approaches. The far-field diffraction pattern of the vortex light will vary with the topological charge as it passes through either a triangular or square aperture. The aperture method was initially proposed by Hickmann et al in 2010 for the detection of the topological charge of vortex beams [23]. The far-field diffraction pattern displays a truncated optical lattice relative to the aperture, based on the topological charge of the incident beam. The far-field lattice array, arranged in a triangular configuration with N representing the points on each side, yields a topological charge of $l = N-1$ for the incident vortex beam. The direction of the diffraction pattern is determined by the positive or negative topological charge l. A annular triangular aperture will also result in a triangular lattice arrangement of the far-field diffraction pattern of the vortex beam. Additionally, enhancing the clarity of the diffraction intensity distribution is observed by increasing the inner/outer apertures ratio [25]. The lattice array formed by the square aperture is square, and it is possible to detect the magnitude of the topological charge in a similar manner. However, determining the sign of the topological charge is unattainable [26]. The topological charge value has a distinct correlation with the outer points of the pattern. Specifically, an even topological charge value is represented by $l = 2N-2$, while an odd topological charge value is denoted as $l = 2N-1$.

A cylindrical lens-based mode converter can realize the mode conversion between HG beams and LG beams [30, 31]. That is, the cylindrical lens can be used to detect vortex beams. After passing through a cylindrical lens, the vortex beam can become a light field distribution with multiple separated dark spots, the number of which is exactly equal to the topological charge l of the incident beam. Compared with other methods of astigmatic transforms, the transform with a cylindrical lens proves to be the most effective approach in determining topological charge, with the experimental detection of topological charge up to 100.

The l-defective fork-shaped computer-generated hologram can be used to produce vortex beams, and when the incident vortex beam carries topological charge $-l$, it is converted into the fundamental mode Gaussian beam after passing through the fork grating, which can also be used as a means of detecting topological charge. However, the l-defective fork grating can only detect vortex beams with topological charge $-l$. For vortex beams with different topological charges, multiple holograms must be used for detection. The problem can be solved by superimposing multiple holograms based on a single structure. This more complex hologram based on the Daman grating allows for the measurement of multiple topological charges through a single structure, but the multiple output modes affect efficiency [29]. The gradually changing period grating can detect topological charge in the case where the incident vortex light is not aligned [28]. The number of dark stripes in the first-order diffraction pattern corresponding to size of the topological charge, the direction of which is opposite to the sign. However, the diffraction efficiency does not exceed 10%. The annular grating can measure vortex beams with fractional topological charges, and compared to other gratings, it can achieve high efficiency and does not require alignment for detection [32].

3.2 Geometric coordinate transformation method

Geometric coordinate transformation can effectively separate OAM modes and detect vortex beams [33, 34]. This method involves converting OAM into linear momentum by utilizing diffractive optical elements to transform the helical phase associated with the topological charge into a transverse phase gradient. The input vortex beams with different topological charges are then focused onto distinct lateral positions using lenses. To achieve this, three SLMs are employed: one for generating vortex beams, another for converting the azimuth position in Cartesian coordinates to the transverse position in log-polar coordinates, and the third for correcting any phase distortion caused by the geometric coordinate transformation. Subsequently, the lenses are employed to precisely focus input vortex beams with varying topological charges onto distinct lateral positions. Furthermore, this approach enables identification of multiple superimposed vortex beams.

The performance of a coordinate transformation along the azimuthal angle of the vortex beam, however, leads to a finite length of the field distribution. This makes the method insufficient in terms of resolution for the topological charge of the vortex beam, as the focused spot of vortex beams with adjacent topological charges interfere with each other. Converting the vortex beam into a slanted plane wave mode by cutting off its light field along a spiral path can effectively solve this problem [35], as shown in figure 3.3. Theoretically, the phase excursion along a spiral in the wavefront of an optical vortex is unlimited. As a result, this novel optical transformation exhibits higher resolution in detecting topological charges.

The log-polar coordinate transformation method necessitates precise alignment of two phase elements for mode separation. Furthermore, there have been further

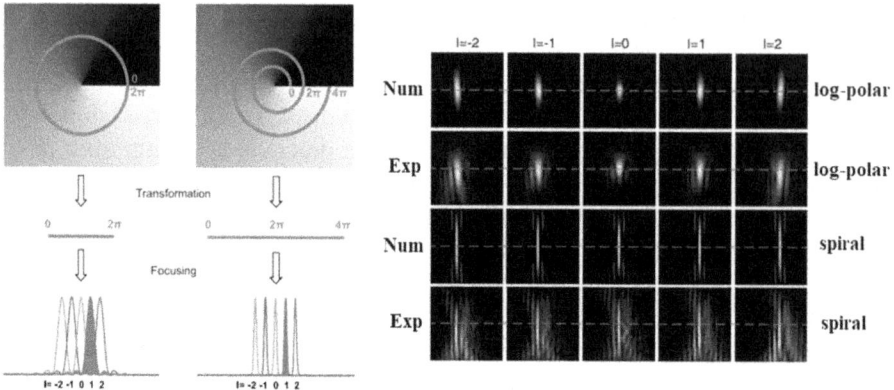

Figure 3.3. Schematic diagram and results of log-polar transformation and spiral transformation for OAM modes sorting. Reprinted figure with permission from [35]. Copyright (2018) by the American Physical Society.

investigations into the miniaturization of topological charge detection systems [36–38], such as metasurfaces [39, 40]. Different vortices can be transformed into focused modes with distinct azimuthal coordinates in the transverse plane through photon momentum transformation based on a single azimuthal-quadratic phase metasurface [41]. This design enables OAM measurement in a large modal space, facilitating SAM and OAM sorting, distinguishing vector vortex beams, and detecting stacked vortex beams with a specific interval step.

3.3 Deep learning method

Traditional detection methods often involve large optical elements and complex optical paths. In order to address these challenges, novel measurement methods have been proposed, including the use of deep learning and surface plasmon polaritons. These innovative approaches offer advantages including low device processing complexity, absence of redundant optical devices, and high speed.

In the recent past, image processing has witnessed widespread utilization of deep learning due to its strong aptitude in analyzing data and processing information [42]. Deep learning allows a computational model composed of multiple processing layers to learn data representations with multiple levels of abstraction. A standard neural network consists of convolutional layers, pooling layers, and fully connected layers. The role of the convolutional layer is to identify local combinations of features from the previous layer, while the pooling layer combines semantically similar features into a cohesive representation. Lastly, the fully connected layer is employed for classification objectives.

The existence of atmospheric turbulence will cause the wavefront distortion of the transmitted beam in free space. The existence of this problem seriously affects OAM-based communication. Recent advancements in deep learning technology have introduced a promising approach for detecting vortex beams using convolutional neural networks (CNNs) [43–46]. A high-precision multi-layer representation learning method can identify the topological charge of multiple vortex beams in different atmospheric turbulence [44]. Modulated vortex beams carrying different topological charges are transmitted through a channel affected by atmospheric turbulence, and the intensity images generated at the receiving end are collected and resized. Subsequently, feature extraction is performed using convolutional layers followed by dimension compression through pooling layers. Finally, output generation is achieved using a fully connected layer. Subsequently, a dedicated network consisting of six layers of neural networks was proposed, which could effectively distinguish vortex beams with different topological charges under different turbulence intensities and transmission distances [45]. Additionally, there is an approach that combines the CNN with Gerchberg–Saxton (GS). This combined method achieves higher accuracy and enhances recognition performance [43]. The GS algorithm is initially employed to retrieve the intensity distribution of the vortex beam, followed by the utilization of CNN for the classification of multiple LG beams. Moreover, a hybrid interference-CNN technique has been introduced to improve the recognition performance of multi-vortex beams under varying turbulence levels [46]. Compared to traditional CNN-based methods, this interference-CNN scheme can also determine the sign of topological charge while maintaining feasibility when utilizing different vortex beams with a radial index other than zero ($p{\neq}0$).

The topological charge recognition accuracy based on deep learning can reach up to 0.01 level. By changing the phase, it is possible to produce different intensity patterns for vortex beams with the same topological charge, which can be used for neural network training [47] (as shown in figure 3.4). Deep learning cannot only achieve high-precision topological charge recognition, but also large topological charge detection. By adding Gaussian white noise and multiplicative noise, vortex beam images can be simulated under non-ideal conditions. Using it for neural network training, Knutson *et al* achieved recognition of topological charge up to 100 [48]. In addition, trained CNN networks can also recognize vector vortex beams [49].

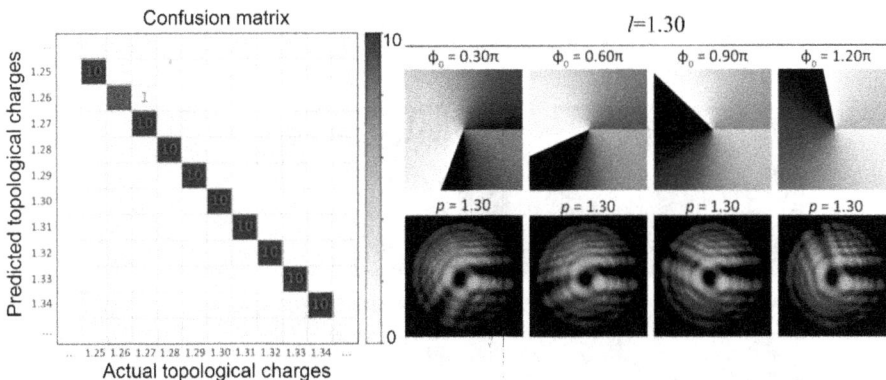

Figure 3.4. Quantitative analysis of CNN and the recognized OAM modes with fractional topological charge. Confusion matrix from $l = 1.25$ to $l = 1.34$. First row: phase pictures uploaded on the SLM. Second row: the intensity distributions of vortex modes recorded by the CMOS camera. Reprinted figure with permission from [47]. Copyright (2019) by the American Physical Society.

3.4 Surface plasmon polaritons method

Unlike traditional methods of phase accumulation for optical field control, SPPs can break the diffraction limit and have nanoscale optical manipulation capabilities, which offer unique advantages in detecting the topological charge of vortex light [50–54]. In 2012, Capasso *et al* used a holographic grating to focus SPPs excited by a vortex beam, thereby detecting the topological charge of vortex beam [50]. However, the fixed grating structure could only detect specific vortex light beams. Later, Mei *et al* used a semi-ring plasmonic grating structure to achieve the identification of multiple topological charges. When the topological charge of the vortex beam is matched with the on-chip plasmonic grating, the excited SPPs will focus at different positions, with a resolution of 100 nm [51]. However, this structure design requires the incident light beam to be circularly polarized vortex light.

In addition to detecting the topological charge of vortex light through the focusing position of surface plasmon, it can also be detected by the relative distance between multiple spots [53]. When a vortex light is incident on a ribbon grating, the SPPs excited by it propagate along two branches, and the splitting angle is related to

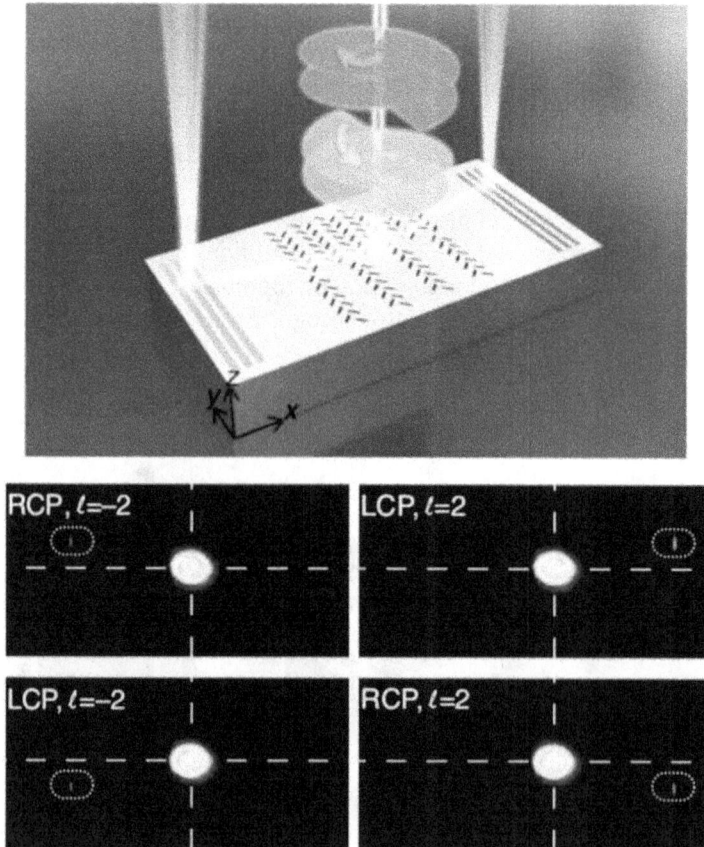

Figure 3.5. OAM pattern detection method based on plasma spin-Hall nanograting. Reproduced from [54]. CC BY 4.0.

the topological charge. This design does not require precise alignment of the incident light, but cannot detect the sign of the topological charge. By using a composite plasmonic grating structure composed of two different period gratings, the vortex light with different sign of topological charge can be coupled to different directions to excite SPPs. It is worth noting that the on-chip nanograting mentioned above can only detect the OAM state, but cannot detect the SAM state of the incident light.

The replacement of the slit in the composite plasmonic grating with a spin-Hall unit results in a plasmonic spin-Hall nanograting, which enables simultaneous detection of both polarization and phase singularity of an incident cylindrical vortex vector beam, as illustrated in figure 3.5 [54]. The SPPs excited by the vortex beams with different SAM and OAM are propagated to the four quadrants, and the chiral response and topological charge of the incident light can be distinguished by the position of the focal spot. These on-chip nanogratings hold significant potential for diverse applications, particularly in integrated photonic circuits.

References

[1] Yang Y, Ren Y-X, Chen M, Arita Y and Rosales-Guzmán C 2021 Optical trapping with structured light: a review *Adv. Photon.* **3** 034001

[2] Friese M E J, Nieminen T A, Heckenberg N R and Rubinsztein-Dunlop H 1998 Optical alignment and spinning of laser-trapped microscopic particles *Nature* **394** 348–50

[3] Mair A, Vaziri A, Weihs G and A Z 2001 Entanglement of the orbital angular momentum states of photons *Nature* **412** 313–16

[4] Molina-Terriza G, Torres J P and Torner L 2007 Twisted photons *Nat. Phys.* **3** 305–10

[5] Heckenberg N R, McDuff R, Smith C P and G W A 1992 Generation of optical phase singularities by computer-generated holograms *Opt. Lett.* **17** 221–23

[6] Vickers J, Burch M, Vyas R and Singh S 2008 Phase and interference properties of optical vortex beams *J. Opt. Soc. Am.* A **25** 823–27

[7] Huang H *et al* 2013 Phase-shift interference-based wavefront characterization for orbital angular momentum modes *Opt. Lett.* **38** 2348–50

[8] Harris M, Hill C A, Tapster P R and Vaughan J M 1994 Laser modes with helical wave fronts *Phys. Rev.* A **49** 3119–22

[9] Chen L, Zhang W, Lu Q and Lin X 2013 Making and identifying optical superpositions of high orbital angular momenta *Phys. Rev.* A **88** 053831

[10] Sztul H I and Alfano R R 2006 Double-slit interference with Laguerre–Gaussian beams *Opt. Lett.* **31** 999–1001

[11] Emile O and Emile J 2014 Young's double-slit interference pattern from a twisted beam *Appl. Phys.* B **117** 487–91

[12] Chen T, Lu X, Zeng J, Wang Z, Zhang H, Zhao C, Hoenders B J and Cai Y 2020 Young's double-slit experiment with a partially coherent vortex beam *Opt. Express* **28** 38106–14

[13] Qi W-R, Liu R, Kong L-J, Wang Z-X, Huang S-Y, Tu C, Li Y and Wang H-T 2020 Double-slit interference of single twisted photons *Chin. Opt. Lett.* **18** 102601

[14] Zhou H, Shi L, Zhang X and Dong J 2014 Dynamic interferometry measurement of orbital angular momentum of light *Opt. Lett.* **39** 6058–61

[15] Fu D, Chen D, Liu R, Wang Y, Gao H, Li F and Zhang P 2015 Probing the topological charge of a vortex beam with dynamic angular double slits *Opt. Lett.* **40** 788–91

[16] Malik M, Murugkar S, Leach J and Boyd R W 2012 Measurement of the orbital-angular-momentum spectrum of fields with partial angular coherence using double-angular-slit interference *Phys. Rev.* A **86** 063806

[17] Berkhout G C and Beijersbergen M W 2008 Method for probing the orbital angular momentum of optical vortices in electromagnetic waves from astronomical objects *Phys. Rev. Lett.* **101** 100801

[18] Zhao Q, Dong M, Bai Y and Yang Y 2020 Measuring high orbital angular momentum of vortex beams with an improved multipoint interferometer *Photonics Res* **8** 745–49

[19] Courtial J, Dholakia K, Robertson D A, Allen L and Padgett M J 1998 Measurement of the rotational frequency shift imparted to a rotating light beam possessing orbital angular momentum *Phys. Rev. Lett.* **80** 3217–19

[20] Leach J, Padgett M J, Barnett S M, Franke-Arnold S and Courtial J 2002 Measuring the orbital angular momentum of a single photon *Phys. Rev. Lett.* **88** 257901

[21] Slussarenko S, D'Ambrosio V, Piccirillo B, Marrucci L and Santamato E 2010 The polarizing Sagnac interferometer: a tool for light orbital angular momentum sorting and spin–orbit photon processing *Opt. Express* **18** 27205–16

[22] Zhang W, Qi Q, Zhou J and Chen L 2014 Mimicking Faraday rotation to sort the orbital angular momentum of light *Phys. Rev. Lett.* **112** 153601

[23] Hickmann J M, Fonseca E J, Soares W C and Chavez-Cerda S 2010 Unveiling a truncated optical lattice associated with a triangular aperture using light's orbital angular momentum *Phys. Rev. Lett.* **105** 053904

[24] Araujo L E E d and Anderson M E 2011 Measuring vortex charge with a triangular aperture *Opt. Lett.* **36** 787–89

[25] Liu Y, Tao H, Pu J and Lü B 2011 Detecting the topological charge of vortex beams using an annular triangle aperture *Opt. Laser Technol.* **43** 1233–36

[26] Silva J G, Jesus-Silva A J, Alencar M A, Hickmann J M and Fonseca E J 2014 Unveiling square and triangular optical lattices: a comparative study *Opt. Lett.* **39** 949–52

[27] Moreno I, Davis J A, Pascoguin B M L, Mitry M J and Cottrell D M 2009 Vortex sensing diffraction gratings *Opt. Lett.* **34** 2927–29

[28] Dai K, Gao C, Zhong L, Na Q and Wang Q 2015 Measuring OAM states of light beams with gradually-changing-period gratings *Opt. Lett.* **40** 562–65

[29] Gibson G, Courtial J and Padgett M J 2004 Free-space information transfer using light beams carrying orbital angular momentum *Opt. Express* **12** 5448–56

[30] Peng Y, Gan X-T, Ju P, Wang Y-D and Zhao J-L 2015 Measuring topological charges of optical vortices with multi-singularity using a cylindrical lens *Chin. Phys. Lett.* **32** 024201

[31] Kotlyar V V, Kovalev A A and Porfirev A P 2017 Astigmatic transforms of an optical vortex for measurement of its topological charge *Appl. Optics* **56** 4095–104

[32] Zheng S and Wang J 2017 Measuring orbital angular momentum (OAM) states of vortex beams with annular gratings *Sci Rep.* **7** 40781

[33] Berkhout G C, Lavery M P, Courtial J, Beijersbergen M W and Padgett M J 2010 Efficient sorting of orbital angular momentum states of light *Phys. Rev. Lett.* **105** 153601

[34] Berkhout G C G, Lavery M P J, Padgett M J and Beijersbergen M W 2011 Measuring orbital angular momentum superpositions of light by mode transformation *Opt. Lett.* **36** 1863–65

[35] Wen Y, Chremmos I, Chen Y, Zhu J, Zhang Y and Yu S 2018 Spiral transformation for high-resolution and efficient sorting of optical vortex modes *Phys. Rev. Lett.* **120** 193904

[36] Ruffato G, Massari M and Romanato F 2017 Compact sorting of optical vortices by means of diffractive transformation optics *Opt. Lett.* **42** 551–54

[37] Ruffato G, Massari M, Girardi M, Parisi G, Zontini M and Romanato F 2019 Non-paraxial design and fabrication of a compact OAM sorter in the telecom infrared *Opt. Express* **27** 24123–34

[38] Lightman S, Hurvitz G, Gvishi R and Arie A 2017 Miniature wide-spectrum mode sorter for vortex beams produced by 3D laser printing *Optica* **4** 605–10

[39] Fu Y *et al* 2019 Measuring phase and polarization singularities of light using spin-multiplexing metasurfaces *Nanoscale* **11** 18303–10

[40] Zhang S *et al* 2020 Broadband detection of multiple spin and orbital angular momenta via dielectric metasurface *Laser Photonics Rev.* **14** 2000062

[41] Guo Y, Zhang S, Pu M, He Q, Jin J, Xu M, Zhang Y, Gao P and Luo X 2021 Spin-decoupled metasurface for simultaneous detection of spin and orbital angular momenta via momentum transformation *Light-Sci. Appl.* **10** 63

[42] LeCun Y, Bengio Y and Hinton G 2015 Deep learning *Nature* **521** 436–44

[43] Dedo M I, Wang Z, Guo K and Guo Z 2020 OAM mode recognition based on joint scheme of combining the Gerchberg–Saxton (GS) algorithm and convolutional neural network (CNN) *Opt. Commun.* **456** 124696

[44] Li J, Zhang M, Wang D, Wu S and Zhan Y 2018 Joint atmospheric turbulence detection and adaptive demodulation technique using the CNN for the OAM-FSO communication *Opt. Express* **26** 10494–508

[45] Wang Z, Dedo M I, Guo K, Zhou K, Shen F, Sun Y, Liu S and Guo Z 2019 Efficient recognition of the propagated orbital angular momentum modes in turbulences with the convolutional neural network *IEEE Photonics J.* **11** 1–14

[46] Fu X, Bai Y and Yang Y 2021 Measuring OAM by the hybrid scheme of interference and convolutional neural network *Opt. Eng.* **60** 064109

[47] Liu Z, Yan S, Liu H and Chen X 2019 Superhigh-resolution recognition of optical vortex modes assisted by a deep-learning method *Phys. Rev. Lett.* **123** 183902

[48] Knutson E M, Lohani S, Danacı O, Huver S D and Glasser R T 2016 Deep learning as a tool to distinguish between high orbital angular momentum optical modes *Opt. Photon. Inform. Process.* X **9970** 997013

[49] Giordani T, Suprano A, Polino E, Acanfora F, Innocenti L, Ferraro A, Paternostro M, Spagnolo N and Sciarrino F 2020 Machine learning-based classification of vector vortex beams *Phys. Rev. Lett.* **124** 160401

[50] Genevet P, Lin J, Kats M A and Capasso F 2012 Holographic detection of the orbital angular momentum of light with plasmonic photodiodes *Nat. Commun.* **3** 1278

[51] Mei S *et al* 2016 On-chip discrimination of orbital angular momentum of light with plasmonic nanoslits *Nanoscale* **8** 2227–33

[52] Chen J, Li T, Wang S and Zhu S 2017 Multiplexed holograms by surface plasmon propagation and polarized scattering *Nano Lett.* **17** 5051–55

[53] Chen J, Chen X, Li T and Zhu S 2018 On-chip detection of orbital angular momentum beam by plasmonic nanogratings *Laser Photonics Rev.* **12** 1700331

[54] Feng F, Si G, Min C, Yuan X and Somekh M 2020 On-chip plasmonic spin-Hall nanograting for simultaneously detecting phase and polarization singularities *Light-Sci. Appl.* **9** 95

IOP Publishing

Optical Vortices

Fundamentals and applications

Yuanjie Yang, Yu-Xuan Ren and Carmelo Rosales-Guzmán

Chapter 4

Partially coherent vortex beams

The previous chapters are limited to fully coherent vortex beams. Indeed, typically the field may be partially spatially coherent under usual circumstances [1]. Thus, partially coherent vortex beams have attracted much attention and exhibit some unique characteristics, e.g., lower beam scintillation and beam wander during propagation in random media, less image noise, better self-reconstruction ability, and are expected to be useful in free-space optical communications, optical micromanipulations, optical imaging and information transfer [2–5]. In this chapter, the fundamentals and applications of partially coherent vortex beams will be discussed.

4.1 Orbital angular momentum of partially coherent beams

Orbital angular momentum (OAM) carried by the vortex beam provides powerful capabilities for widespread applications [6]. Such attention has naturally shifted to the study of the OAM of partially coherent beams. In 2001, Serna *et al* extended the definition of the OAM established for coherent beams to partially coherent beams, derived the orbital angular momentum of partially coherent beams based on the Wigner distribution function [7]. From the theoretic deduction results, the expression of the OAM of a coherent beam through a z-plane, which is also suitable for partially coherent beams, is given by:

$$\bar{J}_z^{\,L} = \left(\frac{I}{c}\right)(\langle xv \rangle - \langle yu \rangle), \tag{4.1}$$

where $(\langle xv \rangle - \langle yu \rangle)$ is related to the phase spatial structure. $\langle xv \rangle$ $\langle yu \rangle$ are the second-order irradiance moments of the beam, which are defined in terms of the Wigner distribution function $h(x, y, u, v)$ of the partially coherent beams, namely:

$$\langle xv \rangle = \left(\frac{1}{I}\right)\int\int\int\int xvh(x,\ y,\ u,\ v)dxdydudv, \tag{4.2}$$

doi:10.1088/978-0-7503-5844-6ch4

$$\langle yu \rangle = \left(\frac{1}{I}\right) \int \int \int \int yuh(x, y, u, v)dxdydudv, \tag{4.3}$$

where u and v represent angles of propagation.

Taking the Gaussian–Schell Model (GSM) beam as a particular case, the relation between the OAM and the twist parameter is analyzed. Following equation (4.2) and reference [8], the expression of the OAM of GSM fields is given by:

$$\bar{J_z} = \left(\frac{I}{2c}\right)\{[T\mathbf{r}(\mathbf{R}^{-1}\sigma_I{}^2 - \sigma_I{}^2\mathbf{R}^{-1})\mathbf{J}] - 2\tau T\mathbf{r}(\sigma_I{}^2)\}, \tag{4.4}$$

where

$$\mathbf{J} = \begin{bmatrix} 0 & 1 \\ -1 & 0 \end{bmatrix}, \tag{4.5}$$

σ_I is beam width, σ_g is transverse coherence width, $\mathbf{R}$ is curvature radius, and a scalar τ is the twisted phase parameter.

One can get the following conclusion from equation (4.4): $\mathbf{J}_z$ is independent of the coherence properties given by σ_g. σ_I and $\mathbf{R}$ are two contributions to $\mathbf{J}_z$. One can find three different beams by choosing these parameters, such as beams that lend both contributions to the angular momentum (for example, the pseudoaligned simple astigmatic beams) or beams that fulfill $\mathbf{J}_z \neq 0$ but with $\tau = 0$ (for example, coherent Gaussian beams whose intensity profiles rotate in free propagation) or beams (for example, the family of twisted GSM beams) for which twist and OAM are equivalent.

4.2 Singularities in partially coherent vortex beams

Singularity means the point in the field has a zero amplitude with indefinite phase. Partially coherent beams, however, are known to have no region of zero intensity and no well-defined phase, and consequently do not typically possess optical phase singularities. In 2003, the correlation function of partially coherent fields was found to contain analogous singularities, which are named 'correlation singularities' [8]. Subsequently, the vortex nature of partially coherent singular beams with a separable phase was exhibited, and further research confirmed the existence of a hidden correlation singularity (or coherence vortex) [9–14]. Since then, much attention has been given to the vortices of correlation functions.

Correlation singularity is defined as the point in the field whose cross-spectral density (CSD) or degree of coherence equals to zero while the corresponding phase is indefinite. Such vortices are pairs of points (r_1 and r_2) at which the spectral degree of coherence of the field vanishes, i.e., where

$$\mu(\mathbf{r}_1, \mathbf{r}_2) = \frac{W(\mathbf{r}_1, \mathbf{r}_2)}{\sqrt{S(\mathbf{r}_1)}\sqrt{S(\mathbf{r}_2)}} = 0, \tag{4.6}$$

In other words, coherence singularities exist at locations where the real and imaginary parts of $W(r_1, r_2)$ are simultaneously zero:

$$\mathrm{Re}[W(r_1, r_2)] = 0, \tag{4.7}$$

$$\mathrm{Im}[W(r_1, r_2)] = 0 \tag{4.8}$$

Here $W(r_1, r_2)$ denotes the CSD which is used to characterize the statistic properties of the partially coherent beam, but at which the spectral density (often referred to as intensity), $S(r_i) = W(r_i, r_i)$ of the field is nonzero. Re and Im denote taking the real and imaginary parts, respectively.

In 2004, Palacios *et al* verified the existence of a robust correlation singularity in experiment [15]. As the coherence length decreases, the dark vortex core disappears gradually while a ring dislocation appears in the cross-correlation function, and with the decrease of coherence length, the radial size of ring dislocation increases and becomes more recognizable. In addition to the ring dislocation form, the coherence singularity also has a vortex form. The two forms can be transformed into each other by choosing the suitable observation point position (x, y).

Coherence vortices and optical vortices have similarities and differences [16–18]. Both are associated with phase singularities. In the case of the fully coherent limit, coherence vortices are transformed into optical vortices. The optical vortex is the phase of complex amplitude, while the coherence vortex is the phase of the CSD. Therefore, the optical vortex has null intensity, but the intensity of the coherence vortex is not necessarily null. Furthermore, optical vortices are 'real' and have OAM and are independent of the observation point, while coherence vortices are 'virtual' and are dependent of the observation point.

4.3 Measuring the topological charge of partially coherent vortex beams

4.3.1 Partially coherent Laguerre–Gaussian beams

Standard Laguerre–Gaussian (LG) beams are the most frequently encountered in laser optics, material processing and atomic optics, the most extensively studied in the literature and have been widely applied in optical communications, atom trapping, atom interferometers, etc [19]. Simultaneously, partially coherent standard LG beams have attracted enormous attention. Taking the partially coherent standard Laguerre–Gaussian (LG$_{p\ell}$) beam as a representative, where p is radial index mode and l is topological charge, measuring its topological charge (l) is of typical significance.

(1) **For the case of $p = 0$**

A method that observes cross-correlation function (CCF) to measure the topological charge was proposed [20]. The CCF is defined as the mutual coherence function of two points (x, y) and $(-x, -y)$. The far-field CCF after the beam propagates a distance z is expressed as $\chi(\rho') = \Gamma'(\rho_1', -\rho_1')$, where $\Gamma'(\rho_1, \rho_2)$ is the mutual coherence function at far-field position z, and is given by:

$$\Gamma(\rho_1, \rho_2, z) = \left(\frac{1}{\lambda z}\right)^2 \iint \iint d\rho'_1 d\rho'_2 \Gamma(\rho'_1, \rho'_2, 0)$$

$$\times \exp\left(-\frac{2\pi i}{\lambda z}(\rho'_1 \cdot \rho_1 - \rho'_2 \cdot \rho_2)\right)$$

(4.9)

where $\Gamma(\rho_1', \rho_2', 0)$ is the mutual coherence function in the source.

Figure 4.1 illustrates although the dark vortex core of intensity fills with diffuse light with coherence decreases, the far-field CCF appears as interesting bright and dark rings and the radial size of ring dislocation increases gradually. Indeed, the far-field CCF maintains the striking robust attribute of the vortex (singularity). Moreover, it is noted that the number of the dark rings present is three, the very value of the original topological charge of the field [20].

Figure 4.2, further, demonstrates a definitive linkage between the azimuthal index of an optical vortex and the far-field CCF for the partial and low-coherence cases. It can be concluded that the number of dark ring dislocations (N) in the far-field CCF is equal to the topological charge of the partially coherent optical vortex, i.e., $N = |\ell|$ [20]. This finding has been verified by experiments.

(2) **For the case of $p \neq 0$**

When the radial mode index p is considered, i.e., when $p \neq 0$, there exists a definitive general relation between the CCF and the radial and azimuthal mode indices (p, ℓ) [21, 22]. Figure 4.3 demonstrates theoretical and experimental results of the far-field CCF of partially coherent vortex beams

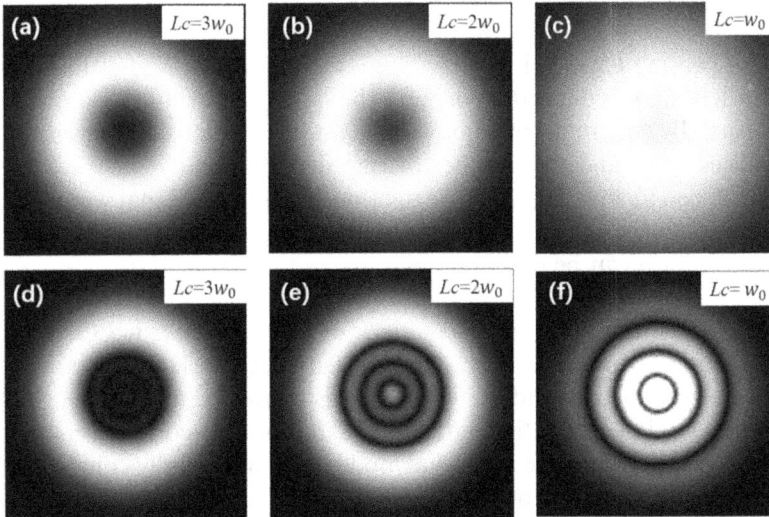

Figure 4.1. Far-field intensity (a–c) and CCF (d–f) for vortex beam $l = 3$ for different coherence length.

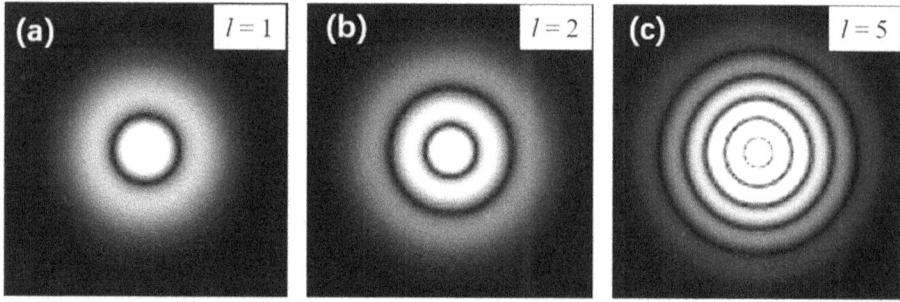

Figure 4.2. Far-field CCF for different topological charges.

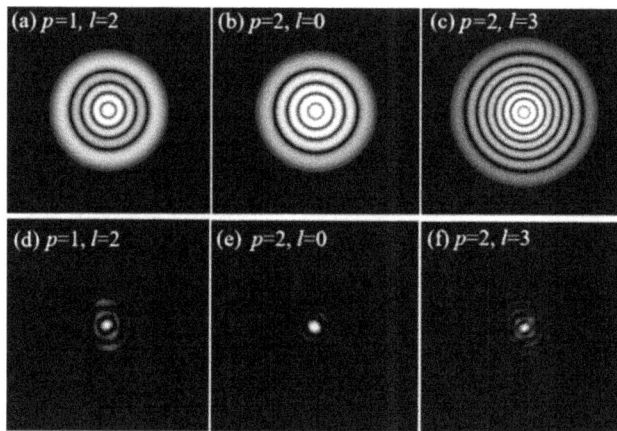

Figure 4.3. Theoretical (a–c) (Reproduced from [21] © IOP Publishing and Deutsche Physikalische Gesellschaft. CC BY 3.0) and experimental (d–f) (Reprinted from [22], with the permission from AIP Publishing) results of the far-field CCF of partially coherent vortex beams with varying l and p.

with varying l and p. It is shown that the number of ring dislocations in the far-field CCF is dependent on both the radial and azimuthal mode indices of a partially coherent optical vortex beam the relationship between the mode indices (p, ℓ) and the number of ring dislocations (N) in the CCF can be concluded as $N = 2p + |\ell|$. The relationship is implied that we can determine the topological charge ℓ of a partially coherent $LG_{p\ell}$ beam so long as the parameter p can be determined.

Usually, the radial index mode p of a fully coherent LG beam can be measured by the dark rings of the intensity pattern. However, for the partial coherence case, the radial mode index is concealed at the intensity pattern. A new kind of correlation singularity called double-correlation function was proposed to determine the radial mode index of partially coherent LG beam [23]. Double-correlation function (DCF) is defined as the mutual coherence function of two points (x, y) and $(2x, 2y)$. Unlike the cross-correlation

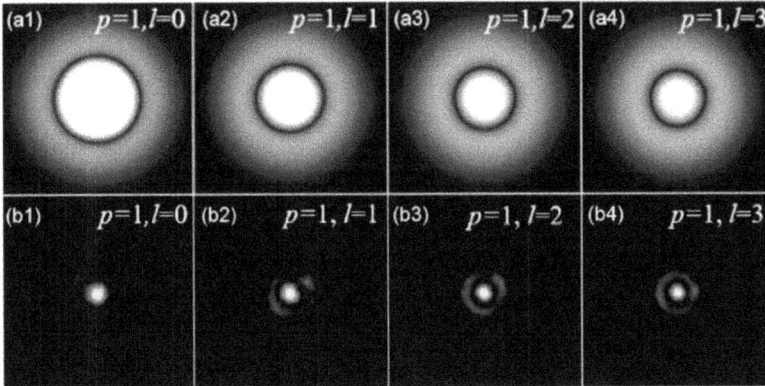

Figure 4.4. Far-field DCF of a partially coherent LG beam with $p = 1$ and varying topological charge. (a) Simulation (Reproduced from [23]. © 2016 IOP Publishing Ltd. All rights reserved) and the corresponding experimental results (Reprinted with permission from [24]. © 2017 Chinese Laser Press).

singularity, which depends on both radial and azimuthal mode indices, the double-correlation singularity is dependent on the radial mode index only, which can be used to measure the radial mode index.

Figure 4.4 shows the theoretical and experimental results of far-field DCF of partially coherent LG beam with $p = 1$ and varying topological charges [23, 24]. We can see that with the increasing azimuthal mode index, the radius of the dark ring decreases, but the number of dark ring dislocations remains the same. Even though there is no vortex, dark ring dislocation still exists. It is implied that the number of dark ring dislocations in the far-field DCF is independent of the azimuthal index ℓ, furthermore, a definitive relationship can be deduced, namely that the number of dark ring dislocations M is equal to p, i.e., $M = p$.

Thus, a method to measure the azimuthal and radial mode indices of a partially coherent vortex beam can be concluded. First, the number of dark ring dislocations in the far-field DCF should be counted. According to the relationship $M = p$, we can retrieve the radial mode index p, Then we can determine the azimuthal mode index (topological charge) by observing CCF, recalling the relationship between the number of dark ring dislocations in the far-field CCF and the radial and azimuthal indices $N = 2p + |\ell|$. We can measure the radial and azimuthal mode indices of partially coherent beams step by step.

4.3.2 Partially coherent elegant Laguerre–Gaussian beams

Elegant Laguerre–Gaussian (LG) was proposed by Takenaka *et al* as an extension of a standard LG beam. Both standard LG modes and elegant LG modes satisfy the paraxial wave equation, while the elegant LG modes have a more symmetrical form [23]. Partially coherent elegant LG beams possess quite different and unique properties, compared to partially coherent standard LG beams. The previously mentioned method of measuring the topological charge is not valid for partially coherent elegant LG beams. Therefore, a new method of using a complex degree of coherence (CDOC) to measure its topological charge of such beams has been recently proposed [25]. CDOC is defined as a normalized mutual coherence function, which is given by:

$$\mu(r_1, r_2, z) = \frac{\Gamma(r_1, r_2, z)}{\Gamma(r_1, r_1, z)\Gamma(r_2, r_2, z)} \quad (4.10)$$

where $\Gamma(r_1, r_2, z)$ is the mutual coherence function at the propagation distance z, $\Gamma(r_i, r_i, z)$ is intensity at the location r_i.

Figure 4.5 shows that the far-field CDOC between two points (x_1, y_1) and $(0, 0)$ of partially coherent elegant LG beam (coherence length $\delta = w_0$) with $p = 1$ for different values of l and p. One can see that the radial mode index doesn't affect the number of the ring dislocations in the far-field CDOC. In other words, the number of ring dislocations is dependent on the topological charge l only, which is much different from the case of partially coherent standard LG beams [26].

Figure 4.6 shows that theoretical and experimental results of the far-field CDOC between two points (x_1, y_1) and $(0, 0)$ of a partially coherent elegant LG beam with $p = 1$ and varying topological charges. It can be concluded that the number of ring dislocations in the focal-field CDOC is equal to the topological charge. Therefore, the topological charge of the partially coherent elegant LG can be measured by observing its CDOC in the far field [26].

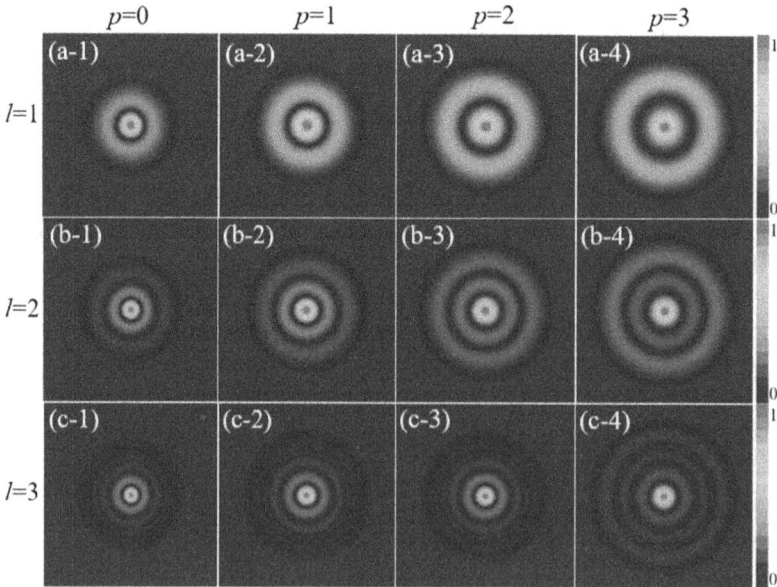

Figure 4.5. The far-field CDOC of partially coherent LG beam with $p = 1$ for different values of l and p. Reprinted with permission from [26].© 2018 Optical Society of America under the terms of the OSA Open Access Publishing Agreement.

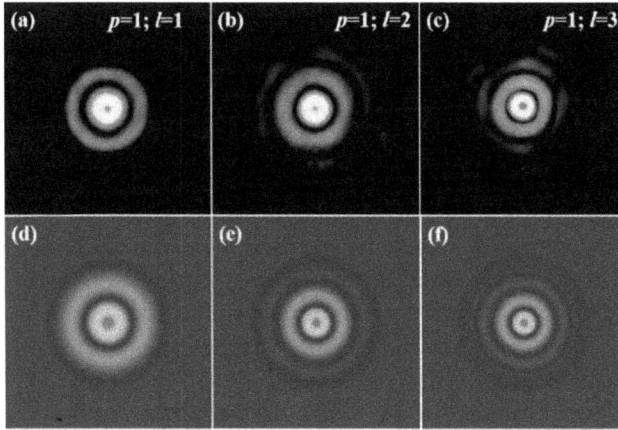

Figure 4.6. Experimental results (a) and the corresponding theoretical results (b) of the far-field CDOC of partially coherent elegant LG beam with $p = 1$ and varying topological charges. Reprinted with permission from [26]. © 2018 Optical Society of America under the terms of the OSA Open Access Publishing Agreement.

4.4 Applications of partially coherent vortex beams

4.4.1 Optical communication

Optical communication is one of the most popular applications of partially coherent beams. Using partially coherent beams can be used to improve communication efficiency. It is well known that the scintillation and phase distortion of the beam can be caused by random fluctuation of turbulence [27–31]. The vortex phase embedded in the partially coherent beam can improve the beam quality. In other words, partially coherent vortex beams are more resistant to distortions caused by a turbulent atmosphere. The partially coherent vortex beam has the lowest on-axis scintillation index compared with other types of lights. In addition, it is found that using partially coherent radially polarized vortex beams and increasing topological charge can further reduce their on-axis scintillation index. On the other hand, it is found that increasing the channel capacity can improve the efficiency of optical communication. However, turbulence, which often occurs in transmission, will affect the crosstalk between OAM modes during optical transmission, resulting in reduced information capacity. A method to reduce the crosstalk effect by using partially coherent vortex beams is proposed to solve this problem effectively. In addition, the study found that reducing beam coherence can reduce the effects of turbulence. Therefore, partially coherent vortex beams have received more and more attention in the field of optical communication.

4.4.2 Optical manipulation

The optical manipulation of micro-particles, including trapping, rotating and guiding these objects, is based on the radiation force generated by a strongly focused beam of light. The radiation force is affected by the intensity distribution, polarization and coherence of the incident beam. Compared with the coherent vortex field, the partially coherent vortex beam shows more freedom in the control of coherence length and topological charge. Therefore, the radiant force can be modulated more easily with partially coherent vortex beams, which may lead to the development of new optical manipulation techniques and improved capture efficiency. For example, Zhao and Cai found that two Rayleigh particles could be captured using a focused partially coherent elegant LG beam of the appropriate mode order by varying the spatial coherence width [32]. Researchers also demonstrated that other partially coherent vortex beams have the similar ability. In addition, researchers found that the OAM of partially coherent vortex beams formed by superimposing mutually incoherent vortex modes can trap particles by exploiting their coherence. The results show that even with a small number of modes, OAM can be controlled by changing the beam parameters, including controlling the radial distribution of OAM. These promising properties of partially coherent vortex beams are beneficial to optical manipulation and trapping.

4.4.3 Imaging

Improving image resolution is of great significance in the fields of biology, medicine and image processing. Resolution is usually limited by the diffraction properties of light. Recently, researchers have found that using unconventional partially coherent beams can improve image resolution [33]. Specifically, the use of partially coherent beams can not only reduce the noise of image spots and crosstalk between objects, but also improve the phase contrast of objects. In addition, correlations in the light field are known to have a profound effect on image contrast for features near the resolution limit. In simple video microscopy, the contrast of horizontal separation features can be improved by using partially correlated azimuth vortex illumination. Therefore, partially coherent vortex beams have potential applications in metrology, microscopy and lithography.

4.4.4 Other promising applications

Based on the special properties of partially coherent vortex beams, the researchers also propose scenarios with potential applications. For example, based on the fact that coherent vortices can be recovered during speckle imaging, the researchers propose a new way of information transmission [34]. Researchers have proposed a new and simpler method to detect spatial coherence of fields based on the link between vortex behavior and coherence length [35].

References

[1] Dong M, Zhao C L, Cai Y J *et al* 2021 Partially coherent vortex beams: fundamentals and applications *Sci. China: Phys. Mech. Astron.* **64** 224201

[2] Gbur G and Wolf E 2002 Spreading of partially coherent beams in random media *J. Opt. Soc. Am.* A **19** 1592–8

[3] Cai Y J and He S 2006 Propagation of a partially coherent twisted anisotropic GSM beam in turbulent atmosphere *Appl. Phys. Lett.* **89** 041117

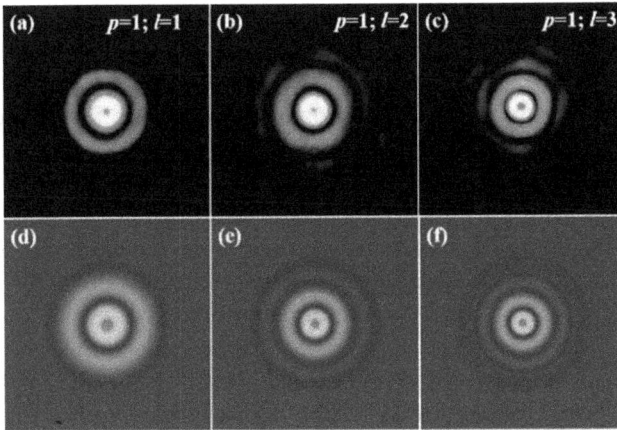

Figure 4.6. Experimental results (a) and the corresponding theoretical results (b) of the far-field CDOC of partially coherent elegant LG beam with $p = 1$ and varying topological charges. Reprinted with permission from [26]. © 2018 Optical Society of America under the terms of the OSA Open Access Publishing Agreement.

4.4 Applications of partially coherent vortex beams

4.4.1 Optical communication

Optical communication is one of the most popular applications of partially coherent beams. Using partially coherent beams can be used to improve communication efficiency. It is well known that the scintillation and phase distortion of the beam can be caused by random fluctuation of turbulence [27–31]. The vortex phase embedded in the partially coherent beam can improve the beam quality. In other words, partially coherent vortex beams are more resistant to distortions caused by a turbulent atmosphere. The partially coherent vortex beam has the lowest on-axis scintillation index compared with other types of lights. In addition, it is found that using partially coherent radially polarized vortex beams and increasing topological charge can further reduce their on-axis scintillation index. On the other hand, it is found that increasing the channel capacity can improve the efficiency of optical communication. However, turbulence, which often occurs in transmission, will affect the crosstalk between OAM modes during optical transmission, resulting in reduced information capacity. A method to reduce the crosstalk effect by using partially coherent vortex beams is proposed to solve this problem effectively. In addition, the study found that reducing beam coherence can reduce the effects of turbulence. Therefore, partially coherent vortex beams have received more and more attention in the field of optical communication.

4.4.2 Optical manipulation

The optical manipulation of micro-particles, including trapping, rotating and guiding these objects, is based on the radiation force generated by a strongly focused beam of light. The radiation force is affected by the intensity distribution, polarization and coherence of the incident beam. Compared with the coherent vortex field, the partially coherent vortex beam shows more freedom in the control of coherence length and topological charge. Therefore, the radiant force can be modulated more easily with partially coherent vortex beams, which may lead to the development of new optical manipulation techniques and improved capture efficiency. For example, Zhao and Cai found that two Rayleigh particles could be captured using a focused partially coherent elegant LG beam of the appropriate mode order by varying the spatial coherence width [32]. Researchers also demonstrated that other partially coherent vortex beams have the similar ability. In addition, researchers found that the OAM of partially coherent vortex beams formed by superimposing mutually incoherent vortex modes can trap particles by exploiting their coherence. The results show that even with a small number of modes, OAM can be controlled by changing the beam parameters, including controlling the radial distribution of OAM. These promising properties of partially coherent vortex beams are beneficial to optical manipulation and trapping.

4.4.3 Imaging

Improving image resolution is of great significance in the fields of biology, medicine and image processing. Resolution is usually limited by the diffraction properties of light. Recently, researchers have found that using unconventional partially coherent beams can improve image resolution [33]. Specifically, the use of partially coherent beams can not only reduce the noise of image spots and crosstalk between objects, but also improve the phase contrast of objects. In addition, correlations in the light field are known to have a profound effect on image contrast for features near the resolution limit. In simple video microscopy, the contrast of horizontal separation features can be improved by using partially correlated azimuth vortex illumination. Therefore, partially coherent vortex beams have potential applications in metrology, microscopy and lithography.

4.4.4 Other promising applications

Based on the special properties of partially coherent vortex beams, the researchers also propose scenarios with potential applications. For example, based on the fact that coherent vortices can be recovered during speckle imaging, the researchers propose a new way of information transmission [34]. Researchers have proposed a new and simpler method to detect spatial coherence of fields based on the link between vortex behavior and coherence length [35].

References

[1] Dong M, Zhao C L, Cai Y J *et al* 2021 Partially coherent vortex beams: fundamentals and applications *Sci. China: Phys. Mech. Astron.* **64** 224201
[2] Gbur G and Wolf E 2002 Spreading of partially coherent beams in random media *J. Opt. Soc. Am.* A **19** 1592–8
[3] Cai Y J and He S 2006 Propagation of a partially coherent twisted anisotropic GSM beam in turbulent atmosphere *Appl. Phys. Lett.* **89** 041117

[4] Li J H, Yang A L and Lü B D 2008 A comparative study of the beam-width spreading of partially coherent Hermite-sinh-Gaussian beams in atmospheric turbulence *J. Opt. Soc. Am. A* **25** 2670–9

[5] Wang T, Pu J X and Chen Z Y 2008 Propagation of partially coherent vortex beams in a turbulent atmosphere *Opt. Eng.* **47** 6002

[6] Allen L, Beijersbergen M W, Spreeuw R J C and Woerdman J P 1992 Orbital angular momentum of light and the transformation of Laguerre–Gaussian laser modes *Phys. Rev. A* **45** 8185–9

[7] Serna J and Movilla J M 2001 Orbital angular momentum of partially coherent beams *Opt. Lett.* **26** 405–7

[8] Schouten H F, Gbur G, Visser T D and Wolf E 2003 Phase singularities of the coherence functions in Young's interference pattern *Opt. Lett.* **28** 968–70

[9] Gbur G and Visser T D 2003 Coherence vortices in partially coherent beams *Opt. Commun.* **222** 117–25

[10] Gbur G, Visser T D and Wolf E 2004 Hidden' singularities in partially coherent wavefields *J. Opt. A* **6** 68579–5

[11] Wang W, Duan Z H, Hanson S G, Miyamoto Y and Takeda M 2006 Experimental study of coherence vortices: local properties of phase singularities in a spatial coherence function *Phys. Rev. Lett.* **96** 073902

[12] Gori F, Santarsiero M, Borghi R and Vicalvi S 1998 Partially coherent sources with helicoidal modes *J. Mod. Opt.* **45** 539–54

[13] Boggatyryova V G, Felde V C, Polyanskii P V, Ponomarenko S A, Soskin M S and Wolf E 2003 Partially coherent vortex beams with a separable phase *Opt. Lett.* **28** 878–80

[14] Ponomarenko S A 2001 A class of partially coherent beams carrying optical vortices *J. Opt. Soc. Am. A* **18** 150–6

[15] Palacios D M, Maleev I D, Marathay A S and Swartzlander G A 2004 Spatial correlation singularity of a vortex field *Phys. Rev. Lett.* **92** 143905

[16] Gbur G and Swartzlander G 2008 Complete transverse representation of a correlation singularity of a partially coherent field *J. Opt. Soc. Am. B* **25** 1422–9

[17] Gbur G 2008 Optical and coherence vortices and their relationships *Proc. SPIE* **7008** 70080N

[18] Gbur G and Visser T D 2006 Phase singularities and coherence vortices in linear optical systems *Opt. Commun.* **259** 428–35

[19] Wang F, Cai Y and Korotkova O 2009 Partially coherent standard and elegant Laguerre–Gaussian beams of all orders *Opt. Express* **17** 22366–79

[20] Yang Y J, Mazilu M and Dholakia K 2012 Measuring the orbital angular momentum of partially coherent optical vortices through singularities in their cross-spectral density functions *Opt. Lett.* **37** 4949–51

[21] Yang Y, Chen M, Mazilu M, Mourka A, Liu Y and Dholakia K 2013 Effect of the radial and azimuthal mode indices of a partially coherent vortex field upon a spatial correlation singularity *New J. Phys.* **15** 113053

[22] Liu R F, Wang F R, Chen D X *et al* 2016 Measuring mode indices of a partially coherent vortex beam with Hanbury Brown and Twiss type experiment *Appl. Phys. Lett.* **108** 051107

[23] Yang Y and Liu Y 2016 Measuring azimuthal and radial mode indices of a partially coherent vortex field *J. Opt.* **18** 015604

[24] Liu X L, Wu T F, Liu L, Zhao C L and Cai Y J 2017 Experimental determination of the azimuthal and radial mode orders of a partially coherent LG_{pl} beam (invited paper) *Chin. Opt. Lett.* **15** 030002

[25] Wang F, Cai Y J, Eyyuboğlu H T and Baykal. Y 2010 Average intensity and spreading of partially coherent standard and elegant Laguerre–Gaussian beams in turbulent atmosphere *Prog. Electromagn. Res.* **103** 33–56

[26] Dong M, Lu X Y, Zhao C L, Cai Y J and Yang Y J 2018 Measuring topological charge of partially coherent elegant Laguerre–Gaussian beam *Opt. Express* **26** 33035–43

[27] Cheng M J, Guo L X, Li J T, Huang Q Q, Cheng Q and Zhang D 2006 Propagation of an optical vortex carried by a partially coherent Laguerre–Gaussian beam in turbulent ocean *Appl. Opt.* **55** 4642–8

[28] Liu D J, Luo X X, Yin H M, Wang G Q and Wang Y C 2017 Effect of optical system and turbulent atmosphere on the average intensity of partially coherent flat-topped vortex hollow beam *Optik* **130** 227–36

[29] Li Y X, Cui Z W, Han Y P and Hui Y F 2019 Channel capacity of orbital-angular-momentum-based wireless communication systems with partially coherent elegant Laguerre–Gaussian beams in oceanic turbulence *J. Opt. Soc. Am.* A **36** 471–7

[30] Peng J, Zhang L, Zhang K C and Ma J X 2018 Channel capacity of OAM based FSO communication systems with partially coherent Bessel–Gaussian beams in anisotropic turbulence *Opt. Commun.* **418** 32–6

[31] Liu X L, Shen Y, L L, Wang F and Cai Y J 2013 Experimental demonstration of vortex phase-induced reduction in scintillation of a partially coherent beam *Opt. Lett.* **38** 5323–6

[32] Zhao C L, Wang F, Dong Y, Han Y J and Cai Y J 2012 Effect of spatial coherence on determining the topological charge of a vortex beam *Appl. Phys. Lett.* **101** 261104

[33] Brown D P and Brown T G 2008 Partially correlated azimuthal vortex illumination: coherence and correlation measurements and effects in imaging *Opt. Express* **16** 20418–26

[34] Alves Cleberson R, Jesus-Silva Alcenísio J and Fonseca Eduardo. J S 2015 Characterizing coherence vortices through geometry *Opt. Lett.* **40** 2747–50

[35] Gu Y L and Gbur G 2009 Topological reactions of optical correlation vortices *Opt. Commun.* **282** 709–16

IOP Publishing

Optical Vortices
Fundamentals and applications
Yuanjie Yang, Yu-Xuan Ren and Carmelo Rosales-Guzmán

Chapter 5

Vector vortex beams

Vector vortex modes, characterized by a non-homogeneous polarization distribution, are the result of a nonseparable superposition of the spatial and polarization degrees of freedom [1–5]. Even though historically, both degrees of freedom were considered independently, since their inception in the decade of 1970, vector vortex modes have drawn significant attention, in part due to their potential applications, spanning fields such as classical and quantum communications, optical metrology, optical manipulations, amongst others [6–20]. With the sake of simplifying the understanding of this chapter, we will start from the very basic, the Helmholtz equation, which sets the basis for any light wave treatment. This will follow a complete description of polarization, including some of the mathematical tools that have been introduced to simplify its understanding. Afterwards, we will outline some of the basic concepts of vector beams followed by some of the modern approaches for their generation and characterization. Finally, we will describe some of the applications that rely on the use of vector beams.

5.1 Basic concepts

To begin with, it is worth mentioning that the electric (**E**) and magnetic (**B**) fields of an electromagnetic wave are vector quantities with both a magnitude and a direction, and therefore, light itself is vectorial. Further, while **E** is perpendicular to both **B** and the wave vector, **k**, the components of $\mathbf{E} = (E_x, E_y)$ in the transverse plane may be selected independently of one another. For example, for a travelling electromagnetic plane wave in the z direction, Maxwell's equations restrict $E_x = cB_y$ and $E_y = -cB_x$, but place no restrictions on the relation between E_x and E_y. The addition of these components with various amplitude and phase weightings gives rise to the concept of polarization. In addition, the magnitude of **E** at each point in the plane determines the transverse spatial mode, the pattern of light, with an intensity proportional to $|E|^2$.

doi:10.1088/978-0-7503-5844-6ch5

5.1.1 The vectorial nature of light

Polarization is one of the most salient properties of light and has been known for several centuries. Historically, the discovery of polarization dates back to the year 1670, credited to Erasmus Bartholinus. He observed that when a ray of natural light passes though a calcite crystal, it splits into two new rays that are equally intense. This discovery was shortly followed by the realization by Huygens that these two rays carry orthogonal polarizations. However, it was only in 1803 when Young demonstrated that light vibrates in the plane perpendicular to the direction of propagation that this was associated with the transverse nature of light. The series of discoveries that followed quickly led to a full understanding of polarization. It suffices to say that it is now clear that the state of polarization is related to the time variations of the electric field in the plane perpendicular to that of propagation. Moreover, if the electric field varies randomly in time then the wave is said to be unpolarized. On the contrary, if it varies in a preferable direction, it is said to be polarized. In the following section we will provide a mathematical description of polarization, starting with the wave equation.

The time-dependent vectorial wave equation for the electric field in free space is given by

$$\nabla^2 \mathbf{E}(\mathbf{r},\, t) - \varepsilon\mu\frac{\partial^2 \mathbf{E}(\mathbf{r},\, t)}{\partial t^2} = 0, \tag{5.1}$$

known as the Helmholtz equation. Here, $\mathbf{E}(\mathbf{r},\, t)$ is the electric field, $\mathbf{r}$ is the position vector, while ε and μ are the dielectric permittivity and magnetic permeability, respectively. Further, ∇^2 is the Laplacian operator, which involves the second partial derivatives, and takes the form of the coordinates systems in which the wave equation is described. For a monochromatic light beam of frequency ω, the electric field can be written as $\mathbf{E}(\mathbf{r},\, t) = \text{Re}\{\mathbf{E}_0(\mathbf{r})\exp[i(\mathbf{k} \cdot \mathbf{r} - \omega t)]\}$, $\mathbf{k}$ being the wave vector, whose magnitude is given by $|\mathbf{k}| = 2\pi/\lambda$. The vector quantity $\mathbf{E}_0(\mathbf{r})$ is given by

$$\mathbf{E}_0(\mathbf{r}) = [E_1\hat{e}_1 + E_2\hat{e}_2 + E_3\hat{e}_3]u(\mathbf{r}), \tag{5.2}$$

where the complex coefficients $E_1(\mathbf{r})$, $E_2(\mathbf{r})$ and $E_3(\mathbf{r})$ represent the amplitude of the electric field along the direction given by the unitary vectors $\hat{e}_1$, $\hat{e}_2$ and $\hat{e}_3$, associated to the state of polarization of the electric field. The function $u(\mathbf{r})$ denotes the spatial shape of the beam. For example, if $E_1 = 1$ and $E_2 = E_3 = 0$, then the entire spatial mode, $u(\mathbf{r})$, is linearly polarized in the $\hat{e}_1$ direction.

5.1.1.1 Representation of polarization using Jones vectors
The different polarization states of light can be used represented and manipulated mathematically through the use of vectors and the tools of vectorial algebra. Such representation was first proposed by R C Jones in 1941 and therefore the states of polarization represented in this way are known as Jones vectors [21]. In this representation, each polarization state is represented by a vector of the form

$$\mathbf{E}(\mathbf{r}) = \begin{bmatrix} E_1 \\ E_2 \\ E_3 \end{bmatrix} u(\mathbf{r}), \tag{5.3}$$

In the paraxial approximation and using the Cartesian coordinates (x, y, z), the *transverse* mode of the light beam propagating along the z can be conveniently represented in a normalized vector form as

$$\mathbf{E} = \frac{1}{\sqrt{E_x^2 + E_y^2}} \begin{bmatrix} E_x \\ E_y e^{i\delta} \end{bmatrix}, \tag{5.4}$$

where E_x and E_y represent the polarization components along the x and y directions, respectively. In this representation, all the linear polarization states can be obtained by setting $E_x = \cos\theta$, $E_y = \sin\theta$ and $\delta = 0$ as

$$\mathbf{E} = \begin{bmatrix} \cos\theta \\ \sin\theta \end{bmatrix}. \tag{5.5}$$

In particular, the two orthogonal polarization states, horizontal (H) and vertical(V), are obtained by setting $\theta = 0$ and $\theta = \pi/2$, respectively. In addition, the diagonal (D), and antidiagonal (A) polarization states, which are also orthogonal, are obtained by setting $\theta = \pi/4$ and $\theta = -\pi/4$, respectively. The corresponding Jones vector representation of these four linearly polarized states is

$$H = \begin{bmatrix} 1 \\ 0 \end{bmatrix}, \quad V = \begin{bmatrix} 0 \\ 1 \end{bmatrix} \quad D = \frac{1}{\sqrt{2}} \begin{bmatrix} 1 \\ 1 \end{bmatrix}, \quad \text{and } A = \frac{1}{\sqrt{2}} \begin{bmatrix} 1 \\ -1 \end{bmatrix} \tag{5.6}$$

Two other polarization states which are of interest are the right (R) and left (L) circular polarizations, which are obtained by an equally weighted superposition of horizontal and vertical polarization and a relative phase delay of $\delta = \pm\pi/2$, that is, $E_x = E_y = E$. Their respective Jones vector representation is

$$R = \frac{1}{\sqrt{2}} \begin{bmatrix} 1 \\ i \end{bmatrix} \quad \text{and } L = \frac{1}{\sqrt{2}} \begin{bmatrix} 1 \\ -i \end{bmatrix}. \tag{5.7}$$

From the above equations it is easy to see that any polarization state can be represented as a linear combination of two orthogonal polarization states. For example, right and left-handed polarization can be represented as linear combinations of horizontal and vertical polarization states. Similarly, horizontal and vertical polarization can be written as a linear combination of diagonal and antidiagonal polarization. For the sake of clarity these four examples are written below:

$$R = \frac{H - iV}{\sqrt{2}}, \quad L = \frac{H + iV}{\sqrt{2}} \quad H = \frac{D + A}{\sqrt{2}}, \quad \text{and} \quad V = \frac{D - A}{\sqrt{2}}. \tag{5.8}$$

The most general state of polarization is the elliptical polarization, which can be represented using Jones vectors as

$$\begin{bmatrix} E_x \\ E_y \end{bmatrix} = \begin{bmatrix} \cos \varepsilon \cos \xi \pm i \sin \varepsilon \sin \xi \\ \sin \varepsilon \cos \xi \mp i \cos \varepsilon \sin \xi \end{bmatrix}, \tag{5.9}$$

where ε is the angle that the major axis of the polarization ellipse forms with the horizontal axis and $\xi = \arctan(E_x/E_y)$ is the ellipticity of the state of polarization. By taking $\varepsilon \in [0, \pi]$ and $\xi \in [-\pi/4, \pi/4]$, all polarization states can be reached.

5.1.1.2 Representation of polarization using Stokes matrices
The Jones representation of polarization facilitates its understanding and mathematical manipulation but it only applies to pure polarization states. Hence, a more general treatment of polarization, which can also be applied to partially polarized states is required. Such treatment, which set the basis for the modern description of polarization, was proposed by Stokes in 1842. Crucially, the Stokes treatment of polarization allows us to experimentally determine the state of polarization of any light beam, polarized or not, by using only intensity measurements. To this end, four quantities known as the Stokes parameters have to be determined, which are determined using a minimum of four intensity measurements, namely:

$$\begin{aligned} S_0 &= I_0 = I_R + I_L = I_H + I_V = I_D + I_A, \\ S_1 &= I_H - I_V = 2I_H - S_0, \\ S_2 &= I_D - I_A = 2I_D - S_0 \\ S_3 &= I_R - I_L = 2I_R - S_0, \end{aligned} \tag{5.10}$$

where $I_0 = I_R + I_L = I_H + I_V = I_D + I_A$ is the total intensity of the given optical field and I_H, I_V, I_A and I_D are the measured intensities of the transmitted electric field through a linear polarizer orientated at 0, $\pi/2$, $\pi/4$ and $3\pi/4$, respectively. The parameters I_L and I_R are the corresponding intensity components of the right- and left-handed polarization components, respectively. In this way, the first Stokes parameter (I_0) accounts for the total intensity of the field, the second quantifies the predominance of horizontal polarization over vertical, the third the predominance of diagonal over antidiagonal and the fourth the predominance of right over left circular polarization. In other words, if $S_1 > 0$ the beam's polarization is closer to horizontal than vertical, and so on.

An alternative representation of the Stokes parameters can be given in terms of the components of the electric field, E_y and E_y, as,

$$\begin{aligned} S_0 &= |E_x|^2 + |E_y|^2, \\ S_1 &= |E_x|^2 - |E_y|^2, \\ S_2 &= E_x^* E_y + E_x E_y^* \\ S_3 &= -i(E_x^* E_y + E_x E_y^*), \end{aligned} \tag{5.11}$$

which can be further simplified using the unitary matrix σ_0 and the Pauli spin matrices σ_i, $i = 1, 2, 3$,

$$\sigma_0 = \begin{bmatrix} 1 & 0 \\ 0 & 1 \end{bmatrix}, \quad \sigma_1 = \begin{bmatrix} 0 & 1 \\ 1 & 0 \end{bmatrix}, \quad \sigma_2 = \begin{bmatrix} 0 & -i \\ i & 0 \end{bmatrix} \quad \text{and} \quad \sigma_3 = \begin{bmatrix} 1 & 0 \\ 0 & -1 \end{bmatrix} \quad (5.12)$$

as

$$S_0 = \mathbf{E}^{\dagger}\sigma_0\mathbf{E}, \qquad S_1 = \mathbf{E}^*\sigma_3\mathbf{E}, \qquad S_2 = \mathbf{E}^*\sigma_1\mathbf{E}, \qquad S_3 = \mathbf{E}^*\sigma_2\mathbf{E},$$

As mentioned above, this treatment of polarization also applies to non-polarized light, in general, $S_0^2 \geqslant S_1^2 + S_2^2 + S_3^2$, with the equality holding for polarized light. As a final remark, the four Stokes parameters can be represented as a four entry column vector as

$$\mathbf{S} = \begin{bmatrix} S_0 \\ S_1 \\ S_2 \\ S_3 \end{bmatrix} = \begin{bmatrix} \mathbf{E}^*\sigma_0\mathbf{E} \\ \mathbf{E}^*\sigma_3\mathbf{E} \\ \mathbf{E}^*\sigma_1\mathbf{E} \\ \mathbf{E}^*\sigma_2\mathbf{E} \end{bmatrix}. \qquad (5.13)$$

Common examples of the Stokes representation for unpolarized, linear horizontal and vertical vertical polarization, as well as right- and left-handed circular polarization, are

$$\mathbf{S}_u = \begin{bmatrix} 1 \\ 0 \\ 0 \\ 0 \end{bmatrix}, \quad \mathbf{S}_H = \begin{bmatrix} 1 \\ 1 \\ 0 \\ 0 \end{bmatrix}, \quad \mathbf{S}_V = \begin{bmatrix} 1 \\ -1 \\ 0 \\ 0 \end{bmatrix}, \quad \mathbf{S}_R = \begin{bmatrix} 1 \\ 0 \\ 0 \\ 1 \end{bmatrix}, \quad \mathbf{S}_L = \begin{bmatrix} 1 \\ 0 \\ 0 \\ -1 \end{bmatrix} \quad (5.14)$$

5.1.1.3 The Poincaré sphere for polarization

In 1892, Henry Poincaré, who more than any other saw the physical implications of geometry in polarization, provided an alternative representation of polarization that relied on stereographic projection. In this representation, each point on a plane, representing all the possible polarization states, is mapped in a one-to-one correspondence with all the points of a unit radius sphere, which is nowadays known as the Poincaré sphere.

To start with, let us remember that any state of polarization can be fully determined by two parameters, its ellipticity $\chi \in [0, \pi]$, which is the ratio between the major and minor axis of the polarization ellipse, and the angle of rotation $\psi \in [-\pi/4, \pi/4]$, the angle that the major axis of the polarization ellipse forms with the horizontal axis, whose sign indicates the direction of the polarization rotation, negative for left-handed and positive for right-handed. For the sake of clarity, these parameters of polarization are schematically illustrated in figure 5.1(a).

Crucially, the Stokes parameters can also be written in terms of ψ and χ as [22],

$$\begin{aligned} S_1 &= S_0 \cos(2\chi)\cos(2\psi), \\ S_2 &= S_0 \cos(2\chi)\sin(2\psi), \\ S_3 &= S_0 \sin(2\chi) \end{aligned} \qquad (5.15)$$

from which expressions for ψ and χ in terms of the Stokes parameters can be derived, namely,

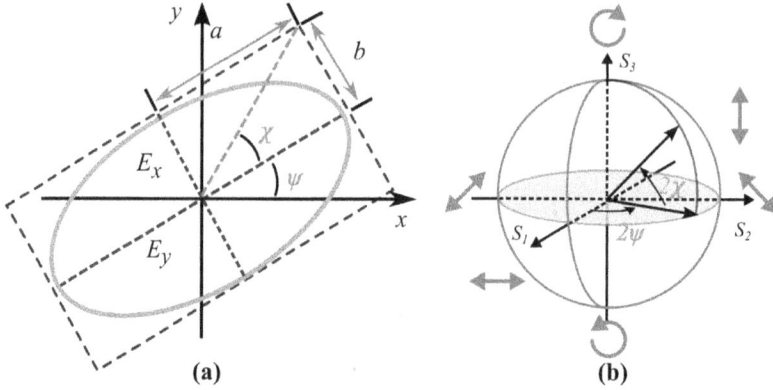

Figure 5.1. Geometric representation of polarization. (a) Polarization ellipse and (b) its representation on the the Poincaré sphere.

$$\tan(2\psi) = \frac{S_2}{S_1} \quad \text{and} \quad \sin(2\chi) = \frac{S_3}{S_0}. \tag{5.16}$$

Equation (5.15) is analogous to the transformation expressions that link the Cartesian coordinates (x, y, z) to the spherical coordinates (r, α, β) via the transformation

$$
\begin{aligned}
x &= r \sin\theta \cos\phi, \\
y &= r \sin\theta \sin\phi, \\
z &= r \cos\theta.
\end{aligned} \tag{5.17}
$$

Here $r \in [0, \infty)$ is the radius of the sphere, $\alpha \in [0, 2\pi]$ is the azimuthal angle in the $x - y$ plane and $\beta \in [0, \pi]$ is the polar angle from the z- axis. The analogy between this transformation and the representation of polarization can be readily seen by setting $\theta = \phi/2 - 2\chi$, $\phi = 2\psi$, $r = S_0$, $x = S_1$, $y = S_2$ and $z = S_3$, which immediately yields equation (5.15). Hence, any state of polarization can be described in the surface of a sphere, which for convenience is set to unitary radius, as schematically shown in figure 5.1(b). In this geometric representation the upper and lower poles represent right and left circular polarization states, corresponding to $\chi = \pi/4$ and $\chi = -\pi/4$, respectively, while points on the equator correspond to linear polarization, for which $\chi = 0$. Diametrically opposite points on the equator correspond to horizontal and vertically polarized light with $\psi = 0$ and $\psi = \pi/2$, respectively. The rest of the points represent elliptic polarization states.

5.2 Definition of vector vortex beams

As mentioned earlier, until very recently, it was common to study light modes with homogeneous polarization distributions, known as scalar beams, but in recent time vector modes started to gain popularity [1]. In fact, vector beams have been known for a long time in the context of conical diffraction [23], singular optics [24, 25], spin–orbit coupling [26–28], and more recently and controversially, classical entanglement [20, 29–34], but the recent ability to easily create such structured light beams in common laboratories has fuelled their use in fundamental and applied studies. Vector modes are easily distinguished from scalar modes when passing through a linear polarizer, which reveals either a changing pattern (vector beam) or an unchanging pattern (scalar beam), as shown in figure 5.2. Further, the overall structure of vector beams may change periodically or quite dramatically during propagation in free space [35, 36], paving the way towards a novel type of vector beams, with new potential applications.

From the mathematical perspective, vector beams are natural solutions to the vectorial Helmholtz equation [37–40], which are often generated as a nonseparable coaxial superposition of the spatial and polarization degrees of freedom. More precisely, as the coaxial superposition of two orthogonal scalar fields bearing orthogonal polarization states. Mathematically, this nonseparable superposition can be written as a weighted addition of the form [4, 7, 41],

$$U(\mathbf{r}) = \cos\left(\frac{\theta}{2}\right) u_R(\mathbf{r}) e^{i\alpha_1} \hat{e}_R + \sin\left(\frac{\theta}{2}\right) u_L(\mathbf{r}) e^{i\alpha_2} \hat{e}_L, \qquad (5.18)$$

where the unitary vectors $\hat{e}_R$ and $\hat{e}_L$ represent the right and left circular polarization components, and $u_R(\mathbf{r})$ and $u_L(\mathbf{r})$ are the associated amplitude of the spatial modes with weighting coefficients $\cos\theta$ and $\sin\theta$, respectively. The parameters α_1 and α_2 represent an initial phase term, which in general is not zero. Crucially, even though the polarization degree of freedom is limited to a two-dimensional space, the spatial degree of freedom is unbounded, and $u_R(\mathbf{r})$ and $u_L(\mathbf{r})$ can be any two orthogonal modes from the various sets of solutions of the wave equation. Importantly, the wave equation has been solved in Cartesian, cylindrical, elliptical and parabolic coordinates, amongst others. This paved the path for the generation of Laguerre–, Ince–, Parabolic–and Mathieu–Gauss vector modes, to mention a few [35, 42–45].

With the aim of providing a topical and intuitive approach to unpack the general concepts of vector modes, in what follows and without the loss of generality, we will

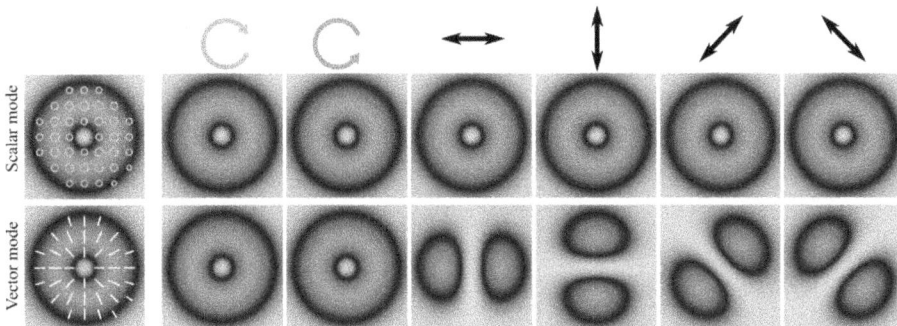

Figure 5.2. Intensity profile and polarization distribution of scalar and vector beams. Scalar beams have homogeneous polarization while vector beams have inhomogeneous polarization. As a consequence, when scalar modes (top row) traverse a linear polarizer oriented at different angles, their spatial profile is unaffected (except for an amplitude factor). On the contrary, when vector modes (bottom row) traverse through the linear polarizer, its intensity profile rotates as the linear polarizer is rotated.

restrict ourselves to the set of Laguerre–Gauss (LG_p^ℓ) vector modes, which have become popular in recent times, as they have found numerous applications. To make this clear, we will rewrite equation (5.18) using the LG_p^ℓ modes as [4, 7, 46, 47]

$$U_{LG}(\mathbf{r}) = \cos\left(\frac{\theta}{2}\right)LG_{p_1}^{\ell_1}\,\hat{e}_R + \sin\left(\frac{\theta}{2}\right)LG_{p_2}^{\ell_2}\,e^{i\alpha}\hat{e}_L, \qquad (5.19)$$

where we have assumed $\alpha_1 = 0$ and $\alpha_2 = \alpha$. The set of vector modes given by equation (5.19), commonly known as Poincaré beams, can be classified into two subsets defined by $p = 0$ and $p \neq 0$. Each of these subsets can be split into two smaller subsets, $\ell_1 = \ell_2$ and $\ell_1 \neq \ell_2$. More precisely, such modes are typically classified by the type of disclinations (line patterns) present in the orientation of the polarization ellipse and can be divided into three morphologies: lemon, star and monstar [4, 48, 49]. The centre of a disclination, commonly known as a C-point, features a singularity of circular polarization with undefined orientation. The strength of the singularity, known as the index of disclination, is measured through the number of rotations in the orientation of the lines per circulation about the singularity and denoted as I_C. The number of radial lines N originating at the singularity is also a signature of the type of disclination, this being a consequence of the rotation of the lines about the singularity. The lemon pattern with only one radial line has a singularity index $I_C = 1/2$, the star pattern with three radial lines has an index $I_C = -1/2$ and the monstar also with three radial lines has and index $I_C = 1/2$ [50–53]. It is noteworthy that the disclination index I_C as well as the number of radial lines can be linked to the topological charges ℓ_1 and ℓ_2 of the LG_0^ℓ modes generating the Poincaré beam through the relation

$$I_C = \frac{\ell_1 - \ell_2}{2}, \qquad N = |2(I_c - 1)| = |(\ell_1 - \ell_2) - 2|. \qquad (5.20)$$

Importantly, high-order disclinations can also be obtained from this equation, giving raise to high-order topological structures that have adopted names such as spiders, flowers or spider webs. In fact, each modal combination can have its own space of high-order polarization disclinations; a full analysis of this lies beyond the scope of this chapter so we refer the reader to references [54–56] for further information on this fascinating topic. Figure 5.3 shows some examples of low- and high-order topological disclinations.

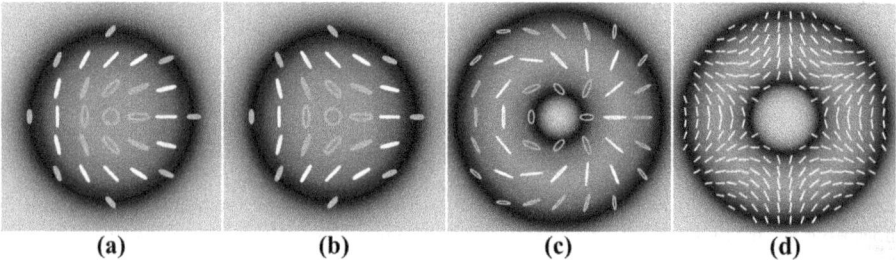

Figure 5.3. Polarization morphology overlapped with the intensity distribution of the lowest- and higher-order polarization disclinations, lemon (a), star (b) and spider (c), web (d), respectively.

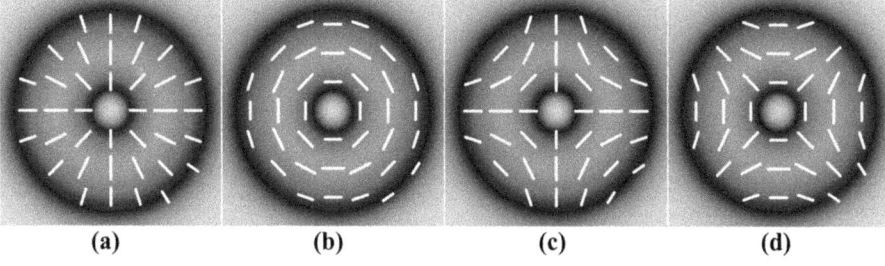

Figure 5.4. Common low order vector fiber modes. (a) TE, (b) TM, (c) HE^e and (d) HE^o.

Amongst all this variety of vector modes, of particular interest are the so-called Bell modes, which consist of four vector modes obtained for $p = 0$, $\ell = \pm 1$, and $\alpha = \{0, \pi\}$, namely

$$TE(\mathbf{r}) = \frac{1}{\sqrt{2}}(e^{i\varphi}\hat{e}_R + e^{-i\varphi}\hat{e}_L)u(\mathbf{r}),$$

$$TM(\mathbf{r}) = \frac{1}{\sqrt{2}}(e^{i\varphi}\hat{e}_R - e^{-i\varphi}\hat{e}_L)u(\mathbf{r}),$$

$$HE^o(\mathbf{r}) = \frac{1}{\sqrt{2}}(e^{i\varphi}\hat{e}_L + e^{-i\varphi}\hat{e}_R)u(\mathbf{r}),$$

$$HE^e(\mathbf{r}) = \frac{1}{\sqrt{2}}(e^{i\varphi}\hat{e}_L - e^{-i\varphi}\hat{e}_R)u(\mathbf{r}),$$

(5.21)

where $u(\mathbf{r})$ carries the radial amplitude information of the LG_p^ℓ modes. The four modes, which are illustrated in figure 5.4, can propagate in both free space and optical fibers. The modes TE and TM, feature a radial and azimuthal polarization, which are shown in figures 5.4(a) and 5.4(b), respectively. The modes HE^o and HE^e feature a hybrid polarization distribution, which are shown in figures 5.4(c) and 5.4(d), respectively.

5.2.1 The higher-order Poincaré sphere

In the previous section we described a geometric way to visualise polarization in the so-called Poincaré sphere. Crucially, such formalism can be extended to represent vector vortex modes in a higher-order Poincaré Sphere (HOPS) [57, 58] and its more general version, the hybrid-order Poincaré sphere [59]. In this representation, a new set of higher-order Stokes parameters are defined, which takes the form

$$S_0^{\ell_1,2} = |\Psi_R^{\ell_1}|^2 + |\Psi_L^{\ell_2}|^2$$

$$S_1^{\ell_1,2} = 2|\Psi_R^{\ell_1}||\Psi_L^{\ell_2}|\cos(\delta_2 - \delta_1)$$

$$S_2^{\ell_1,2} = 2|\Psi_R^{\ell_1}||\Psi_L^{\ell_2}|\sin(\delta_2 - \delta_1)$$

$$S_3^{\ell_1,2} = |\Psi_R^{\ell_1}|^2 - |\Psi_L^{\ell_2}|^2.$$

(5.22)

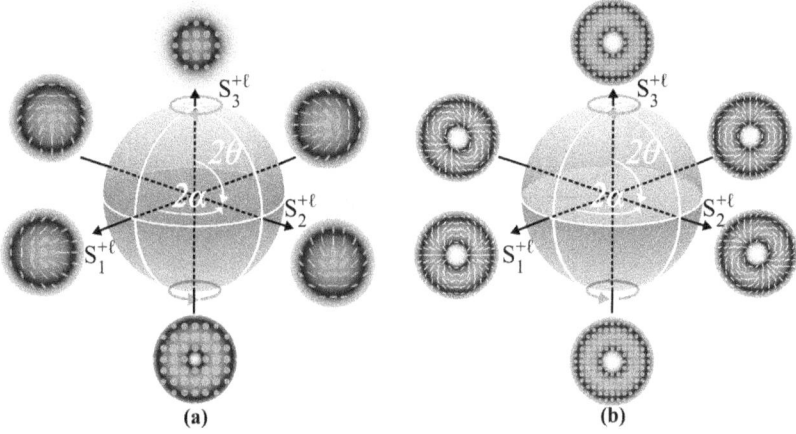

Figure 5.5. Higher-order Poincaré sphere representing vector vortex modes, for (a) $\ell_1 = 0, \ell_2 = 1, p = 0$ and (b) $|\ell_1| = 2, |\ell_2| = -2, p = 0$. Here orange and green ellipses represent right- and left-handed circular polarization, respectively, whereas white segments represent linear polarization.

Here, $\Psi_R^{\ell_1}$ and $\Psi_L^{\ell_2}$ are two orthogonal scalar modes with corresponding right- and left-handed circular polarization. Notice the dependence of these parameters on the topological charges ℓ_1 and ℓ_2. In a similar way to the Poincaré sphere for polarization, the higher-order Stokes parameters defined above enable the representation of vector modes. Here, scalar states with right- and left-handed circular polarization are represented on the poles and nonseparable states everywhere else, with maximally nonseparable states on the equator. In general, any vector mode is represented as a point on the HOPS with coordinates (α, θ) defined in terms of the Stokes parameters as

$$\alpha = \frac{1}{2} \arctan\left(\frac{S_2^{\ell_{1,2}}}{S_1^{\ell_{1,2}}}\right) \qquad \text{and} \qquad \theta = \frac{1}{2}\arcsin\left(\frac{S_3^{\ell_{1,2}}}{S_0^{\ell_{1,2}}}\right). \tag{5.23}$$

where $\theta \in [0, \pi]$ and $\alpha \in [0, 2\pi]$. Optical vortices with homogeneous right or left states of polarization are located at the upper and lower poles, where ϕ takes the values 0 and π, respectively. The equator, where $\phi = \pi/2$, is occupied by equally weighted left-right circular polarized vector modes with opposite topological charges. For the sake of clarity, figure 5.5(a) illustrates the HOPS for $|\ell_1| = 0$, $|\ell_2| = 1$ and $p = 0$. In a similar way, figure 5.5(b) shows the HOPS for the parameters $|\ell_1| = 2, |\ell_2| = -2$ and $p = 0$.

5.3 Generation of vector vortex beams

Given the mathematical description of a vector beam, its generation process relies on the ability to tailor (or structure) light in its various degrees of freedom and to recombine them as a nonseparable superposition. Of particular interest are vector beams generated as a nonseparable superposition of orthogonal spatial modes with orthogonal polarization states. The means and methods to achieve this have evolved quite dramatically over the years, from bulky static fashions, towards more versatile and compact solutions, some of which have easy integration into small-scale devices. Many of these techniques can be implemented either by dynamic or geometric phase modulation. The former relies on the coaxial superposition of two coherent orthogonal spatial modes bearing orthogonal polarization, commonly achieved by interferometric means. Here the Sagnac and Mac-Zehnder interpreters have become popular, even though in recent time the Michelson interferometer has also been used [60]. In essence an input beam is split into two beams propagating along different paths where the polarization and spatial degrees of freedom of each beam can be easily manipulated in an independent way, prior to their coaxial recombination. Furthermore, the state of polarization can be manipulated through polarization elements as well as phase retarders, such as quarter and half wave plates. In contrast, there are several ways to manipulate the spatial degree of freedom, that is, phase and amplitude, such as spiral phase plates (SPPs), conical lenses (also known as axicons, fork holograms and more recently, computer-controlled devices). Examples of the latter are the liquid crystal spatial light modulators (SLMs) and the digital micro-mirror devices (DMDs), which provide almost unlimited flexibility for the generation of a wide variety of spatial modes. By contrast the geometric phase control is based on the direct transformation from spin-to-orbital angular momentum using an inhomogeneous and anisotropic material as the intermediary, a process known as spin-to-orbital conversion, achieved through the use of liquid crystal birefringent material or subwavelength gratings with periods smaller than the wavelength of the input light beam. This spatially-varying polarization transformation induces an optical phase delay linked to the geometry of the subwavelength structure, which is responsible for the conversion of arbitrary homogeneously polarized states of light into vector beams. One of the drawbacks of the geometric phase control is that they cannot be reconfigured, once fabricated they can only generate one particular optical field of specific intensity profile, polarization distribution and wavelength. In what follows we will describe both the geometric and dynamic phase control in more detail.

5.3.1 Dynamic phase control

Perhaps the most direct, but not necessarily the easiest, way to generate vector modes relies on the coaxial superposition of two optical fields by interferometric means, previously tailored in both degrees of freedom, spatial and polarization. Some of the first approaches relied on the use of optical elements such as SPPs, which transforms a Gaussian beam into so-called vortex beams, light beams endowed with orbital angular momentum [61–63]. Crucially, recent technological developments have provided us with computer-controlled devices that are more flexible and versatile, allowing for the almost unlimited generation of vector modes. There are mainly two types of devices, phase-only SLMs and DMDs. The former are polarization-dependent devices based on the liquid crystal technology that operate either by reflection or transmission and can be addressed with grayscale images that modulate the phase of the input beam [64]. The latter employs a completely different technology initially developed for projection systems, which is still in use at present time: the DMD technology. A DMD consists of an array of hundreds of thousands of micron-sized mirrors, each of which can be set to an 'Off' or 'On' state, by tilting the micromirror at two different angles. Therefore, DMDs reshape light using digital binary holograms [43, 65–67]. By way of example, figure 5.6 illustrates both types of holograms (grayscale and binary). First in figure 5.6(a) we show two types of grayscale holograms phase only and complex-amplitude modulation, by contrast, in figure 5.6(b) we show a binary hologram.

As a first example for vector beam generation using the dynamic phase control we will explain a polarization-sensitive device that does not require the input beam to be split [68–71]. For the sake of clarity, we will assume that liquid crystal SLMs are birefringent and therefore they can only modulate light polarized along the extraordinary axis, leaving light polarized along the ordinary axis unaffected. This polarization-sensitive property will be revisited in chapter 8, where it will be explained in more detail. As such, when an input light beam impinges on the SLM with diagonal polarization, only one polarization component is modulated, typically horizontal, reflecting the vertical exactly as it entered. This property is commonly considered a drawback, since in order to achieve the maximum modulation efficiency, it is necessary to ensure the input beam impinges with horizontal polarization. Nonetheless, it can been exploited favourably to generate vector beams. The experimental implementation is schematically shown in figure 5.7, whereby a light beam with diagonal polarization (D-pol) impinges on first section of an SLM, addressed with the hologram shown as an inset on the bottom of the image (H_1), which correspond to an optical vortex of topological charge $\ell = 1$. It is worth noting that the displayed holograms do not contain a linear grating and therefore, modulated and unmodulated light propagate along the same axis. As a result of the SLM configuration, the resulting beam contains a nonseparable superposition of two orthogonal modes with orthogonal polarization. More precisely, the input Gaussian beam carrying vertical polarization and the modulated light, in this case an optical vortex, carrying horizontal polarization. Afterwards, both polarization components are rotated 90° to an antidiagonal polarization state (A-pol) and the beam is reflected back to the second section of the SLM where the hologram shown

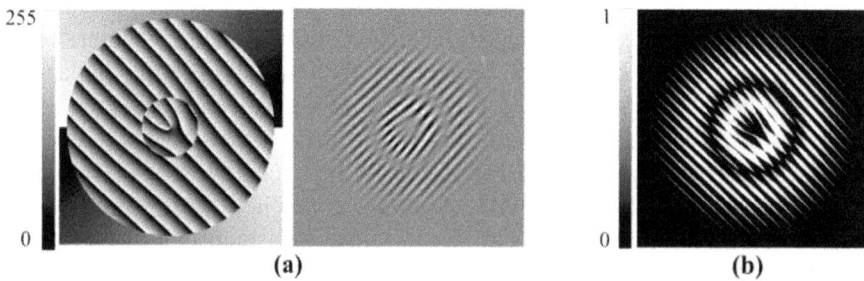

Figure 5.6. (a) Grayscale holograms encoded on liquid crystal spatial light modulators. On the left it is shown as a phase-only hologram, whereas on the right it is a complex-amplitude one. (b) Binary holograms encoded on digital micromirror devices.

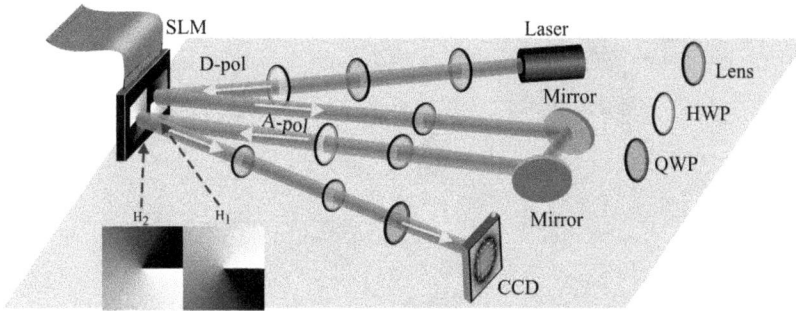

Figure 5.7. Generation of vector vortex modes with an SLM. Here, a vector beam is generated via a double pass of a single beam through two different sections of an SLM. In a first pass, the beam impinges with diagonal polarization (D-pol). Given the configuration of the SLM, only the horizontal polarization component will be modulated, leaving the vertical intact. The polarization of the resulting beam is then rotated to an orthogonal polarization sate, that is antidiagonal polarization (A-pol) and reflected back to the second section of the SLM, where now the other polarization state will get the modulation encoded on the SLM, generating in this way the desired vector beam.

on the bottom of the image is displayed (H_2). This time the previously generated light beam will pass through the SLM without any alteration as it now carries vertical polarization. On the contrary, the other component, which is still a Gaussian beam, will be transformed according to the hologram H_2, in this case also an optical vortex with opposite topological charge $\ell = -1$. The resulting beam is precisely a nonseparable superposition of two orthogonal spatial modes with orthogonal polarization. Lastly, a quarter-wave plate can be inserted to transform the vector beam from the linear polarization basis to the circular. For this technique it is crucial that the modulated and unmodulated light travel together all the time. Hence, in principle it cannot be combined with complex-amplitude modulation, which relies on the spatial separation of these two. Nonetheless a recent technique, which is an advanced version of this, was proposed, capable of generating arbitrary vector beams combining the technique explained above with complex-amplitude modulation [72].

Alternative approaches, which also rely on manipulating the transverse profiles of both polarization components independently, have been implemented using interferometric arrays containing one or two SLMs [42, 62, 69, 70, 73–82], most of which follow a principle similar to the one reported by Neil *et al* [83]. A variation of this approach based on a Sagnac interferometer is illustrated in figure 5.8. Here a horizontally polarized laser beam, collimated and expanded to fully cover the liquid crystal screen of the SLM, is split into two independent beams, by digitally splitting the SLM's screen into two independent screens. Each section is addressed with an independent digital hologram that generates, in the first diffraction order, two independent optical fields. Hence, while one side of the SLM encodes one mode, the other encodes a mode orthogonal to this, by way of examples, the holograms labelled as H_1 and H_2 are shown at the bottom of figure 5.8. Each hologram is superimposed with a linear grating to separate the different diffraction orders and to

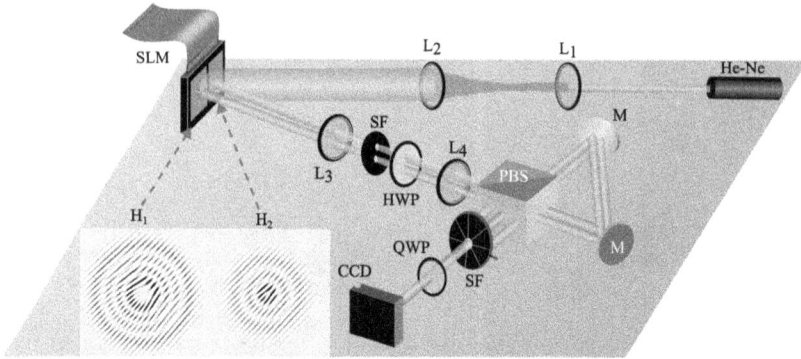

Figure 5.8. Generation of vector beams through a modified Sagnac interferometer. Here an expanded and collimated Gaussian beam impinges onto the screen of an SLM, whereby it is split into two orthogonal spatial modes, which enter a triangular Sagnac interferometer formed by a polarising beam splitter (PBS) and two mirrors. Prior to entering the interferometer, the polarization of both beams is rotated to diagonal so that the PBS splits each beam into its vertical and horizontal components, which travels along opposite directions inside the interferometer. After one round, the beams exit through one of the faces of the PBS, where two orthogonal spatial modes with orthogonal polarization are coaxially recombined to generate the desired vector mode. Here L_1 to L_4 are lenses; HWP and QWP are a half- and quarter-wave plate, respectively; M represents the mirrors; SF represents spatial filters and CCD stands for charge-coupled device camera. H_1 and H_2 are the corresponding holograms encoded on both sections of the SLM.

filter the first diffraction using a telescope formed by two lenses and a spatial filter located in the focal plane of the first lens. The two optical fields are then redirected to a common-path triangular Sagnac interferometer comprising a polarising beam splitter (PBS) and two mirrors. Prior to entering the interferometer, the two beams are rotated to a diagonal polarization state. In this way, when the two beams enter the interferometer after traversing the PBS, each of them is separated into two new beams with orthogonal linear polarization states (horizontal and vertical) travelling along opposite optical paths. After a round trip, all four beams exit the interferometer from the opposite side of the PBS, two with horizontal polarization and two with vertical. Finally, the horizontal polarization component of one of the beams is aligned coaxially with the vertical polarization component of the other beam. To improve the coaxial superposition, fine tuning is performed digitally via the period of the linear gratings encoded on each digital hologram. The use of SLMs enable the simultaneous generation of over 100 scalar modes using a single hologram [84–86], and similarly to generate multiple vector vortex beams [86].

Digital micromirror devices represent an alternative and powerful tool for the generation of vector modes. Amongst their various advantages, we can highlight their polarization-independent property, which allows us to generate vector modes in a more compact fashion [35, 43, 45, 87]. Other advantages related to the use of DMDs are related to their high refresh rates, low cost and wavelength independence; this, however, is at the cost of a low efficiency. At this point, it is worth emphasising that the DMD technology has been around for several decades and is the main component of projection systems. Hence, they were not invented as light modulation systems, nonetheless, in the last ten years or so, they have gained popularity in

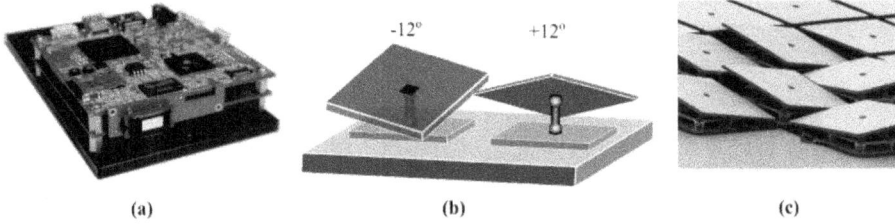

Figure 5.9. (a) Illustrative example of a digital micromirror device after removing the projection optics to expose the screen where the digital holograms are displayed. The screen of the DMD is composed of over two million micromirrors (depending on the model), each of which can be tilted to $-12°$ or $+12°$. (c) A section of a DMD observed through a scanning electron microscope.

structured light laboratories. For illustrative purposes, figure 5.9(a) shows a schematic illustration of the DLP3000-C300REF, whereby the optics have been already removed to expose the DMD screen, where the digital holograms are displayed. To understand how DMDs are used to generate vector modes, it is worth emphasising that a DMD consists of an array of millions of micron-sized mirrors ($\approx 8\mu$ m in size), each of which can be turned to an 'Off' or 'On' state by tilting it $-12°$ or $+12°$, respectively, as illustrated schematically in figure 5.9(b). For completeness, figure 5.9(c) shows a section of a DMD observed through a scanning electron microscope, where some the micromirrors can be seen in either of these two directions. Hence, when the DMD is properly aligned, each mirror in the 'On' state reflects light in the desired direction, contributing to the generation of the light beam encoded on the digital hologram [65, 88–90].

Since each mirror can only be in two states, the DMD has to be addressed with binary holograms, which are appropriately designed to perform full complex-amplitude modulation. The methods to shape both the amplitude and phase of a complex light field $u(x, y) = A(x, y)\exp[i\phi(x, y)]$, with amplitude $A(x, y)$ and phase $\phi(x, y)$, via binary-amplitude holograms were developed in the 1960s [91]. In particular, one of the most known methods, which was proposed by Lee, consists of creating a periodic binary-amplitude grating, where the phase and amplitude information is encoded in the first diffraction order. The mathematical expression for the binary grating has the form [92, 93],

$$T(x, y) = \frac{1}{2} + \frac{1}{2}\text{sgn}\{\cos[p(x, y)] + \cos[\pi q(x, y)]\}, \qquad (5.24)$$

where sgn $\{\cdot\}$ is the sign function that reflects the binary-amplitude modulation and forces all arguments to either a value of 0 or 1. Hence, to generate the complex field $u(x, y) = A(x, y)\phi(x, y)$, the functions $p(x, y)$ and $q(x, y)$ are defined in terms of the amplitude and phase information as

$$\begin{aligned} q(x, y) &= \arcsin(A(x, y)/A_{max})/\pi, \\ p(x, y) &= \phi(x, y) + 2\pi(\nu x + \eta y), \end{aligned} \qquad (5.25)$$

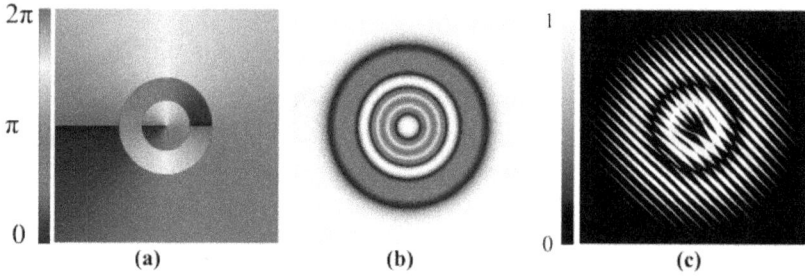

Figure 5.10. The phase (a) and amplitude (b) information of a complex light beam, in this case a Laguerre–Gaussian beam of parameters $p = 1$ and $\ell = 2$ are captured by a binary hologram (c), which is displayed as an image on the screen of the DMD.

respectively. Here the term A_{max} represents the maximum amplitude value that normalizes $A(x, y)$. By way of example, figure 5.10(a) shows the phase distribution $\phi(x, y)$ of a Laguerre–Gauss beam of parameters $p = 1$ and $\ell = 2$, along with its amplitude distribution $A(x, y)$, which is shown in figure 5.10(b). The corresponding binary hologram for this particular example is also shown in figure 5.10(c) where the amplitude distribution of the LG beam is mimicked by the micromirrors in the 'On' state. In this figure the mirrors in the 'Off' state correspond to black colour, which are associated to a 0 value, whereas those in the 'On' state correspond to white colour and take the value 1. Notice how the spatial variations of the amplitude $A(x, y)$ and phase $\phi(x, y)$ of the complex light field are achieved locally through local variations of the width of the apertures apertures (the local duty cycle of the grating). Similarly, the phase is modulated through local variations of the lateral position of the apertures.

In addition, the term $2\pi(\nu x + \eta y)$ corresponds to a linear grating, which controls the separation and angle of the multiple diffraction orders as a function of the parameters. By way of example, figure 5.11(a)–(c) illustrates the effect of such a binary periodic grating for different values of ν and η, taking the particular case of an LG beam of parameters $p = 1$ and $\ell = 1$. First, figure 5.11(a) illustrates the effect of a horizontal grating (top), for which $\eta = 0$ and separates the beams along the vertical direction, as schematically illustrated on the bottom of the same figure. In figure 5.11(b) shows the case of a horizontal ($\nu = 0$), which separates the beams along the horizontal direction. Figure 5.11(c) illustrates the case of a diagonal grating oriented at $45°$ ($\nu = \eta$), which separate the beams along the diagonal directions. Finally, figure 5.11(d) illustrates the effect of increasing the values of ν and η, which increases the separation of the modes.

To generate complex vector modes of arbitrary spatial distributions, several methods have been proposed, but perhaps one of the first approaches that took full advantage of the polarization-insensitive is the one described in [43], which is schematically represented in figure 5.12(a). To begin with, a horizontally polarized laser beam ($\lambda = 523$ nm) is collimated and expanded by lenses L_1 ($f_1 = 20$ mm) and L_2 ($f_2 = 200$ mm). Afterwards, the polarization state of the expanded beam is rotated to the diagonal polarization state with the help of a half-wave plate (HWP)

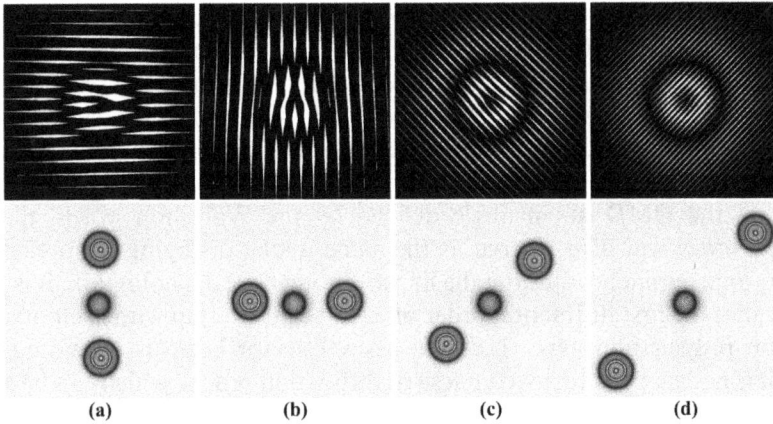

Figure 5.11. (a) A horizontal grating (top) separates the generated modes along the vertical direction. (b) By contrast, a vertical grating separates the modes along the horizontal direction. (c) A diagonal grating separates the generated beams in the orthogonal direction to that of the grating. (d) An increase in the grating period increases the separation of the generated beams.

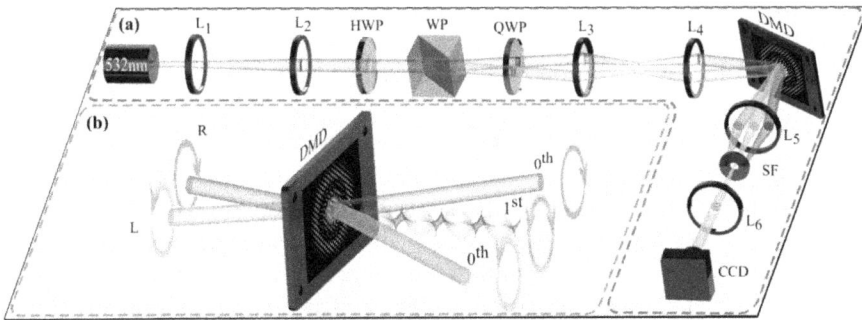

Figure 5.12. (a) Schematic representation of the experimental setup for the polarization-insensitive generation of arbitrary vector mode based on DMDs. (b) The concept of how the linear diffraction grating affects the angle of the first diffraction order. Here, the multiplexing hologram encoded on the DMD is the superposition of two independent holograms with unique spatial carrier frequencies. Such parameters need to be carefully picked to guarantee the overlap of the first diffraction order from each beam along the same propagation path. Adapted from [43] CC BY 4.0.

at 22.5°. A Wollaston prism (WP) subsequently separates the beam into its horizontal and vertical polarization components, both of which are transformed to the circular polarization basis $(\hat{l}, \hat{r})$ by means of a quarter-wave plate (QWP). A 4f imaging system composed of lenses L_3 and L_4 ($f_3 = f_4 = 200$ mm) redirects these two beams towards the centre of a DMD where they impinge at slightly different angles ($\approx 1.5°$) but exactly at the same spatial location, the centre of the hologram. A multiplexed binary-amplitude hologram based on the spatial random multiplexing method is encoded on the DMD. Such multiplex holograms consist of the superposition of two individual holograms, overlapped with a linear diffraction grating. The period of the diffraction grating is carefully chosen to ensure the overlap of the first diffraction order of each beam along a common propagation path, where the desired complex vector field $\vec{u}(\vec{r})$ is generated. This process can be explained in a more detailed way using the schematic representation shown in figure 5.12(b), where for the sake of clarity the DMD is represented as a transmission device. Here, the two input beams, which carry orthogonal circular polarization, impinge in the centre of the displayed binary hologram. As illustrated and given that the input beams impinge on the DMD at an angle defined by the Wallaston prism, the zeroth diffraction order will also emerge at the same angle, diverging from each other. Hence, by appropriately adjusting the linear gratings of each hologram, it is possible to ensure that the first diffraction order of each beam overlap with each other along a common propagation axis where the desired vector beam is generated. In this scheme, it is necessary to remove undesired diffraction orders, which can be achieved by placing a spatial filter (SF) in the far-field plane of a telescope, in this case formed by lenses L_5 and L_6. For the sake of clarity, figures 5.12(a) and (b) show only the first and zeroth diffracting orders but many other are present. It is worth mentioning that this generation device enables the generation of arbitrary vector modes with almost unlimited spatial and polarization distributions, and therefore it has been used to generated vector beams in non conventional coordinate systems [35, 44, 45, 94]. An alternative approach based on a modified Sagnac interferometer has also been proposed recently [67, 87].

To illustrate the superposition process described before for the generation of the digital hologram displayed on the DMD, figure 5.13 shows a particular example whereby two LG beams are used to generate a vector beam. Figure 5.13(a) shows the binary hologram that generates the LG beam with an intensity distribution shown on the top-right corner of the same figure. Similarly, figure 5.13(b) shows the binary hologram of the LG mode whose intensity profile is shown also in the top-right corner. Notice that both linear gratings are oriented at orthogonal directions, to ensure that the first diffraction order of each generated beam coincide along the same propagation axis. Finally, figure 5.13(c) shows the resulting multiplexed hologram (left), which generates the vector beams illustrated on the right of the

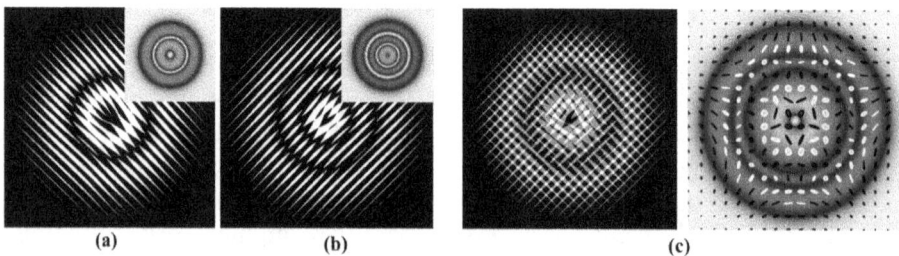

(a)   (b)   (c)

Figure 5.13. Schematic representation of the superposition of two binary holograms to generate a multiplexed hologram, which is displayed on the DMD to generate a vector beam. (a) Binary hologram that generates the LG beam shown on the top-right corner. (b) Binary hologram of the LG mode whose intensity profile is shown also in the top-right corner. (c) Resulting multiplexed hologram (left), which generates the vector beam shown on the right, where the polarization distribution of the beam is overlapped with its intensity profile.

same figure, where the polarization distribution of the beam is overlapped with its intensity profile.

As a closing remark for this section, it is worth mentioning that while SLMs and DMDs represent the most flexible and versatile devices to generate arbitrary vector modes, SLMs are an expensive solution while DMDs have poor diffraction efficiency. Hence, alternative tools are desirable to produce vector light fields, and diffractive optical element (DOEs) represent an alternative solution to this end [95, 96]. Importantly, they have higher diffraction efficiency when compared to DMDs, are much cheaper than SLMs, and are very compact, making them suitable for small-scale integration. However, this comes at the cost of versatility since DOE are manufactured to produce a unique optical field.

5.3.2 Geometric phase control

The geometric phase control involves a process known as spin-to-orbital conversion (SOC) of angular momentum, where a direct transformation from spin-to-orbital angular momentum takes place, with matter being the intermediary [97]. Two phenomena are key in this process; first, in an optically anisotropic media only the spin angular momentum of light is transferred to matter, and second, in an inhomogeneous isotropic transparent media only an orbital angular momentum interaction takes place. Crucially, these two mechanisms of light–matter interactions are not independent in a material which is both inhomogeneous and anisotropic and under certain conditions the exchange of spin affects the direction (sign) of the exchange of orbital angular momentum. Furthermore, under specific geometrical conditions both exchanges remain always exactly opposite to each other, resulting in a zero transfer of angular momentum from light to matter. A widely known optical element based on this optical process is the q-plate [98–101]. A q-plate is essentially a slab of a liquid crystal with a uniform birefringent phase retardation δ and a transverse optical axis pattern with a nonzero topological charge that can be dynamically aligned with an external electric, magnetic or optical field. The distribution of the pattern is described by the number of cyclic rotations (q) around the centre of the plate. For example, when a circularly polarized Gaussian beam traverses a QP, a helical beam of topological charge $\ell = \pm 2q$ is generated at the output, with its sign determined by the input polarization state. More precisely, given an input (left-) right-circularly polarized incident on such an optical element, the result is a total polarization conversion to an output beam that is (right-) left-circularly polarized and carries a topological charge $\ell = 2q$, as illustrated schematically in figure 5.14(a). Since their invention, q-plates have simplified the generation of vector beams, whereby control of the q value allows one to generate a wide variety of vector vortex beams. q-plates of values 0.5, 1.5 and 3 are also shown in the bottom panels of figure 5.14(a). Crucially, for a linearly polarized beam, a superposition of left and right circular polarizations, the output is precisely a vector beam. Unfortunately, q-plates also face some limitations; on the one hand, the spin–orbit coupling is only applicable for input circular polarization states, and on the other, the output topological charge on the left- and right-circular polarization states are

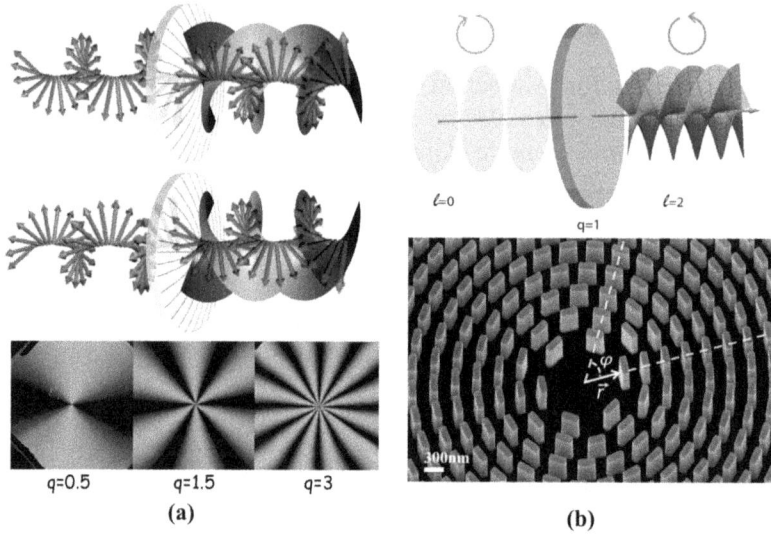

Figure 5.14. Generation of vector beams using the geometric phase. (a) In the top panel, an input (left-) right-circularly polarized incident on a q-plate is converted to the opposite circular polarization and acquires a topological charge $\ell = 2q$, depending on the specific value of the q-plate, examples of which are shown on the bottom panel (adapted from [101]). (b) The top panel illustrates that a j-plate converts arbitrary polarization states to arbitrary output states of total angular momentum $J = s + \ell$ using subwavelength gratings with periods smaller than the wavelength of the input light, as shown on the bottom panel (adapted from [105]).

constrained to be conjugate of each other, namely, $\ell = \pm 2q$. Hence, alternative approaches to the liquid crystal-based q-plate have been proposed, for example Zhao *et al* proposed a metamaterial q-plate, for broadband generation of vector beams [102]. The reported metamaterial q-plates were fabricated on a micrometre-size gold film, and consisted of two concentric rings with nanometric-size apertures with height 200nm. Zhao *et al* achieved an operational bandwidth from 1000 nm to 2500 nm, a significant range for transmission through free-space and optical fibers. A similar but more powerful proposal, which relies on the use of metamaterial subwavelength structures than allow quasi-continuous control of spatial phase shifts, birefringence and the orientation of the fast axis. Such a device has the capability to perform arbitrary spin–orbit coupling independent of the input polarization state [103–105]. This novel device have been termed J-plate, to emphasise their ability to convert arbitrary polarization states to arbitrary output states of total angular momentum, $J = s + \ell$, where s is the spin angular momentum, related to the state of circular polarization. The principle behind the J-plate relies on the use of subwavelength gratings with periods smaller that the wavelength of the input light, as shown in figure 5.14(b). Under this condition, each groove acts as a polarizer with a variant transmission axis across the transverse profile of the beam. This spatially-varying polarization transformation induces an optical phase delay linked to the geometry of the subwavelength structure, a manifestation of the geometric phase. The fact that it is metamaterial-based also makes it attractive for integration onto small-scale devices. For example, interesting solutions have been proposed to generate vector beams directly from a laser cavity making use of metamaterials [106] or saturable absorbers [107], with advances mostly in fiber lasers [108–114]. Importantly, in recent times the use of intra-cavity geometric phase elements for spin–orbit coupling, resulting in direct generation from a laser cavity of vector beams was reported, see for example [115, 116].

5.4 Characterization of vector beams

Another important aspect of complex light fields is their characterization, which provides useful information about their purity or spatial polarization distribution, a challenging task due to the nonseparable coupling between the spatial and polarization DoFs. A well-known technique, aiming at the reconstruction of their entire transverse polarization, is Stokes polarimetry, which comprises a series of intensity measurements [22, 117]. A more modern approach, which exploits their similarity with quantum entangled states, utilises well-established tools from quantum mechanics, namely, *concurrence* (*C*) [118]. Such a measure, which for vector modes is called vector quality factor (VQF), measures the degree of coupling between the spatial and polarization DoF assigning a number in the range [0, 1], 0 to scalar modes with a null degree of coupling and 1 to vector modes with a maximum degree of coupling [34, 119–123]. This technique was first implemented with SLMs requiring the spatial separation of both polarization components, followed by the projection of each beam onto a series of spatial filters encoded on the SLM. It is worth noting that the polarization-insensitivity property of DMDs enables the real-time and all-digital Stokes polarimetry to reconstruct the transverse polarization distribution as well as a simplified implementation of the VQF measurement [124–126]. In what follows, we will provide some of the details related to these techniques.

5.4.1 Stokes polarimetry

Stokes polarimetry represents the most powerful intensity-based technique for reconstructing the transverse polarization distribution of homogeneous and non-homogeneous light fields. Here, a minimum of four intensity measurements enclosed in a set of four quantities known as Stokes parameters is required, from which the polarization distribution can be reconstructed [22]. The relation between the intensities and the Stokes parameters is given by

$$S_0 = I_0, \quad S_1 = 2I_H - S_0, \quad S_2 = 2I_D - S_0, \quad S_3 = 2I_R - S_0, \qquad (5.26)$$

where I_0 is the total intensity of the given optical field. I_H and I_D are the measured intensities of the field after a linear polarizer orientated at 0 and $\pi/4$, respectively. I_R is the intensity acquired after the combination of a QWP at $\pi/4$ and a linear polarizer at $\pi/2$. Traditionally, these intensities are obtained individually, one-by-one, at different times, limiting its performance to light beams with static states of polarization but in recent time the use of DMDs allowed for the real-time reconstruction of the SoP of any light field. The key idea behind this novel technique relies on performing all the intensity measurements simultaneously. To this end, a multiplexed digital hologram is displayed on a DMD, enabling one to split the input beam into four identical copies propagating along different paths. In this way, all the four required intensities can be recorded simultaneously in a single shot with the help of the required optical filters and a CCD [124]. This technique allows us to monitor in real-time the polarization evolution of dynamically-changing vector modes, which is of great relevance in optical metrology applications.

5.4.2 Quantum-like nonseparability of vector beams

Classical mechanics seems to differ quite dramatically from its quantum counterpart, yet many similarities can be found. [5, 127–133]. This is the case of quantum entanglement, which captures the nonseparability property of entangled photons but can also describe the nonseparability of classical systems [5, 17, 30, 134–136]. Nonetheless, a strong difference exists here, while at the quantum level entanglement can exist between systems that are spatially separated from each other (nonlocal) at the classical level, vector beams can only be entangled in a local sense, between the internal degrees of freedom of a system [137–141]. In the language of quantum mechanics we might say that for nonseparable vector beams, the choice of the polarization measurement affects the outcome of the measurement of the spatial mode. For a vector beam, one cannot one cannot follow the evolution of the polarization and spatial mode as independent quantities through an optical system, since what we measure in terms of amplitude and phase very much depends on the choice of polarising optics. This realization makes it possible to measure the classical entanglement, through the degree of nonseparability, of optical modes [20, 120, 142, 143]. The result provides information on whether the mode is purely scalar, purely vector, or somewhere in-between. This requires the introduction of a traditionally quantum toolbox to describe classical light [34, 119]. To see this, let us rewrite equation (5.19) as

$$\Psi = \sqrt{a}\,e^{i\delta}\,LG_0^\ell e^{i\delta}\hat{e}_R + \sqrt{(1-a)}\,e^{-i\delta}\,LG_0^{-\ell}\hat{e}_L. \tag{5.27}$$

From this equation we can see that polarization and spatial degrees of freedom cannot be factorized as a product of two independent terms, in a similar way to two entangled photons. From now on and also to highlight this similarity, we will use notation from quantum mechanics, more precisely, Dirac's notation. Hence equation (5.27) takes the form

$$|\psi\rangle = \sqrt{a}\,|u_R\rangle \otimes |R\rangle + \sqrt{(1-a)}\,|u_L\rangle \otimes |L\rangle, \tag{5.28}$$

where the kets $|u_R\rangle$ and $|u_L\rangle$ represents the spatial modes $LG_0^\ell e^{i\delta}$ and $LG_0^{-\ell}e^{-i\delta}$, weighted by the terms $\sqrt{a}$ and $\sqrt{(1-a)}$, respectively, with $a \in [0, 1]$. The kets $|R\rangle$ and $|L\rangle$, are the right and left circular polarization vectors, respectively. The symbol $\otimes$ denotes the tensor product between the vectors. The weighting factor a controls the degree of classical entanglement of the field $|\psi\rangle$, which can go from purely scalar ($a = 0, 1$) to purely vector ($a = 1/2$), corresponding to completely separable and completely nonseparable, respectively. Given the mathematical similarity between classical and quantum states, it has become possible to apply the formalism of quantum mechanics to measure the degree of nonseparability of a state defined by equation (5.28), in other words [34, 119]. This can be done through the entanglement entropy, which quantifies the quantum entanglement in a pure state of bipartite systems [118]. The entanglement entropy is given by the von Neumann entropy of the reduced density matrix of one of the subsystems, polarization or the spatial degrees of freedom. In the case of the polarization, it is obtained by tracing over the spatial degree of freedom,

$$E(|\psi\rangle) = -\mathrm{Tr}[\rho_p \log(\rho_p)], \qquad \rho_p = \mathrm{Tr}_s[|\psi\rangle\langle\psi|], \tag{5.29}$$

where the reduced density matrix ρ_p is given by

$$\rho_p = \begin{pmatrix} a & \sqrt{a(1-a)}|u_L\rangle\langle u_R| \\ \sqrt{a(1-a)}|u_R\rangle\langle u_L| & (1-a) \end{pmatrix}. \tag{5.30}$$

From the physical perspective, ρ_p quantifies the average polarization of the vector beam. More importantly, it can be determined by measuring the components $s_i = \mathrm{Tr}[\sigma_i \rho_p]$, which correspond to the Stokes parameters, of the Bloch vector $\mathbf{s}$, where $\rho_p = (\mathbb{I} + \sum_i s_i \sigma_i)/2$ ($i = 1, 2, 3$). σ_i are the Pauli operators given by,

$$\begin{aligned} \sigma_1 &= |H\rangle\langle H| - |V\rangle\langle V|, \\ \sigma_2 &= \frac{1}{2}(|H+V\rangle\langle H+V| - |H-V\rangle\langle H-V|), \\ \sigma_3 &= |R\rangle\langle R| - |L\rangle\langle L|. \end{aligned} \tag{5.31}$$

In fact, the reduced density matrix, ρ_p, that depicts the average state of polarization across the transverse plane, resembles the polarization matrix of an incoherent mixture. For a vector beam this matrix results from a coherent mixture of pure states of the form

$$\rho_p = \sum_{i=1}^{3} p_i |P_i\rangle\langle P_i|. \tag{5.32}$$

Hence, classical entanglement can be measured using the von Neumann entropy as

$$E(|\Psi\rangle) = -\left(\frac{1+s}{2}\right)\log\left(\frac{1+s}{2}\right) - \left(\frac{1-s}{2}\right)\log\left(\frac{1-s}{2}\right), \tag{5.33}$$

where s is the length of the Bloch vector and measures the mixedness of ρ_p and thus the vector nature of ρ. Explicitly, s is given by

$$s(\rho_P) = (\mathrm{Tr}[\rho_P^2])^{1/2} = \left(\sum_{i=1}^{3}\langle\sigma_i\rangle\right)^{1/2}, \tag{5.34}$$

where $\langle \sigma_i \rangle$ are the expectation values of the Pauli operators. Finally, the degree of classical entanglement or nonseparability can be measured by computing the real part of the concurrence C or vector quality factor (VQF) as

$$\text{VQF} = \text{Re}(C) = \text{Re}(\sqrt{1 - s^2}). \qquad (5.35)$$

Crucially, the expectation values of the Pauli operators can be expressed in terms of its eigenvectors $|\lambda_0\rangle$ and $|\lambda_1\rangle$, and eigenvalues ± 1 as

$$\langle \sigma_i \rangle = |\lambda_0\rangle\langle\lambda_0| - |\lambda_1\rangle\langle\lambda_1|. \qquad (5.36)$$

and can be determined by projecting the input vector field $|\psi\rangle$ on one degree of freedom, while tracing over the other. In particular, tracing over the polarization degree of freedom allows us to rewrite equation (5.36) as

$$\langle \sigma_i \rangle = \langle \lambda_0, \sigma_R |\psi\rangle|^2 + \langle \lambda_0, \sigma_L |\psi\rangle|^2 - \langle \lambda_1, \sigma_L |\psi\rangle|^2 - \langle \lambda_1, \sigma_L |\psi\rangle|^2, \qquad (5.37)$$

where the eigenvectors $|\lambda_0\rangle$ and $|\lambda_1\rangle$ correspond to spatial modes. Importantly, the expectation values can be measured experimentally from a series of twelve projective on-axis intensity measurements I_{ij} ($i \in [1, 2]$ and $j \in [1, 6]$), six spatial projections for each polarization constituting the vector mode. More explicitly, in terms of the intensities I_{ij}, the expectation values $\langle \sigma_i \rangle$ take the form

$$\langle \sigma_1 \rangle = [I_{13} + I_{23}] - [I_{15} + I_{25}],$$
$$\langle \sigma_2 \rangle = [I_{14} + I_{24}] - [I_{16} + I_{26}], \qquad (5.38)$$
$$\langle \sigma_3 \rangle = [I_{11} + I_{21}] - [I_{12} + I_{22}].$$

For the sake of clarity, figure 5.15 shows a schematic representation of the intensities I_{ij} and how each of them are obtained by projecting a given vector mode over the two degrees of freedom, spatial mode and polarization. The example is given for a vector vortex mode defined as the superposition of two optical vortices of topological charges $\ell = +1$ and $\ell = -1$, which in Dirac's notation can be written as

$$|\psi\rangle = |+1\rangle|R\rangle + |-1\rangle|L\rangle \qquad (5.39)$$

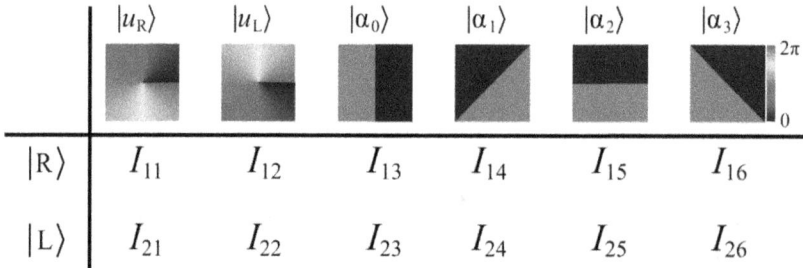

	$|u_R\rangle$	$|u_L\rangle$	$|\alpha_0\rangle$	$|\alpha_1\rangle$	$|\alpha_2\rangle$	$|\alpha_3\rangle$
$|R\rangle$	I_{11}	I_{12}	I_{13}	I_{14}	I_{15}	I_{16}
$|L\rangle$	I_{21}	I_{22}	I_{23}	I_{24}	I_{25}	I_{26}

Figure 5.15. Required intensity measurements to determine the vector quality factor of a vector beam. Here, $|u_R\rangle$ and $|u_L\rangle$ represent the spatial modes that constitute the vector mode and $\alpha_j = j\pi/2$ and $j = 0, 1, 2, 3$ represents an intermodal phase for the superposition of both modes. In addition $R\rangle$ and $|L\rangle$ represent the right-handed and left-handed circular polarization.

Here, $|\alpha_j\rangle \equiv (|u_R\rangle + \exp(i\alpha_j)|u_L\rangle)/\sqrt{2}$, with $\alpha_j = j\pi/2$ and $j = 0, 1, 2, 3$, is an intramodal phase of the superposition of both modes. To determine these intensities experimentally, the two orthogonal polarization states $|R\rangle$ and $|L\rangle$ are first spatially separated using polarising optical elements, afterwards they are projected onto the spatial degree of freedom given by the spatial modes $|u_R\rangle$, $|u_L\rangle$ and $|\alpha_j\rangle$, which are encoded on an SLM. The intensities I_{ij} can then be measured as the on-axis values of the far-field intensity, obtained in the focal plane of a lens [119]. Crucially, all intensities can be measured simultaneously by displaying on the SLM a multiplexed hologram containing all the spatial projections, as conceptually illustrated in figure 5.16. Here, the unknown vector mode is first separated into its two orthogonal polarization components, for example with a polarization grating, each of which are redirected to two independent sections of an SLM where two independent multiplexed holograms are displayed to perform the spatial projection. The resulting beams are then transmitted through a lens and sent to a CCD camera placed at the focal plane of the lens, where all the intensities are captured. Each of the intensities I_{ij} is previously associated to a specific location in the far field. Notice that additional polarization optics are required in order to meet the polarization requirements of the SLM.

For the sake of clarity, figure 5.17(a) shows the measured far-field intensities for the pure vector mode previously introduced. The on-axis intensities for each of the twelve intensity patterns is then measured and all normalized to one. Notice that it is required to previously determine the on-axis spatial coordinates of each beam, which can be performed through a calibration image containing only Gaussian beams, where the on-axis corresponds to the maximum intensity value. All the normalized intensities are shown in figure 5.17(b) and labelled with their corresponding I_{ij} for easy identification. Finally, the VQF can be computed easily from these intensity values, according to equation (5.35).

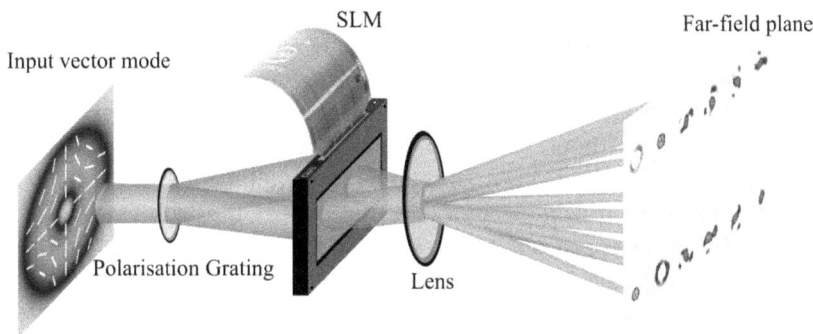

Figure 5.16. Schematic representation of the experimental setup to determine the VQF of a given vector mode in a single frame. Here the unknown vector mode is first separated into its two polarization components, each of which is sent through an independent section of an SLM. Each section of the SLM is addressed with a multiplexed hologram containing the six spatial projections required to determine the 12 intensity measurements. The far field is implemented with a lens placed a focal distance away from the SLM and the intensities are measured with a CCD camera located in the focal plane of the lens.

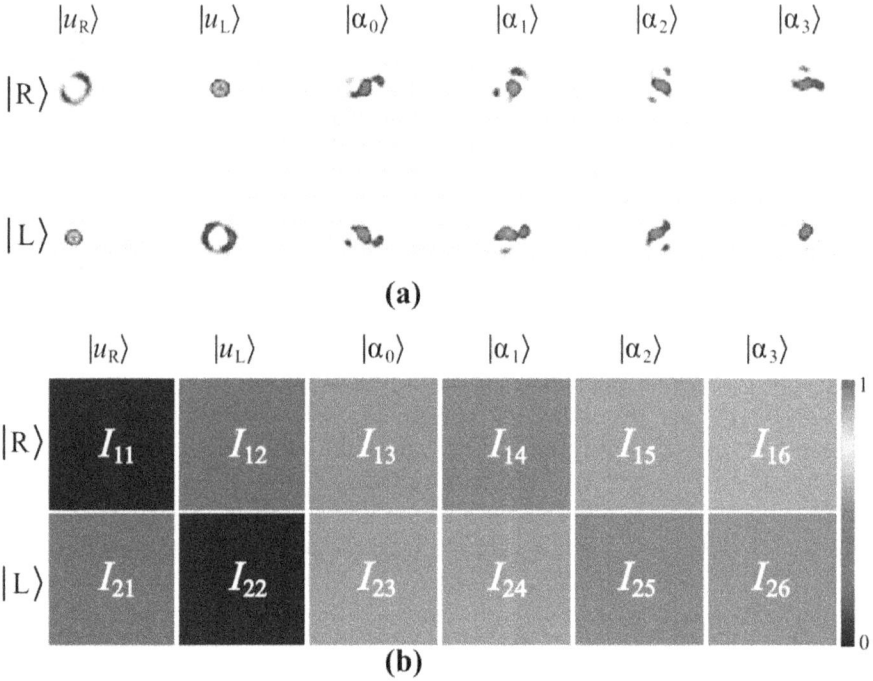

$$|u_R\rangle \quad |u_L\rangle \quad |\alpha_0\rangle \quad |\alpha_1\rangle \quad |\alpha_2\rangle \quad |\alpha_3\rangle$$

$$|R\rangle$$

$$|L\rangle$$

(a)

$$|u_R\rangle \quad |u_L\rangle \quad |\alpha_0\rangle \quad |\alpha_1\rangle \quad |\alpha_2\rangle \quad |\alpha_3\rangle$$

$|R\rangle$	I_{11}	I_{12}	I_{13}	I_{14}	I_{15}	I_{16}
$|L\rangle$	I_{21}	I_{22}	I_{23}	I_{24}	I_{25}	I_{26}

(b)

Figure 5.17. (a) Example of the typical far-field intensities obtained in a single shot through spatial multiplexing. (b) Normalized on-axis intensity values where, according to the colour code, blue is associated with 0, red with 1 and the darkest green with 0.5.

5.4.2.1 Vector quality factor with DMDs

Importantly, when tracing over the spatial degree of freedom, the projective measurements I_{ij} can be reduced by 25%, as demonstrated in [126]. In this approach the VQF is measured by projecting a given vector mode directly on the spatial basis, encoded as binary holograms on a DMD. The resulting mode is then passed through a series of polarization filters that projects the beam onto the polarization DoF, enabling a reduction in the number of required measurements from twelve to eight. This is in fact the optimal number of required measurements, and there is no way to reduce this number without losing any information. Crucially, a multiplexed hologram displayed on the DMD facilitates the simultaneous measurement of all the required intensities in a single shot, as conceptually illustrated in figure 5.18.

Here, a vector mode is projected onto a digital hologram displayed on the DMD. The hologram consist of a series of eight multiplexed holograms, each with unique spatial frequency, so that the outcomes are directed along independent trajectories. By way of example, if we use again the mode $\psi = \cos\theta|\ell\rangle|R\rangle + \sin\theta|-\ell\rangle|L\rangle$, four holograms perform the $|+\ell\rangle$ projections and the other four the $|-\ell\rangle$ projections. Thereafter, each beams is passed through a series of polarization filters to perform the polarization, namely, onto the $|R\rangle, |L\rangle, |H\rangle$ and $|D\rangle$ polarization components. All beams are then passed through a lens to obtain the far field of each beam, where the on-axis intensity values are then measured to compute the expectation values $\langle\sigma_1\rangle$,

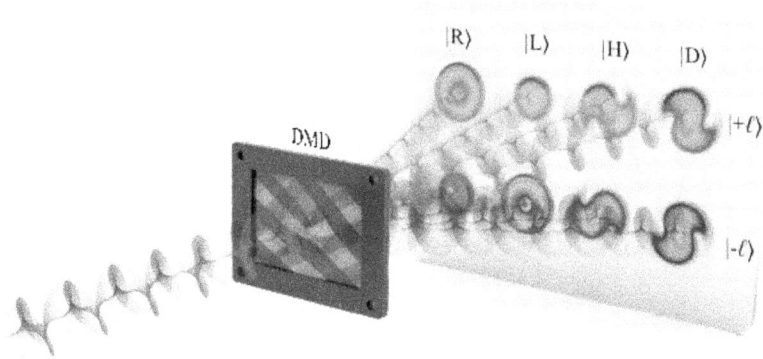

Figure 5.18. The VQF of vector modes is measured by projecting onto the OAM basis encoded on a DMD and tracing over the polarization DOF. For this, eight intensity measurements are required, as shown in figure 5.19. Importantly, all intensities can be measured simultaneously using a multiplexing approach. Reprinted from [126], with permission from AIP Publishing.

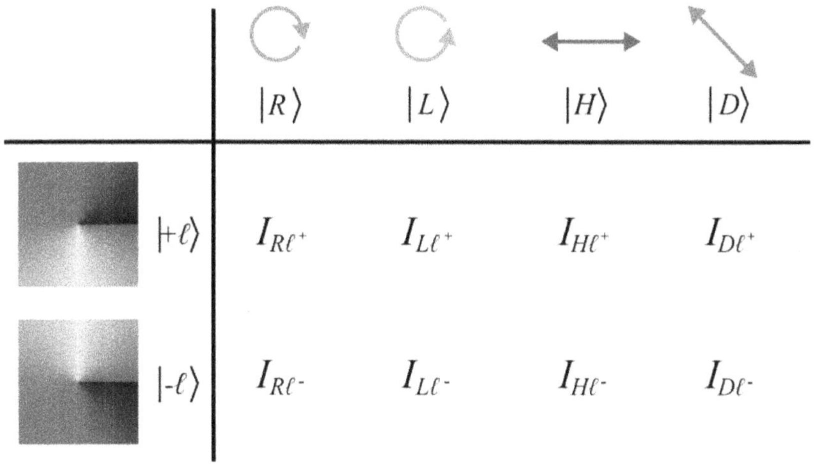

Figure 5.19. Required intensities to determine the vector quality factor by projecting over the spatial mode and tracing over the polarization degree of freedom.

$\langle\sigma_2\rangle$ and $\langle\sigma_3\rangle$, from which the VQF can be finally computed. For the sake of clarity, the required projections are shown in figure 5.19. Here, for example, $I_{H\ell^+}$ represents the intensity after projecting the vector mode on the $+\ell$ OAM phase filter and passing it through a H polarization filter (figure 5.20).

Explicitly, the expectation values $\langle\sigma_i\rangle$ will take the form

$$\begin{aligned}
\langle\sigma_1\rangle &= 2(I_{H\ell^+} + I_{H\ell^-}) - (I_{\ell^+} + I_{\ell^-}),\\
\langle\sigma_2\rangle &= 2(I_{D\ell^+} + I_{D\ell^-}) - (I_{\ell^+} + I_{\ell^-}),\\
\langle\sigma_3\rangle &= 2(I_{R\ell^+} + I_{R\ell^-}) - (I_{\ell^+} + I_{\ell^-}),
\end{aligned} \tag{5.40}$$

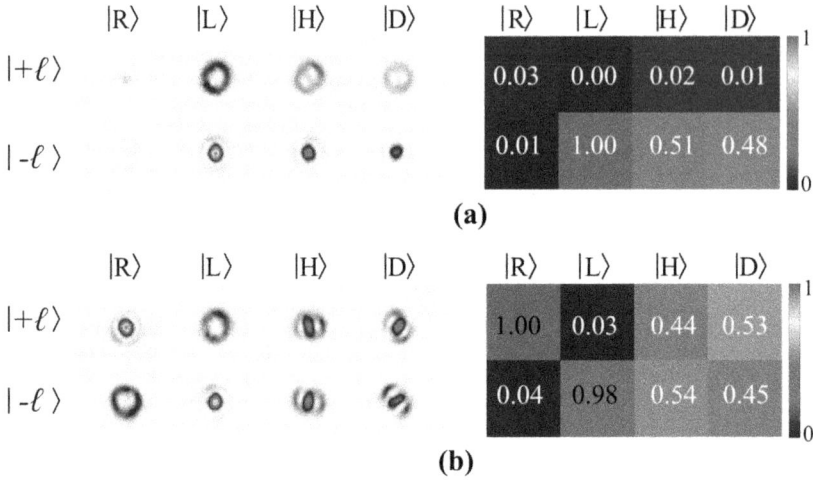

	$\lvert R\rangle$	$\lvert L\rangle$	$\lvert H\rangle$	$\lvert D\rangle$		$\lvert R\rangle$	$\lvert L\rangle$	$\lvert H\rangle$	$\lvert D\rangle$
$\lvert +\ell\rangle$						0.03	0.00	0.02	0.01
$\lvert -\ell\rangle$						0.01	1.00	0.51	0.48

(a)

	$\lvert R\rangle$	$\lvert L\rangle$	$\lvert H\rangle$	$\lvert D\rangle$		$\lvert R\rangle$	$\lvert L\rangle$	$\lvert H\rangle$	$\lvert D\rangle$
$\lvert +\ell\rangle$						1.00	0.03	0.44	0.53
$\lvert -\ell\rangle$						0.04	0.98	0.54	0.45

(b)

Figure 5.20. Far-field Intensity distribution (left) and corresponding normalized on-axis intensity (right) for a scalar (a) and a vector (b) beam.

where $\quad I_{\ell^+} = I_{R\ell^+} + I_{L\ell^+} = I_{H\ell^+} + I_{V\ell^+} = I_{A\ell^+} + I_{D\ell^+} \quad$ and $\quad I_{\ell^-} = I_{R\ell^-} + I_{L\ell^-} = I_{H\ell^-} + I_{V\ell^-} = I_{A\ell^-} + I_{D\ell^-}$.

As a final comment, an all-digital approach based on a DMD and capable of performing the required polarization projections for measuring the Stokes parameters has been also proposed. The DMD realizes mode projections onto various polarization states, which in turn enables us to measure all four intensities by addressing the DMD with appropriate holograms. Once the Stokes parameters are measured, the degree of nonseparability of the beam, its state of polarization and its intermodal phase are also determined [125].

References

[1] Rosales-Guzmán C, Ndagano B and Forbes A 2018 A review of complex vector light fields and their applications *J. Opt.* **20** 123001

[2] Rubinsztein-Dunlop H, Forbes A, Berry M V, Dennis M R, Andrews D L, Mansuripur M, Denz C, Alpmann C, Banzer P and Bauer T 2017 Roadmap on structured light *J. Opt.* **19** 013001

[3] Forbes A, de Oliveira M and Dennis M R 2021 Structured light *Nat. Photon.* **15** 253–62

[4] Galvez E J, Khadka S, Schubert W H and Nomoto S 2012 Poincaré beam patterns produced by nonseparable superpositions of Laguerre–Gauss and polarization modes of light *Appl. Opt.* **51** 2925–34

[5] Shen Y and Rosales-Guzmán C 2022 Nonseparable states of light: From quantum to classical *Laser Photon. Rev.* **16** 2100533

[6] Zhan Q 2009 Cylindrical vector beams: from mathematical concepts to applications *Adv. Opt. Photon.* **1** 1–57

[7] Chen S, Zhou X, Liu Y, Ling X, Luo H and Wen S 2014 Generation of arbitrary cylindrical vector beams on the higher order Poincaré sphere *Opt. Lett.* **39** 5274–6

[8] Yang Y, Ren Y, Chen M, Arita Y and Rosales-Guzmán C 2021 Optical trapping with structured light: a review *Adv. Photon.* **3** 034001

[9] Donato M G, Patti F, Saija R, Iatì M A, Gucciardi P G, Pedaci F, Strangi G and Maragò O M 2021 Improved backscattering detection in photonic force microscopy near dielectric surfaces with cylindrical vector beams *J. Quant. Spectrosc. Radiat. Transfer* **258** 107381

[10] Hu X B, Zhao B, Zhu Z H, Gao W and Rosales-Guzmán C 2019 In situ detection of a cooperative target's longitudinal and angular speed using structured light *Opt. Lett.* **44** 3070–3

[11] Kraus M, Ahmed M A, Michalowski A, Voss A, Weber R and Graf G 2010 Microdrilling in steel using ultrashort pulsed laser beams with radial and azimuthal polarization *Opt. Express* **18** 22305

[12] Lou K, Qian S-X, Ren Z-C, Tu C, Li Y and Wang H T 2013 Femtosecond laser processing by using patterned vector optical fields *Sci. Rep.* **3** 2281

[13] Danilov P A *et al* 2018 Polarization-selective excitation of dye luminescence on a gold film by structured ultrashort laser pulses *JETP Lett.* **107** 15–8

[14] Syubaev S A *et al* 2019 Plasmonic nanolenses produced by cylindrical vector beam printing for sensing applications *Sci. Rep.* **9** 19750

[15] Khonina S N, Ustinov A V and Degtyarev S A 2018 Inverse energy flux of focused radially polarized optical beams *Phys. Rev.* A **98** 043823

[16] Kotlyar V V, Stafeev S S and Kovalev A A 2019 Reverse and toroidal flux of light fields with both phase and polarization higher-order singularities in the sharp focus area *Opt. Express* **27** 16689

[17] Töppel F, Aiello A, Marquardt C, Giacobino E and Leuchs G 2014 Classical entanglement in polarization metrology *New J. Phys.* **16** 073019

[18] Berg-Johansen S, Töppel F, Stiller B, Banzer P, Ornigotti M, Giacobino E, Leuchs G, Aiello A and Marquardt C 2015 Classically entangled optical beams for high-speed kinematic sensing *Optica* **2** 864–8

[19] Otte E, Nape I, Rosales-Guzmán C, Denz C, Forbes A and Ndagano B 2020 High-dimensional cryptography with spatial modes of light: tutorial *J. Opt. Soc. Am.* B **37** A309–323

[20] Ndagano B, Perez-Garcia B, Roux F S, McLaren M, Rosales-Guzmán C, Zhang Y, Mouane O, Hernandez-Aranda R I, Konrad T and Forbes A 2017 Characterizing quantum channels with non-separable states of classical light *Nat. Phys.* **13** 397–402

[21] Jones R C 1941 A new calculus for the treatment of optical systems i. Description and discussion of the calculus *J. Opt. Soc. Am.* **31** 488–93

[22] Goldstein D H 2011 *Polarized Light* (Boca Raton, FL: CRC Press) https://doi.org/10.1201/b10436

[23] Berry M and Jeffrey M 2007 Conical diffraction: Hamilton's diabolical point at the heart of crystal optics *Prog. Opt.* **50** 13–50

[24] Soskin M and Vasnetsov M 2001 Singular optics *Prog. Opt.* **42** 219–76

[25] Dennis M 2002 Polarization singularities in paraxial vector fields: morphology and statistics *Opt. Commun.* **213** 201–21

[26] Bliokh K Y, Rodríguez-Fortuño F, Nori F and Zayats A V 2015 Spin–orbit interactions of light *Nat. Photon.* **9** 796–808

[27] Cardano F and Marrucci L 2015 Spin-orbit photonics *Nat. Photon.* **9** 776–8

[28] Bliokh K Y and Nori F 2015 Transverse and longitudinal angular momenta of light *Phys. Rep.* **592** 1–38

[29] Freund I 2002 Polarization singularity indices in Gaussian laser beams *Opt. Commun.* **201** 251–70

[30] Aiello A, Töppel F, Marquardt C, Giacobino E and Leuchs G 2015 Quantum-like nonseparable structures in optical beams *New J. Phys.* **17** 043024

[31] Kagalwala K H, Di Giuseppe G, Abouraddy A F and Saleh B E 2013 Bell's measure in classical optical coherence *Nat. Photon.* **7** 72–8

[32] Qian X F and Eberly J 2011 Entanglement and classical polarization states *Opt. Lett.* **36** 4110–2

[33] Simon B N, Simon S, Gori F, Santarsiero M, Borghi R, Mukunda N and Simon R 2010 Nonquantum entanglement resolves a basic issue in polarization optics *Phys. Rev. Lett.* **104** 023901

[34] McLaren M, Konrad T and Forbes A 2015 Measuring the nonseparability of vector vortex beams *Phys. Rev.* A **92** 023833

[35] Hu X, Perez-Garcia B, Rodríguez-Fajardo V, Hernandez-Aranda R, Forbes A and Rosales-Guzmán C 2021 Free-space local non-separability dynamics of vector modes *Photon. Res.* **9** 439–45

[36] Zhong R Y, Zhu Z H, Wu H J, Rosales-Guzmán C, Song S W and Shi B S 2021 Gouy-phase-mediated propagation variations and revivals of transverse structure in vectorially structured light *Phys. Rev.* A **103** 053520

[37] Hall D G 1996 Vector-beam solutions of Maxwell's wave equation *Opt. Lett.* **21** 9–11

[38] Jordan R H and Hall D G 1994 Free-space azimuthal paraxial wave equation: the azimuthal Bessel-Gauss beam solution *Opt. Lett.* **19** 427–9

[39] Greene P L and Hall D G 1998 Properties and diffraction of vector Bessel-Gauss beams *J. Opt. Soc. Am.* A **15** 3020–7

[40] Bandres M A and Gutiérrez-Vega J C 2005 Vector Helmholtz-Gauss and vector Laplace-Gauss beams *Opt. Lett.* **30** 2155–7

[41] Beckley A M, Brown T G and Alonso M A 2010 Full Poincaré beams *Opt. Express* **18** 10777–85

[42] Maurer C, Jesacher A, Fürhapter S, Bernet S and Ritsch-Marte M 2007 Tailoring of arbitrary optical vector beams *New J. Phys.* **9** 78

[43] Rosales-Guzmán C, Hu X B, Selyem A, Moreno-Acosta P, Franke-Arnold S, Ramos-Garcia R and Forbes A 2020 Polarization-insensitive generation of complex vector modes from a digital micromirror device *Sci. Rep.* **10** 10434

[44] Li Y, Hu X B, Perez-Garcia B, Bo-Zhao W, Gao Z H Z and Rosales-Guzmán C 2020 Classically entangled Ince-Gaussian modes *Appl. Phys. Lett.* **116** 221105

[45] Bo Z, Valeria R F, Xiao-Bo H, Raul I H A, Benjamin P G and Carmelo R G 2022 Parabolic-accelerating vector waves *Nanophotonics* **11** 681–8

[46] Otte E, Tekce K and Denz C 2017 Tailored intensity landscapes by tight focusing of singular vector beams *Opt. Express* **25** 20194–201

[47] Vyas S, Kozawa Y and Sato S 2013 Polarization singularities in superposition of vector beams *Opt. Express* **21** 8972–86

[48] Berry M V and Hannay J H 1977 Umbilic points on Gaussian random surfaces *J. Phys. A: Math. Gen.* **10** 1809

[49] Nye J F 1983 Lines of circular polarization in electromagnetic wave fields *Proc. R. Soc.* **389** 279–90

[50] Kumar V and Viswanathan N K 2014 Topological structures in vector-vortex beam fields *J. Opt. Soc. Am.* B **31** A40–45

[51] Kumar V and Viswanathan N K 2013 Topological structures in the Poynting vector field: an experimental realization *Opt. Lett.* **38** 3886–9

[52] Kumar V, Philip G M and Viswanathan N K 2013 Formation and morphological transformation of polarization singularities: hunting the monstar *J. Opt.* **15** 044027

[53] Boscain U, Sacchelli L and Sigalotti M 2016 Generic singularities of line fields on 2D manifolds *Differ. Geom. Appl.* **49** 326–50

[54] Galvez E J, Rojec B L, Kumar V and Viswanathan N K 2014 Generation of isolated asymmetric umbilics in light's polarization *Phys. Rev.* **89** 031801

[55] Khajavi B and Galvez E J 2016 High-order disclinations in space-variant polarization *J. Opt.* **18** 084003

[56] Dennis M R 2008 Polarization singularity anisotropy: determining monstardom *Opt. Lett.* **33** 2572–4

[57] Milione G, Sztul H I, Nolan D A and Alfano R R 2011 Higher-order Poincaré sphere, Stokes parameters, and the angular momentum of light *Phys. Rev. Lett.* **107** 053601

[58] Holleczek A, Aiello A, Gabriel C, Marquardt C and Leuchs G 2011 Classical and quantum properties of cylindrically polarized states of light *Opt. Express* **19** 9714–36

[59] Yi X, Liu Y, Ling X, Zhou X, Ke Y, Luo H, Wen S and Fan D 2015 Hybrid-order Poincaré sphere *Phys. Rev.* **91** 023801

[60] Román-Valenzuela T, Rodríguez-Fajardo V, Bo-hu X and Rosales-Guzmán C 2024 Generation of cylindrical vector modes via astigmatic mode conversion *Opt. Lett.* **49** 2910–3

[61] Tidwell S C, Ford D H and Kimura W D 1990 Generating radially polarized beams interferometrically *Appl. Opt.* **29** 2234–9

[62] Niziev V G, Chang R S and Nesterov A V 2006 Generation of inhomogeneously polarized laser beams by use of a sagnac interferometer *Appl. Opt.* **45** 8393–9

[63] Passilly N, Treussart F, Hierle R, de Saint Denis R, Aït-Ameur K and Roch J F 2005 Simple interferometric technique for generation of a radially polarized light beam *J. Opt. Soc. Am.* A **22** 984

[64] Rosales-Guzmán C and Forbes A 2017 *How to Shape Light with Spatial Light Modulators* (Bellingham, WA: SPIE Press) https://doi.org/10.1117/3.2281295

[65] Scholes S, Kara R, Pinnell J, Rodríguez-Fajardo V and Forbes A 2019 Structured light with digital micromirror devices: a guide to best practice *Opt. Eng.* **59** 1–12

[66] Hu X B and Rosales-Guzmán C 2022 Generation and characterization of complex vector modes with digital micromirror devices: a tutorial *J. Opt.* **24** 034001

[67] Hu X B, Ma S Y and Rosales-Guzmán C 2021 High-speed generation of singular beams through random spatial multiplexing *J. Opt.* **23** 044002

[68] Otte E, Tekce K, Lamping S, Ravoo B J and Denz C 2019 Polarization nano-tomography of tightly focused light landscapes by self-assembled monolayers *Nat. Commun.* **10** 4308

[69] Rong Z Y, Han Y J, Wang S Z and Guo C S 2014 Generation of arbitrary vector beams with cascaded liquid crystal spatial light modulators *Opt. Express* **22** 1636

[70] Moreno I, Davis J A, Hernandez T M, Cottrell D M and Sand D 2012 Complete polarization control of light from a liquid crystal spatial light modulator *Opt. Express* **20** 364–76

[71] Alpmann C, Schlickriede C, Otte E and Denz C 2017 Dynamic modulation of Poincaré beams *Sci. Rep.* **7** 8076

[72] Rodríguez-Fajardo V, Arvizu F, Daza-Salgado D, Perez-Garcia B and Rosales-Guzmán C 2024 On-axis complex-amplitude modulation for the generation of super-stable vector modes *J. Opt.* **26** 065606

[73] Rosales-Guzmán C, Bhebhe N and Forbes A 2017 Simultaneous generation of multiple vector beams on a single SLM *Opt. Express* **25** 25697–706

[74] Liu S, Qi S, Zhang Y, Li P, Wu D, Han L and Zhao J 2018 Highly efficient generation of arbitrary vector beams with tunable polarization, phase, and amplitude *Photon. Res.* **6** 228–33

[75] Mendoza-Hernández J, Ferrer-Garcia M F, Rojas-Santana J A and Lopez-Mago D 2019 Cylindrical vector beam generator using a two-element interferometer *Opt. Express* **27** 31810–9

[76] Mitchell K J, Radwell N, Franke-Arnold S, Padgett M J and Phillips D B 2017 Polarization structuring of broadband light *Opt. Express* **25** 25079–89

[77] Wang X L, Ding J, Ni W J, Guo C S and Wang H T 2007 Generation of arbitrary vector beams with a spatial light modulator and a common path interferometric arrangement *Opt. Lett.* **32** 3549–51

[78] Chen H, Hao J, Zhang B F, Xu J, Ding J and Wang H T 2011 Generation of vector beam with space-variant distribution of both polarization and phase *Opt. Lett.* **36** 3179–81

[79] Liu S, Li P, Peng T and Zhao J 2012 Generation of arbitrary spatially variant polarization beams with a trapezoid Sagnac interferometer *Opt. Express* **20** 21715–21

[80] Maluenda D, Juvells I, Martínez-Herrero R and Carnicer A 2013 Reconfigurable beams with arbitrary polarization and shape distributions at a given plane *Opt. Express* **21** 5432–9

[81] Bashkansky M, Park D and Fatemi F K 2010 Azimuthally and radially polarized light with a nematic SLM *Opt. Express* **18** 212–7

[82] Tripathi S and Toussaint K C 2012 Versatile generation of optical vector fields and vector beams using a non-interferometric approach *Opt. Express* **20** 10788–95

[83] Neil M A A, Massoumian F, Juskaitis R and Wilson T 2002 Method for the generation of arbitrary complex vector wave fronts *Opt. Lett.* **27** 1929–31

[84] Trichili A, Rosales-Guzmán C, Dudley A, Ndagano B, Ben Salem A, Zghal M and Forbes A 2016 Optical communication beyond orbital angular momentum *Sci. Rep.* **6** 27674

[85] Rosales-Guzmán C, Bhebhe N, Mahonisi N and Forbes A 2017 Multiplexing 200 spatial modes with a single hologram *J. Opt.* **19** 113501

[86] Rosales-Guzmán C, Bhebhe N and Forbes A 2017 Simultaneous generation of multiple vector beams on a single SLM *Opt. Express* **25** 25697–706

[87] Perez-Garcia B, Mecillas-Hernández F I and Rosales-Guzmán C 2022 Highly-stable generation of vector beams through a common-path interferometer and a DMD *J. Opt.* **24** 074007

[88] Cox M A and Drozdov A V 2021 Converting a Texas instruments DLP4710 DLP evaluation module into a spatial light modulator *Appl. Opt.* **60** 465–9

[89] Rodenburg B, Mirhosseini M, Magaña-Loaiza O S and Boyd R W 2014 Experimental generation of an optical field with arbitrary spatial coherence properties *J. Opt. Soc. Am.* B **31** A51–5

[90] Gong L, Ren Y, Liu W, Wang M, Zhong M, Wang Z and Li Y 2014 Generation of cylindrically polarized vector vortex beams with digital micromirror device *J. Appl. Phys.* **116** 183105

[91] Brown B R and Lohmann A W 1966 Complex spatial filtering with binary masks *Appl. Opt.* **5** 967

[92] Lee W H 1979 Binary computer-generated holograms *Appl. Opt.* **18** 3661

[93] Lee W H 1974 Binary synthetic holograms *Appl. Opt.* **13** 1677–82

[94] Rosales-Guzmán C, Hu X, Rodríguez-Fajardo V, Hernandez-Aranda R I, Forbes A and Perez-Garcia B 2022 Experimental generation of helical Mathieu-Gauss vector modes *J. Opt.* **23** 034004

[95] Khonina S N and Karpeev S V 2010 Grating-based optical scheme for the universal generation of inhomogeneously polarized laser beams *Appl. Opt.* **49** 1734

[96] Khonina S N, Karpeev S V, Alferov S V and Soifer V A 2015 Generation of cylindrical vector beams of high orders using uniaxial crystals *J. Opt.* **17** 065001

[97] Marrucci L, Manzo C and Paparo D 2006 Optical spin-to-orbital angular momentum conversion in inhomogeneous anisotropic media *Phys. Rev. Lett.* **96** 163905

[98] Marrucci L, Karimi E, Slussarenko S, Piccirillo B, Santamato E, Nagali E and Sciarrino F 2011 Spin-to-orbital conversion of the angular momentum of light and its classical and quantum applications *J. Opt.* **13** 064001

[99] Cardano F, Karimi E, Slussarenko S, Marrucci L, de Lisio C and Santamato E 2012 Polarization pattern of vector vortex beams generated by q-plates with different topological charges *Appl. Opt.* **51** C1–6

[100] Rubano A, Cardano F, Piccirillo B and Marrucci L 2019 Q-plate technology: a progress review [Invited] *J. Opt. Soc. Am.* B **36** D70

[101] Slussarenko S, Murauski A, Du T, Chigrinov V, Marrucci L and Santamato E 2011 Tunable liquid crystal q-plates with arbitrary topological charge *Opt. Express* **19** 4085–90

[102] Zhao Z, Wang J, Li S and Willner A E 2013 Metamaterials-based broadband generation of orbital angular momentum carrying vector beams *Opt. Lett.* **38** 932–4

[103] Devlin R C, Ambrosio A, Rubin N A, Mueller J B and Capasso F 2017 Arbitrary spin-to-orbital angular momentum conversion of light *Science* **358** 896–901

[104] Devlin R C, Khorasaninejad M, Chen W T, Oh J and Capasso F 2016 Broadband high-efficiency dielectric metasurfaces for the visible spectrum *Proc. Natl. Acad. Sci. USA* **113** 10473–8

[105] Devlin R C, Ambrosio A, Wintz D, Oscurato S L, Zhu A Y, Khorasaninejad M, Oh J, Maddalena P and Capasso F 2017 Spin-to-orbital angular momentum conversion in dielectric metasurfaces *Opt. Express* **25** 377–93

[106] Chriki R, Maguid E, Tradonsky C, Kleiner V, Friesem A A, Davidson N and Hasman E 2018 Spin-controlled twisted laser beams: intra-cavity multi-tasking geometric phase metasurfaces *Opt. Express* **26** 905–16

[107] Hong K G, Hung B J and Wei M D 2016 Low threshold of a continuous-wave mode-locked and azimuthally polarized Nd:YVO4 laser with a semiconductor saturable absorber mirror *J. Opt.* **18** 125603

[108] Carrión-Higueras L, Alcusa-Sáez E P, Díez A and Andrés M V 2017 All-fiber laser with intracavity acousto-optic dynamic mode converter for efficient generation of radially polarized cylindrical vector beams *IEEE Photon. J.* **9** 1–7

[109] Mao D, He Z, Lu H, Li M, Zhang W, Cui X, Jiang B and Zhao J 2018 All-fiber radially/azimuthally polarized lasers based on mode coupling of tapered fibers *Opt. Lett.* **43** 1590–3

[110] Huang B, Yi Q, Yang L, Zhao C and Wen S 2018 Controlled higher-order transverse mode conversion from a fiber laser by polarization manipulation *J. Opt.* **20** 024016

[111] Huang B, Wang Q, Jiang G, Yi J, Tang P, Liu J, Zhao C, Luo H and Wen S 2017 Wavelength-locked vectorial fiber laser manipulated by pancharatnam-berry phase *Opt. Express* **25** 30–8

[112] Mao D, Feng T, Zhang W, Lu H, Jiang Y, Li P, Jiang B, Sun Z and Zhao J 2017 Ultrafast all-fiber based cylindrical-vector beam laser *Appl. Phys. Lett.* **110** 021107

[113] Sun B, Wang A, Gu C, Chen G, Xu L, Chung D and Zhan Q 2015 Mode-locked all-fiber laser producing radially polarized rectangular pulses *Opt. Lett.* **40** 1691–4

[114] Lin D and Clarkson W 2015 Polarization-dependent transverse mode selection in an Yb-doped fiber laser *Opt. Lett.* **40** 498–501

[115] Naidoo D, Roux F S, Dudley A, Litvin I, Piccirillo B, Marrucci L and Forbes A 2016 Controlled generation of higher-order poincaré sphere beams from a laser *Nat. Photon.* **10** 327–32

[116] Sroor H, Huang Y W, Sephton B, Naidoo D, Vallés A, Ginis V, Qiu C W, Ambrosio A, Capasso F and Forbes A 2020 High-purity orbital angular momentum states from a visible metasurface laser *Nat. Photon.* **14** 498–503

[117] Singh K, Tabebordbar N, Forbes A and Dudley A 2020 Digital stokes polarimetry and its application to structured light: tutorial *JOSA* A **37** C33–44

[118] Wootters W 2001 Entanglement of formation and concurrence *Quantum Inf. Comput.* **1** 27–44

[119] Ndagano B, Sroor H, McLaren M, Rosales-Guzmán C and Forbes A 2016 Beam quality measure for vector beams *Opt. Lett.* **41** 3407

[120] Otte E, Rosales-Guzmán C, Ndagano B, Denz C and Forbes A 2018 Entanglement beating in free space through spin-orbit coupling *Light: Sci. Appl.* **7** 18009

[121] Bhebhe N, Rosales-Guzman C and Forbes A 2018 Classical and quantum analysis of propagation invariant vector flat-top beams *Appl. Opt.* **57** 5451–8

[122] Selyem A, Rosales-Guzmán C, Croke S, Forbes A and Franke-Arnold S 2019 Basis-independent tomography and nonseparability witnesses of pure complex vectorial light fields by stokes projections *Phys. Rev.* A **100** 063842

[123] Ndagano B, Nape I, Cox M A, Rosales-Guzmán C and Forbes A 2018 Creation and detection of vector vortex modes for classical and quantum communication *J. Light. Technol.* **36** 292–301

[124] Zhao B, Hu X B, Rodríguez-Fajardo V, Zhu Z H, Gao W, Forbes A and Rosales-Guzmán C 2019 Real-time Stokes polarimetry using a digital micromirror device *Opt. Express* **27** 31087–93

[125] Manthalkar A, Nape I, Bordbar N T, Rosales-Guzmán C, Bhattacharya S, Forbes A and Dudley A 2020 All-digital Stokes polarimetry with a digital micromirror device *Opt. Lett.* **45** 2319

[126] Zhao B, Hu X B, Rodríguez-Fajardo V, Forbes A, Gao W, Zhu Z H and Rosales-Guzmán C 2020 Determining the non-separability of vector modes with digital micromirror devices *Appl. Phys. Lett.* **116** 091101

[127] Dragoman D 2002 Phase space correspondence between classical optics and quantum mechanics *Prog. Opt.* **42** 424–86

[128] Steuernagel O 2005 Equivalence between focused paraxial beams and the quantum harmonic oscillator *Am. J. Phys.* **73** 625–9

[129] Paré C, Gagnon L and Bélanger P A 1992 Aspherical laser resonators: an analogy with quantum mechanics *Phys. Rev.* A **46** 4150–60

[130] Eberly J H, Qian X F, Qasimi A A, Ali H, Alonso M A, Gutiérrez-Cuevas R, Little B J, Howell J C, Malhotra T and Vamivakas A N 2016 Quantum and classical optics-emerging links *Phys. Scr.* **91** 063003

[131] Xu D, Gu B, Rui G, Zhan Q and Cui Y 2016 Generation of arbitrary vector fields based on a pair of orthogonal elliptically polarized base vectors *Opt. Express* **24** 4177–86

[132] Sun Y, Song X, Qin H, Zhang X, Yang Z and Zhang X 2015 Non-local classical optical correlation and implementing analogy of quantum teleportation *Sci. Rep.* **5** 9175

[133] Francisco D and Ledesma S 2008 Classical optics analogy of quantum teleportation *J. Opt. Soc. Am.* B **25** 383–90

[134] Spreeuw R J C 1998 A classical analogy of entanglement *Found. Phys.* **28** 361–74

[135] Guzman-Silva D *et al* 2016 Demonstration of local teleportation using classical entanglement *Laser Photon. Rev.* **10** 317–21

[136] Borges C V S, Hor-Meyll M, Huguenin J A O and Khoury A Z 2010 Bell-like inequality for the spin-orbit separability of a laser beam *Phys. Rev.* A **82** 033833

[137] Balthazar W F, Souza C E R, Caetano D P, Galvão E F, Huguenin J A O and Khoury A Z 2016 Tripartite nonseparability in classical optics *Opt. Lett.* **41** 5797–800

[138] Pereira L J, Khoury A Z and Dechoum K 2014 Quantum and classical separability of spin-orbit laser modes *Phys. Rev.* **90** 053842

[139] Luis A 2009 Coherence, polarization, and entanglement for classical light fields *Opt. Commun.* **282** 3665–70

[140] Goldin M A, Francisco D and Ledesma S 2010 Simulating bell inequality violations with classical optics encoded qubits *J. Opt. Soc. Am.* B **27** 779–86

[141] Qian X F and Eberly J H 2011 Entanglement and classical polarization states *Opt. Lett.* **36** 4110–2

[142] Ndagano B, Brüning R, McLaren M, Duparré M and Forbes A 2015 Fiber propagation of vector modes *Opt. Express* **23** 17330–6

[143] Sit A *et al* 2017 High-dimensional intracity quantum cryptography with structured photons *Optica* **4** 1006

IOP Publishing

Optical Vortices
Fundamentals and applications
Yuanjie Yang, Yu-Xuan Ren and Carmelo Rosales-Guzmán

Chapter 6

Spatio-temporal vortex and applications

The optical vortex can also be created in the spatiotemporal domain. In this chapter, we introduce the basic concepts of the spatiotemporal propagation and control of the light, and the sculpturing of the beam into a spatiotemporal vortex. We will also briefly introduce some advances in the dispersive soliton collision, and the spin–orbit interactions.

6.1 Concept of the spatiotemporally structured light

The light wave can be described using the amplitude, phase, polarization, and frequency. In the previous chapters, most of the discussions focus on the spatial modulation of the amplitude, phase or polarization. The electromagnetic waves of light oscillate in time and propagates in space. In contrast to beam diffraction in space, the light beam can also disperse with time. Thus, the laser beam can also be modulated temporally. More generally, the mode-locking would generate a laser pulsed with different temporal period. For instance, the Q-switching generally produces pulses with a period on the order of nanoseconds (ns), while mode-locking can generate picosecond (ps) pulses. The ultrastrong, ultrashort pulses with a femtosecond (fs) period can be produced by chirped pulse amplification.

In contrast to the spatial wavefront shaping discussed in previous chapters on the spatial modulation of either complex amplitude or polarization, the temporal structure of light can also be tailored to arbitrary patterns. In the spatial domain, the amplitude and phase of the beam can be controlled by propagation and geometric phase through spatial light modulators and metasurfaces [1]. The creation, control and detection of beams with OAM have excited a plethora of applications [2], e.g., optical information storage [3], optical communication, superresolution microscopy [4], and optical micro-manipulation. Optical cycles are too fast to allow direct temporal shaping, and temporal light shaping reduces to spatial light shaping of the frequency components [5]. The frequency components

doi:10.1088/978-0-7503-5844-6ch6

are separated using a dispersive element, e.g., a grating, in the reciprocal space to construct the desired temporal pulses.

In this chapter, we start from the temporal analog of a lens, and briefly discuss the common temporal modes produced in the mode-locked fiber laser, followed by recent advances in shaping spatiotemporal vortices. More generally, we introduce the concepts used in temporal control of light and the related applications, for example, temporal focusing, time-stretch, spatio-temporal vortices, and spin–orbit interaction of light.

6.2 Temporal analog of lens

Spatially, an optical lens focuses a beam into a diffraction limited spot, and eventually the beam broadens due to diffraction. However, if the beam propagates and focuses in a self-focusing medium, the beam will be self-trapped in the form of the so-called spatial soliton. In contrast, the light propagating in a high nonlinear fiber keeps a constant temporal width for each pulse in the presence of dispersion and nonlinearity, forming a temporal solition. Additionally, the concept of an optical lens is extended to the temporal domain as a 'time-lens'. The temporal dispersion is in analogy to the spatial diffraction [6].

Light propagation in the dispersive media can be approximated with the slowly varying envelope equations with diffraction. The space-time duality originates from simplified solutions to the general wave-propagation problem [6]. The time variable in the dispersion is in analogy to the transverse space variable in the diffraction [6]. A similar correspondence can be understood through the temporal and spatial Fourier spectra.

An optical fiber mediates a frequency chirp to an input pulse through self-phase modulation (SPM). As a result, the temporal pulse envelope displays a nearly quadratic time-varying phase shift. Such phenomena are in analogy to the action of a thin lens focusing a beam (figure 6.1(b)). The temporal imaging system can perform distortionless compression or expansion of an input optical waveform. The propagating plane envelope reads

$$E(z, t) = u(z, t) \exp\left(j(\omega t - \beta z)\right). \tag{6.1}$$

The beam follows the parabolic equation in a dispersive medium,

$$\frac{\partial u}{\partial \xi} = \frac{j}{2} \frac{\partial^2 \beta}{\partial \omega^2} \frac{\partial^2 u}{\partial \tau^2}, \tag{6.2}$$

where $\tau = (t - t_0) - (z - z_0)/v_g$, and $\xi = z - z_0$ are the traveling-wave coordinates. The general solution to this dispersion problem reads

$$u(\xi, \tau) = \frac{1}{2\pi} \int_{-\infty}^{+\infty} U(0, \omega) \exp\left(-j\frac{\xi}{2}\frac{\partial^2 \beta}{\partial \omega^2}\omega^2\right) \exp\left(j\omega t\right) d\omega. \tag{6.3}$$

Here, $u(\xi, \tau)$ is the inverse Fourier transform of the initial envelope spectrum, $U(0, \omega)$, multiplied by a phase factor quadratic in frequency.

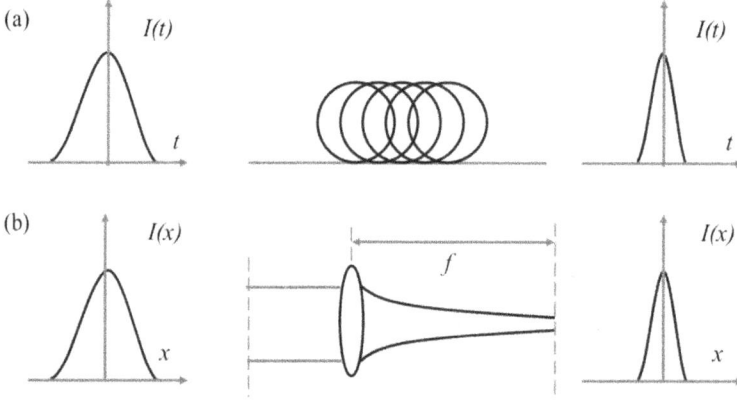

Figure 6.1. Concept of a time lens. (a) Conventional fiber-grating pulse compressor, and (b) its spatial analog, a Fourier transform lens [6].

The time lens performs the multiplication on the envelope time function $u(\xi, \tau)$ by a quadratic phase factor in time. In contradistinction to a spatially thin lens, which performs a phase shift in real space, $\phi(x) = -k_0 x^2/2f$, with f as the focal length, the time lens produces a phase shift, $\phi(t) = -\omega_0 t^2/2f_T$, and the equivalent focal time f_T reads, $f_T = \omega_0/A\omega_m^2$.

There are many ways to achieve a time lens, for instance using a dispersive fiber. The time lens can also be produced by an electro-optic phase modulator with a sinusoidal driving signal of angular frequency ω_m. The phase modulation is quadratic as the input waveform is shorter than $\sim 1/\omega_m$. Accordingly, the pulse accumulates a phase shift via the modulator,

$$\phi(t) = \pm A\left(1 - \frac{\omega_m^2 t^2}{2}\right), \tag{6.4}$$

where A is the peak phase deviation and the $\pm$ accounts for negative (positive) curvature. The instantaneous frequency linearly chirps with a rate of $\partial^2\phi/\partial t^2 = A\omega_m^2$. Combination of the dispersion and quadratic phase modulation results in the output waveform [6],

$$u(\tau, \xi) = \left(\frac{1}{M}\right)^{1/2} \exp\left(-j\frac{A\omega_m^2 \tau^2}{2M}\right) \int_{-\infty}^{+\infty} U(0, \omega)$$

$$\times \exp\left[-j\omega^2\left(\frac{1}{\xi_1 \frac{\partial^2 \beta_1}{\partial \omega^2}} + \frac{1}{\xi_2 \frac{\partial^2 \beta_2}{\partial \omega^2}} - \frac{\omega_0}{f_T}\right)\right] \exp\left(j\omega\frac{\tau}{M}\right) d\omega, \tag{6.5}$$

where $M = 1 - \xi_2(\partial^2\beta_2/\partial\omega^2)A\omega_m^2$ is the magnification and the subscripts 1 and 2 correspond to the object and the image sides, respectively, of the time lens. Let the quadratic phase term in the integrand be zero, resulting the equivalent lens formula in the time-domain,

$$\frac{1}{\xi_1 \frac{\partial^2 \beta_1}{\partial \omega^2}} + \frac{1}{\xi_2 \frac{\partial^2 \beta_2}{\partial \omega^2}} = \frac{\omega_0}{f_T}. \tag{6.6}$$

The similarity between equation (6.6) and its spatial counterpart is striking. In general, the minimum pulse width is $\tau_{min} \approx 1/A\omega_m$. Since the aperture time of the modulator is approximately $1/\omega_m$, then the number of resolvable elements is $N \approx 1/(\omega_m \tau_{min}) = A$.

The time lens bounded by dispersive media performs the distortionless magnification or compression of optical waveforms. With power amplification, the spectrum of an optical pulse can be mapped into a time-domain waveform using the group-velocity dispersion (GVD) and simultaneous amplification, this is coined amplified dispersive Fourier transformation (ADFT) [7]. Such ability can be used to perform time-stretch on optical pulses, and the real-time detection of ultrafast dynamical processes with high sensitivity and resolution. For instance, the 2D image can be detected in a serial time-domain data stream and simultaneously amplifies in the optical domain, as a result, the frame rate goes at least 1000 times faster than the conventional CCD does [8]. The STEAM camera can capture the dynamics of laser ablation and cells in microfluidic flow at a speed on order of m s^{-1} [8]. The temporal resolution is the highest achieved in the observation of the ultrafast microfluidic flow. This capability to develop the microfluidic biochips has the potential to revolutionize cytometry and analysis in molecular biology.

6.3 Spatiotemporal optical vortex

Temporal shaping of the laser pulse includes the use of an acousto-optic modulator loaded with electronic signal. The radio frequency (RF) signal generates acoustic waves in the crystal and forms an erasable optical grating that deflects light. By modulating the RF signal, the deflected light would be controlled by design. For instance, the traditional Ti:sapphire laser generates the mode-locked pulse at a high repetition rate, e.g., 80 MHz. However, in some applications like two photon microscopy, lower repetition rate would be required. The AOM can pick up pulses and lower the repetition rate, e.g., 10 MHz. In general, the AOM could load RF signal with arbitrary shape to control the laser pulses. Traditional optical vortex has a phase singularity, and a central null in intensity. The azimuthal phase resides in transverse spatial dimensions. The spatiotemporal optical vortex (STOV) has a phase wind that resides in the spatiotemporal domain. The STOVs are a new type of mode with optical phase circulation in space-time [9]. The ST vortex has a spiral phase in the meridional plane (x–t plane), and the spiral phase and the zero-intensity center are on the ST plane.

Figure 6.2 shows the spectral support domain for 3D ST wave packets [10]. For a superluminal 3D ST wave packet, shown is the intersection of the light-cone $k_r^2 + k_z^2 = (\omega c)^2$ with a spectral plane that is parallel to the k_r-axis and makes an angle $\theta > 45°$ with respect to the k_z-axis. In (k_x, k_y, ω_c)-space the spectrum is one half of a two-sheet elliptic hyperboloid. The spectral support domain on the light-cone in (k_r, k_z, ω_c)-space is an ellipse for a subluminal ST wave packet with $\theta < 45°$.

a   Superluminal 3D ST wave packets

$(k_r,k_z,\omega/c)$-space $(k_x,k_y,\omega/c)$-space

b   Subluminal 3D ST wave packets

$(k_r,k_z,\omega/c)$-space $(k_x,k_y,\omega/c)$-space

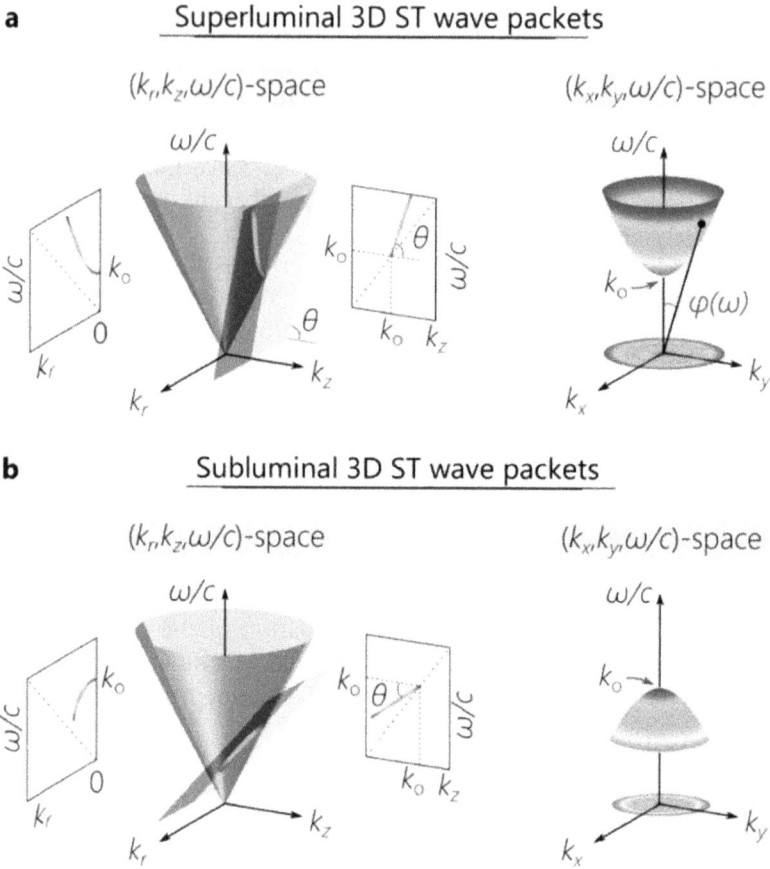

Figure 6.2. The spectral support domain for (a) superluminal and (b) subluminal 3D ST wave packets. Reproduced from [10]. CC BY 4.0.

The well-known optical OAM occurs by applying a spirally varying phase of $\exp(-il\phi)$, where ϕ is the azimuthal angle and l is an integer, i.e., a topological charge l. With such a spiral phase, the optical beam converts into a ring shape, and therefore the energy circles around the phase singularity. The longitudinal OAM can also be embedded in the pulsed vortex beam. A spatial vortex features a spiral phase in the transverse plane with a phase singularity of zero intensity at the centre.

The optical beam with OAM characterizes with phase singularity in the transverse spatial plane. The STOVs have a phase circulation in space-time, and was first demonstrated in self-guiding in material owing to the self-focusing collapse [11]. The linear generation of STOV in free space has also arisen increasing interests. Toroidal STOV is a universal electromagnetic structure that appears from self-focusing collapse of short pulses [11]. The phase winding of STOV resides in the spatiotemporal domain [11]. An electric field component of a simple $|l|$th order line-STOV-carrying pulse at position z along the propagation axis analytically reads

$$E(r_\perp, z, \tau) = a\left(\frac{\tau}{\tau_s} + \text{isgn}(l)\frac{x}{x_s}\right)^{|l|} E_0(r_\perp, z, \tau) = A(x, \tau)e^{il\Phi_{s-t}}E_0(r_\perp, z, \tau), \quad (6.7)$$

where $r_\perp = (x, y)$, $\tau = t - \frac{z}{v_g}$ is a local time coordinate, v_g is the group velocity, τ_s and x_s are temporal and spatial scale of the STOV, $\Phi_{s-t}(x, \tau)$ is the space–time phase circulation in $x - \tau$ space, $l = \pm 1, \pm 2, ...$, $A(x, \tau) = a((\frac{\tau}{\tau_s})^2 + (\frac{x}{x_s})^2)^{\frac{|l|}{2}}$, $a = \sqrt{2}((\frac{x_0}{x_s})^2 + (\frac{\tau_0}{\tau_s})^2)^{-\frac{1}{2}}$ for $l = \pm 1$, E_0 is the STOV-free near-Gaussian pulse input, and x_0 and τ_0 are its spatial and temporal widths. The normalization factor a ensures the energy conservation through $\int d^2r_\perp \, d\tau \, |E|^2 = \int d^2r_\perp \, d\tau \, |E_0|^2$.

In absence of a phase plate, a 50 fs input pulse with a weak parabolic temporal phase can be recovered. Intensity and phase of pulse in the far field of the pulse shaper with $l = 1$ and $l = -1$ spiral phase plates suggest a vortex feature. The STOV displays a new type of optical OAM with vortex phase circulation in space-time. STOV conserves angular momentum in space-time and mediates space-time energy flow within the pulse that can be analyzed with TG-SSSI with enough space- and time-resolution.

6.4 Generation of spatiotemporal optical vortex

The STOV travels along the laser pulse. The global energy flow is a universal phenomenon underlying strong field laser propagation. The STOV was demonstrated by Jhajj et al as a novel type of optical vortex, in all nonlinear optical collapse arrest processes [11]. In contrast to traditional vortex beams with optical phase rotation, the STOV can be embedded in the traveling optical field [11]. STOVs are a fundamental characteristic of strong field laser propagation. The energy flow will be either saddle or spiral around the vortex ring [11].

The formation of STOV comes from the phase slippage between the core and periphery thanks to a nonlinear material response [11]. Depending on the sign of dispersion, the energy flow (Poynting flux) is either saddle or spiral around the vortex ring. Additionally, the STOV ring clearly identifies the spatial locations of the so-called filament 'core' and 'reservoir' [11]. Suppose a 1D plane wave has two regions with a large difference in intensity; the intense half core experiences a larger nonlinear index than the other half (periphery). The two regions will create a phase front shear, when the phase difference suggests a defect and the field envelope appears (figure 6.3). Initially, the phases are aligned ($z = 0$). At $z = z_v$, the null intensity occurs. A continued shear carrying the vortices appears at $z = 2z_v$. The +1 vortex moves to the temporal front, while the −1 vortex moves backwards. The phase winding convention considers a $(\xi - \xi_0) + i(x - x_0)$ winding about a null at $(\xi_0 - x_0)$ to be a +1 STOV. In the third panel, arrows indicate the direction of phase increase; the arrows show that the phase difference between head and tail is ill-defined due to the vorticity of the −1 STOV (figure 6.3).

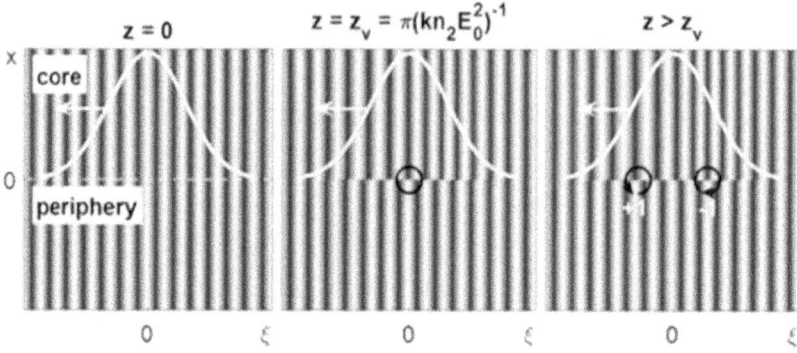

Figure 6.3. Toy model on the birth of vortex pair through spatiotemporal phase shear. The white curve and arrow depict the axial (temporal) intensity and propagation direction, while the 'core' and 'periphery' labels the spatial intensity. Reproduced from [11]. CC BY 3.0.

Freely propagating STOV conserves angular momentum and mediates space-time energy flow inside the pulse [9]. In 2015, the real-time dynamics of a soliton explosion was observed in a fiber laser using the time-stretch dispersive Fourier transform (TS-DFT) [12]. The amplitude and phase of the STOV can be mapped using a TG-SSI technique in single shot. The different vorticities are allocated at different parts of the spectrum such that the short wavelengths of the spectrum are assigned to a spatial mode with vorticity $\ell = +1$ and the long wavelengths with charge $\ell = -1$.

In figure 6.4, row (a) shows the pulse with no phase plate in the pulse shaper [9]. Such far-field output of the shaper is measured by TG-SSI in the witness plate. The temporal leading edge is at $\tau < 0$, and the left column demonstrates $\Delta(x, \tau)$. The fringe passes through a low-pass filter, and generates (x, τ) (the second column), while a high-pass filter leaves the fringe image $f(x, \tau)$ (the third column). The far-right column demonstrates the extracted spatiotemporal phase $\Delta\Phi(x, \tau)$. It is seen that the pulse envelope IS closely agrees with the 50 fs pulse input to the shaper, and that $\Delta\Phi(x,)$ is weakly parabolic in time (small chirp) and relatively flat in space.

One form of line-STOV-carrying pulse can be generated with a spiral phase plate in the pulse shaper. For a $l = 1$ plate, the various extractions from TG-SSI [row (b) of figure 6.4] [9]. The presence of a spatiotemporal phase singularity is evident from the characteristic forked pattern in $f(x, \tau)$. The spatiotemporal envelope $IS(x, \tau)$ and phase $\Delta\Phi(x, \tau)$ of the STOV are displayed in the second and fourth columns, where the pulse appears as an edge-first flying donut with a 2π phase circulation around the phase singularity. A $\tau = -1$ plate generates the opposite spatiotemporal phase circulation [row (c)]. The small insets in (b) and (c) show the near-field intensity envelopes, consisting of two lobes separated by a space–time diagonal.

The magnitude of the transverse OAM carried by the ST vortex can be scalable to a larger value [13]. Suppose $g_R(r)$ represents an optical field in the $k_x - \omega$ domain, where (r, θ) are the polar coordinates with $r = \sqrt{k_x^2 + \omega^2}$ and $\theta = \tan^{-1}(\omega/k_x)$. By

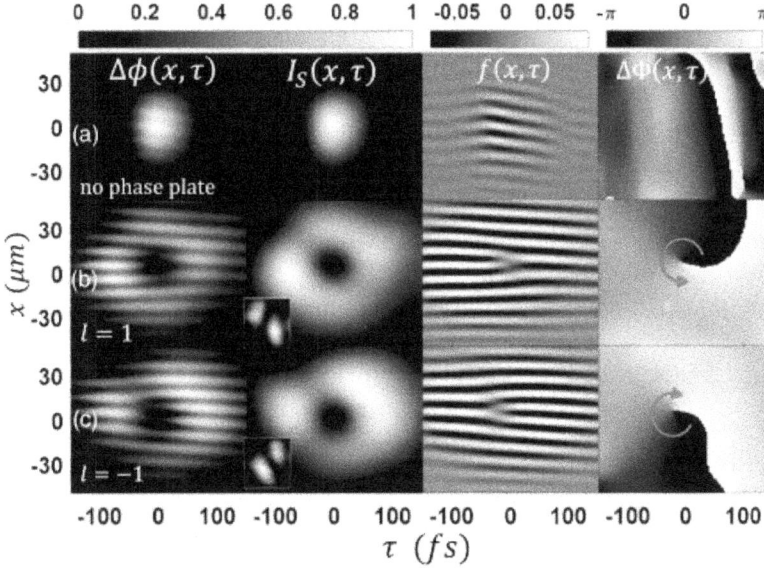

Figure 6.4. Output pulse shape with no phase plate. The 50 fs input pulse, with a weakly parabolic temporal phase is recovered. (b and c) Intensity and phase of pulse in far field of pulse shaper with $l = 1$ and $l = -1$ spiral phase plates. The pulse propagation is from right to left, so the temporal leading edge of the pulse is on the left ($\tau < 0$). The red arrows demonstrate the direction of phase circulation. Adapted with permission from [9]. © 2019 Optical Society of America under the terms of the OSA Open Access Publishing Agreement.

applying a spiral phase of exp $(-il\theta)$ is, a 2D Fourier transform gives the field in the x–t domain [14],

$$G(\rho, \phi) = \mathrm{FT}(g_R(r)e^{-il\theta}) = 2\pi(-i)^l e^{-il\phi} H_l(g_R(r)), \qquad (6.8)$$

where (ρ, ϕ) are the conjugate polar coordinates with $\rho = \sqrt{x^2 + t^2}$ and $\phi = \tan^{-1}(x/t)$, $H_l(g_R(r)) = \int_0^\infty rg_R(r)J_l(2\pi\rho r)dr$ and J_l is the Bessel function of the first kind. The full electric field reads $E(\rho, \phi) = G(\rho, \phi)\exp(ik_z z - i\omega t)$.

Experimentally, an ST vortex can be produced through applying a spiral phase in the spatial frequency-domain. Overlapping of the chirped ST vortex with a short reference pulse ($\sim$90 fs) at a minute angle results in interference fringes that characterize the ST vortex. A diffraction grating combined with a cylindrical lens spatially disperses frequencies akin to a time-frequency Fourier transform, which is implemented by a grating-lens pair. The chirped ST vortex is produced by a spiral phase on the SLM and an inverse Fourier transform using the grating-cylindrical lens pair.

The theoretical intensity profile of the STOV has been experimentally corroborated [13]. An ST hole appears on the 3D intensity profile for the $l = 1$ case owing to the phase singularity. Two ST holes appear in the 3D intensity profile for the $l = 1$ vortex. The $l = 2$ vortex splits into two $l = 1$ vortices; the total topological charge possessed by the wave packet is 2. Even though vortices split, they will act as a single vortex with $l = 2$ with material response times greater than the separation between vortices (~600 fs). The vortex separation will be reduced when the $l = 2$ ST vortex is dechirped to a shorter pulse duration (~100 fs).

6.5 Spatiotemporal solitons and soliton molecules

This section was reproduced with permission from [22], copyright 2020 Chinese Laser Press.

The temporal solitons are naturally structured in response to the fluctuation in polarization and cavity dispersion. The solitons are the consequences of the balance between nonlinearity and dispersion in the nonlinear medium. Solitons form bound states in the form of soliton molecules [15]. Temporal solitons with short pulse widths pose challenges to the detection bandwidth. The dispersive fiber performs the time stretch, and allows the real-time access to both the spectral and temporal dynamics. The time stretch has been applied to explore the explosion dynamics in breathing dissipative solitons [16]. The bound solitons can be closely- and well-separated in ultrafast fiber laser through time-stretch detection.

The soliton explosion dynamics were recorded in a fiber laser by using the time-stretch dispersive Fourier transform (TS-DFT) [12]. Yi and colleagues explored the triggering of soliton explosions in breathing solitons in a bidirectional mode-locking fiber laser [17]. They found that the counterpropagating breathing solitons collide in each round trip. The dissipative solitons stem from the dynamic attractors in dissipative systems, which possess a wide range of nonlinear dynamics, e.g., breathing and chaos [17]. The breathing soliton explosion, as the most fascinating localized nonlinear structures, appears in the soliton buildup process through polarization alteration [16].

Soliton dynamics depend on the net cavity dispersion in fiber lasers. The similarity in spectrum and temporal features appear in the buildup of stationary counterpropagating solitons in fiber laser [18]. In anomalous dispersion regimes, the counterpropagating pulses experience independent buildup features [19]. Yi and colleagues reported on the breathing dissipative soliton explosion in a carbon nanotube (CNT) mode-locked bidirectional ultrafast fiber laser in the net-normal-dispersion regime [17].

The control over the intercavity polarization results in breathing dissipative solitons during the mode-locking buildup. The solitons spectra for both CCW and CW suggest the same center wavelength of 1595 nm but with non-identical bandwidth (18.7 nm for CCW, and 15 nm for the CW) [16]. However, due to the asymmetric pump of the EDF, the CW pulse accumulates higher energy and amplified spontaneous emission (ASE) noise. Figures 6.5(a) and (b) show the temporal evolution along with the CW and CCW directions, respectively. The soliton buildup and explosion behavior take place simultaneously along both directions. The soliton explosion process suggests characteristic temporal shift (figures 6.5(a) and (b)). The temporal separation of counterpropagating solitons remains invariant. The breathing soliton buildup suggests similar spectrum evolution in both directions (figure 6.5(c)). In the soliton explosion, an abrupt spectral collapse appears during the breathing mode-locking, and the periodic spectrum evolution is disrupted in both directions. The duration of the soliton explosion lasts about 900 RTs in both directions before the quasi-stable breathing recovers. The energy evolution suggests similarity for counterpropagating breathing

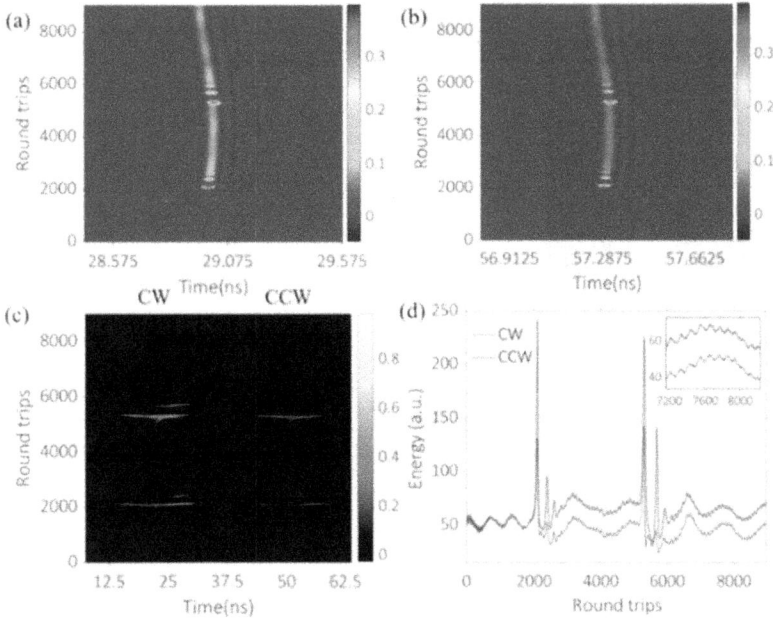

Figure 6.5. Buildup and soliton explosion of breathing soliton. Temporal evolution in (a) CW and (b) CCW directions. (c) Shot-to-shot spectral evolution. (d) Energy evolution. Adapted with permission from [16]. © 2020 Chinese Laser Press.

solitons characterized by the apparent energy oscillation (figure 6.5(d)). The counter-propagating breathing solitons suggests behavior similarity during soliton buildup and explosion thanks to gain/loss modulation.

Moreover, the dissipative soliton suffers from the balance between nonlinearity and dispersion; the existence and stability of dissipative solitons depend on the interplay between gain and loss. In bidirectional ultrafast fiber laser system, lots of interesting phenomena were reported, including the breathing dissipative soliton pairs [17], dissipative soliton molecule switching [20], and reconfigurable soliton molecular complexes [21].

Optical solitons can assemble into stable molecule-like bound states, i.e., soliton molecues. The buildup and dissociation dynamics of soliton molecues was reported in the normal dispersion regime in normal dispersion mode-locked fiber lasers [22]. Under different transmission profiles of the saturable absorber (SA), the soliton pair evolve into a soliton molecule through intense repulsive interaction [22]. The soliton molecule can also dissociate into individual solitons through transient annihilation accompanied by energy transfer, and exhibits distinguishable features according to the time scale. During short-time soliton molecule formation, there drastic changes in DFT spectra exists (figure 6.6(a)). Two pulses directly formed at the initial stage (figure 6.6(b)) with splitting of the single soliton on the dynamics, and the pulses undergo a strong repulsion before the establishment of a stable soliton molecule. First-order field autocorrelation proves the separation evolution between bound solitons and the transient ordering of incoherent dissipative solitons [22].

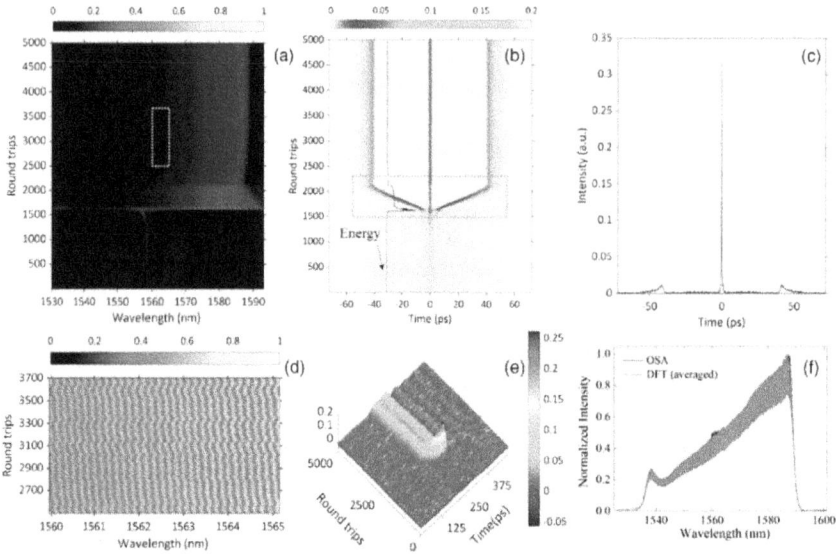

Figure 6.6. Soliton molecule established from background noise. (a) The spectral evolution. (b) The field autocorrelation, the black line: energy evolution. (c) Field autocorrelation trace at RT number of 2500. (d) Zoom-in of dashed rectangle in (a). (e) The corresponding intensity evolution. (f) The agreement between the measured spectra (black) and the average spectrum from 1000 single-shots (red) under stable mode-locking confirms the accuracy and consistency of TS-DFT. Adapted with permission from [22]. © 2020 Optical Society of America under the terms of the OSA Open Access Publishing Agreement.

The Fourier transforms of the single-shot spectra (figure 6.6(a)) produces the field autocorrelation (figure 6.6(b)). Mode-locking begins near RT 1600, and generates two solitons from background noise directly. The initial separation of the two solitons is 4.3 ps. The significant repulsion lasts ≈540 RTs for two solitons, eventually stabilized at RT 2140 with a stable separation of 42 ps. The cross section of the field autocorrelation at the RT 2500 (figure 6.6(c)) implies that the spectrum modulation period is about 0.182 nm, and the two pulses separated with a distance of 42 ps. In contrast, the pulse duration is 0.3 ps, which suggests that the interaction of the soliton molecule originates from weak long-range interactions [22].

Due to the weak long-range interactions of soliton molecules with very weak intensity of sidelobes (figure 6.6(b)), breathing behavior is not apparent. However, the energy oscillation exists in the soliton molecule repulsion (RTs 1700–2100 on the black line in figure 6.6(b)), and the soliton molecule suggests similar breathing. Figure 6.6(e) shows the intensity evolutions. The average spectrum calculated from the 1000 single-shot spectra (red curve) matches well with the spectrum recorded on the OSA (black curve) (figure 6.6(f)).

Furthermore, the soliton molecule could dissociate and transit into an unstable single soliton and annihilate [22]. The DFT spectra suggest evident changes of soliton molecule dissociation (figure 6.7(a)). Figure 6.7(b) shows the field autocorrelation through Fourier transforms of a single-shot spectrum. The spectrum

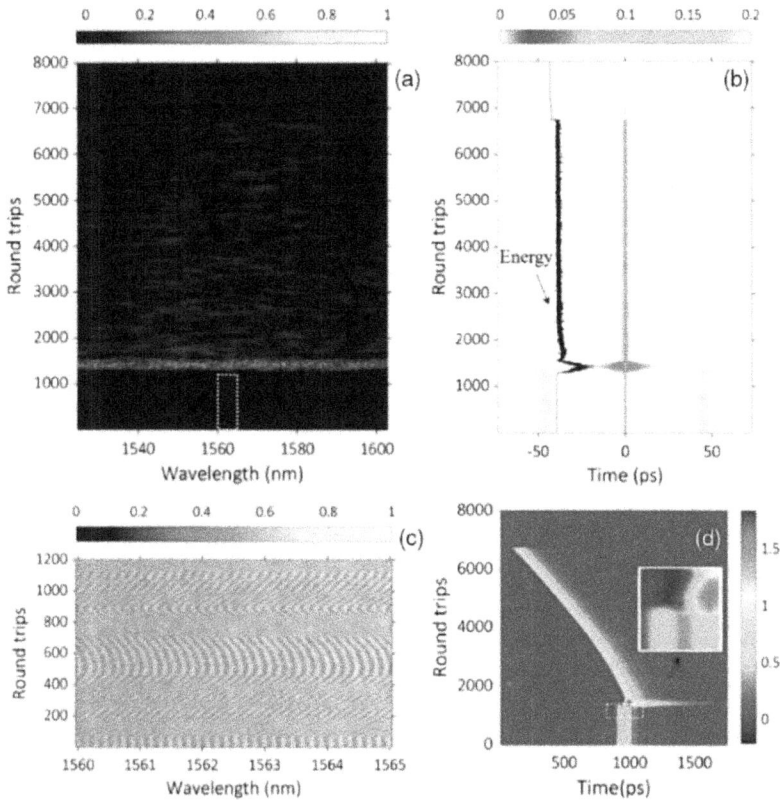

Figure 6.7. Soliton molecule dissociation. (a) The real-time spectral evolution, and the corresponding (b) field autocorrelation. The black line shows the energy evolution. (c) Zoom-in of the dashed rectangle in (a). (d) The temporal intensity evolution. Adapted with permission from [22]. © 2020 Optical Society of America under the terms of the OSA Open Access Publishing Agreement.

between 1560 and 1565 nm (white dashed rectangle in figure 6.7(a)) suggests significant oscillation upon evolution (figure 6.7(c)). From the field autocorrelation and temporal evolution (figures 6.7(b) and (d)), the energy is transferred to the trailing pulse during the soliton molecule dissociation. Then, a single ps pulse with a long soliton tail (~750 ps) forms and becomes a weaker pulse with a shorter soliton tail. Finally, the weaker single soliton annihilates after 5200 RTs. The energy evolution by integrating the spectra over the complete band (black line in figure 6.7(b)) suggests an energy overshoot when the soliton molecule dissociates into a wide pulse.

The different dynamical processes are mainly regulated by the polarization-dependent transmission. In the short-time buildup stage, solition repulsion dominates in the soliton molecule interplay due to large differences in initial pulse intensity and separation. In the long-time buildup stage, the single soliton splits, and the soliton molecules exhibit energy transfer and attraction among solitons in the molecule. These dynamics can obviously differ from net cavity dispersion and solid-state laser systems, i.e., SA mode-locked fiber laser, Ti: sapphire lasers, and microresonators.

6.6 Spin–orbit interaction of spatiotemporal optical vortex

This section was reproduced from [16], copyright 2020 Optical Publishing Group.

Light carries both spin and orbital angular momentum. The topological charge implies that the OAM has the magnitude of $l\hbar$ per photon. Such a beam could also be utilized for the the orbital rotation of the particles [23–25], detection of angular Doppler shift from a rotating surface [26, 27], momentum transmission and quantum communication [28]. The transverse OAM may exhibit a unique spin-Hall effect allowing for an efficient optical diode.

Spatiotemporal analogs of vortex beams, spatiotemporal vortex pulses (STVPs), carry intrinsic OAM transverse (or, generally, tilted) with respect to the propagation direction of the pulse. The dynamical properties of STOV are determined by the polarization and spatial degrees of freedom. A self-consistent full-vector description of optical STOV was proposed to describe the propagation dynamics [29]. The in-plane linear polarization produces a longitudinal field component and a nonzero transverse SAM density. This induces the spin–orbit interaction, such as observable polarization-dependent intensity distributions of STVPs. Importantly, an integral value of the transverse SAM vanishes, while the integral OAM is quantized as $l\hbar$ per photon only for circularly symmetric pulses.

Monochromatic Bessel beams along the z-axis are considered as a superposition of plane waves with frequency $\omega = \omega_0$, wave vectors k distributed within a cone of polar angle $\theta = \theta_0$, and with an azimuthal phase difference $l\phi$ (ϕ is the azimuthal angle in k space) (figure 6.8(a)). In other words, the wave vectors form a circle in the $k_z = k_{\parallel}$ plane, with the center at $(0, 0, k_{\parallel})$ and radius $k_{\perp}$, where $k_{\parallel} = k_0 \cos \theta_0$, $k_{\perp} = k_0 \sin \theta_0$, $k_0 = \omega_0/c$, and c is the light speed. In real space, this superposition results in the scalar wave function $\Psi(r, t) \propto J_l(k_{\perp} r) \exp(ik_{\parallel} z + il\varphi - i\omega_0 t)$, where (r, φ) are the polar coordinates in the (x, y) plane and J_n is the Bessel function of the first kind. Figure 6.8(a) shows the transverse intensity and phase dislocations of such Bessel beam.

A superposition of plane waves with wave vectors distributed over a circle in the $k_y = 0$ plane with the center at $(0, 0, k_0)$ and radius Δk, constructs a Bessel-type STVP with a purely transverse intrinsic OAM (figure 6.8(b)). Using the azimuthal angle ϕ with respect to the center, the real-space wave function is written as a Fourier-type integral:

$$\psi(r, t) \propto \int_0^{2\pi} e^{i[k_0 z + \Delta k \cos \tilde{\phi} z + \Delta k \sin \tilde{\phi} x + l\tilde{\phi} - \omega(\tilde{\phi})t]} d\tilde{\phi}, \qquad (6.9)$$

where $\omega(\tilde{\phi}) = c\sqrt{k_0^2 + \Delta k^2 + 2k_0 \Delta k \cos \tilde{\phi}}$. Parameter Δk determines the degree of paraxiality and monochromaticity of the Bessel STVP. The maximum polar angle of the wave vectors in its spectrum is $\sin \theta_0 = \Delta k / k_0$ (figure 6.10(b)). For $\Delta k \ll k_0$, $\theta_0 \ll 1$, the pulse is near-paraxial and quasi-monochromatic. Under this approximation, $\omega(\tilde{\phi}) \simeq c(k_0 + \Delta k \cos \tilde{\phi})$, and the integral (6.9) results in the analytical Bessel pulse solution,

$$\Psi(r, t) \propto J_l(\tilde{\rho}) \exp(ik_0 \zeta + il\tilde{\varphi}). \qquad (6.10)$$

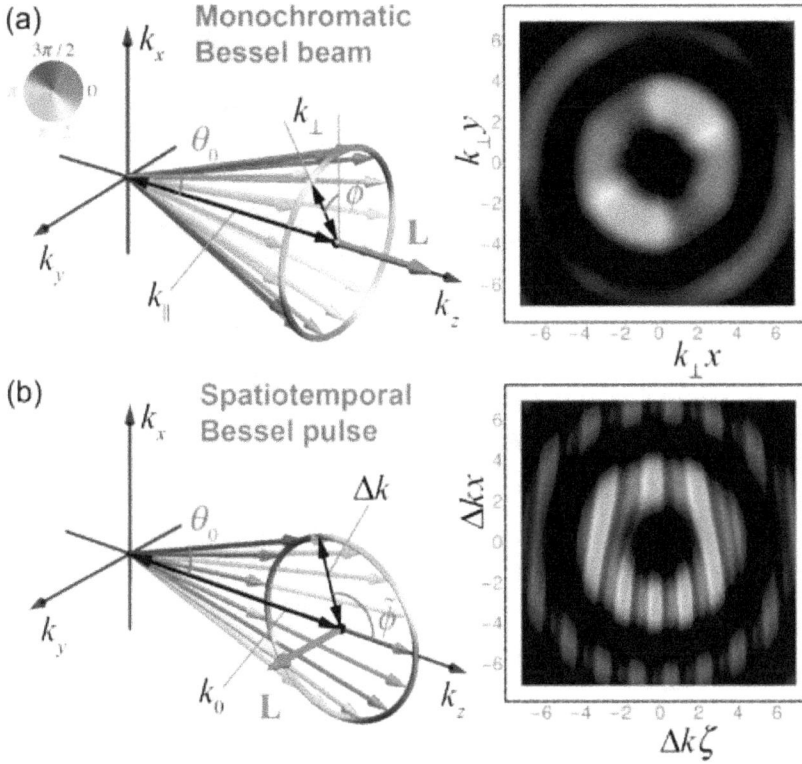

Figure 6.8. The plane-wave spectra (left) and complex amplitude of real-space wave functions $\psi(r, t)$ (right) for (a) the monochromatic Bessel beam with $l = 2$ and (b) spatiotemporal Bessel pulse with $l = 2$. Reprinted with permission from [23]. Copyright (2021) by the American Physical Society.

Here, $\zeta = z - \mathrm{ct} = \tilde{r}\cos\tilde{\varphi}$, $\tilde{\rho} = \Delta k\tilde{r}$, and $(\tilde{r}, \tilde{\varphi})$ are the polar coordinates in the (ζ, x) plane. The intensity $I = |\psi|^2$ and phase $Arg(\psi)$ distributions for the Bessel STVP (figure 6.8(b)). It has typical Bessel intensity profile $I \propto |J_l(\tilde{\rho})|^2$ in the (ζ, x) plane, contains an edge phase dislocation of order l, and propagates along the z-axis.

The nondiffracting solution (6.9) is a result of linear expansion of $\omega(\phi)$ with respect to Δk. The lth order phase dislocation splits into a raw of $|l|$ first-order dislocations oriented diagonally in the (ζ, x) plane. In contrast to the Rayleigh range for spatial diffraction, a typical scale of the temporal diffraction is given by the 'temporal Rayleigh range' $ct_R = k_0/\Delta k^2$. Notably, nondiffracting Bessel-like STVPs can be constructed using the wave vectors distributed along an ellipse in k space, which can be Lorentz transformed to a monochromatic circle. Specifically, the circular spectrum (figure 6.8(b)) should be extended along the k_z axis to become an ellipse with the ratio of semiaxies $\gamma = \sqrt{1 + (k_0/\Delta k)^2}$.

Because of the transversality, the electric field of each plane wave in the vector Bessel STVPs pulse spectrum must be orthogonal to its wave vector k. This determines two basic polarizations in the problem: (1) out-of-plane, E is directed along the y-axis (figure 6.9(a)), and (ii) in-plane, E lies in the (z, x) plane (figure 6.9

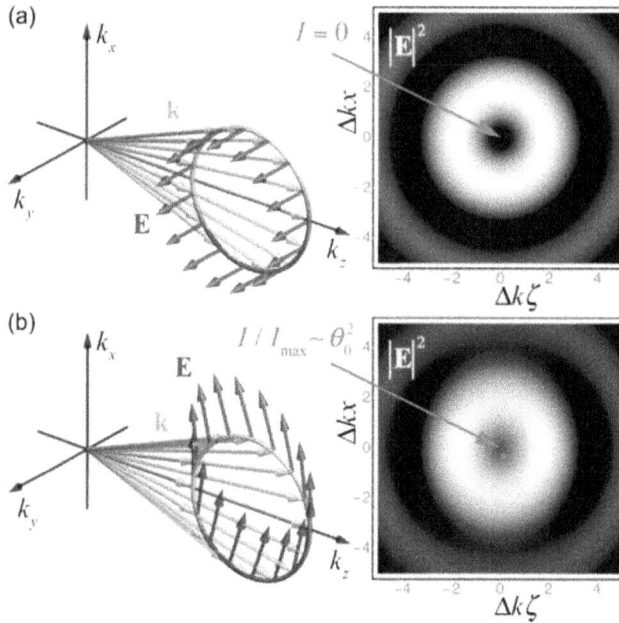

Figure 6.9. Electric fields of plane waves in the spectra of Bessel-type vector STVPs (left) and the corresponding real-space intensity distribution (right) for the (a) out-of-plane, and (b) in-plane polarization. The parameters are $l = 1$ and (b) spatiotemporal Bessel pulse with $\Delta k = 0.7$ for better visibility of nonzero intensity in the center of the pulse in (b). Adapted with permission from [29], Copyright (2021) by the American Physical Society.

(b)). A localized intensity distribution determines the position of a beam or wave packet, whereas the phase gradient describes the wave propagation. For the in-plane polarization, each plane wave in the pulse spectrum has two electric field components, E_x and E_z (figure 6.9(b)). The amplitudes and phases of these components depend on the wave vector k, which signals the spin–orbit interaction.

The spin–orbit interactions (SOIs) of light are striking optical phenomena in which the spin (circular polarization) affects the spatial degrees of freedom of light. The intrinsic SOI of light originates from the fundamental spin properties of Maxwell's equations. The SOI of light covers a broad phenomena in optics, including, the spin-Hall effect in inhomogeneous media and at optical interfaces, spin-dependent effects in nonparaxial fields, spin-controlled shaping of light using metasurfaces, and robust spin-directional coupling via evanescent waves [30]. In paraxial and nonparaxial fields, in both simple optical elements (planar interfaces, lenses, waveguides, anisotropic plates and small particles) and complex nano-structures (photonic crystals, metamaterials and plasmonics structures), the numerous SOI phenomena can be divided into several classes [30]:

(1) a circularly polarized laser beam reflected or refracted at a planar interface experiences a transverse spin-dependent subwaveleneth shift. This offers important evidence of the fundamental quantum and relativistic properties of photons. Supplied with suitable polarimetric tools, it can be employed for precision metrology.

(2) The focusing of circularly polarized light by a high-NA lens, or scattering by a small particle, generates a spin-dependent optical vortex. This is an example of spin-to-orbital angular momentum conversion in nonparaxial fields. Breaking the cylindrical symmetry of a nonparaxial field also produces spin-Hall effect shifts.

(3) A similar spin-to-vortex conversion occurs when a paraxial beam propagates in optical fibers or anisotropic crystals. Proper design of the anisotropic and inhomogeneous structures allows enhancement of the SOI effects and efficient spin-dependent shaping of light. Combining SOI with structured materials provides a versatile platform for optical spin-based elements with desired functionalities.

(4) Any surface or waveguide mode possesses evanescent field tails. Coupling transversely propagating circularly polarized light to these evanescent tails results in a robust spin-controlled unidirection excitation of the surface or waveguide modes. This is a manifestation of the extraordinary transverse spin of evanescent waves associated with the quantum spin-Hall effect of light.

The SOI occurs in the propagation of paraxial light in an inhomogeneous isotropic medium. Light changes its momentum due to refraction or reflection at medium inhomogeneities. However, in traditional geometrical optics in the absence of anisotropy, the trajectory of an optical beam is independent of its polarization. The SOI allows the direct observation of fundamental spin-induced effects in the dynamics of relativistic spinning particles (photons). The SOIs of light are considerably enhanced by material anisotropies and can be artificially designed in optical nanostructures. Moreover, SOI phenomena are usually determined by basic symmetry and are robust with respect to system perturbations.

6.7 Summary

The spatiotemporal vortex offers tremendous freedom to control light propagation feature. The spatiotemporal control of light not only is of fundamental significance, for instance, to study the spin–orbit interaction, second-harmonic generation, parametric downconversion, and propagation in presence of Kerr nonlinearity, but also provides exciting applications to explore ultrafast dynamical processes.

References

[1] Wang B, Liu W, Zhao M, Wang J, Zhang Y, Chen A, Guan F, Liu X, Shi L and Zi J 2020 Generating optical vortex beams by momentum-space polarization vortices centred at bound states in the continuum *Nat. Photon.* **14** 623–8

[2] Shen Y, Nape I, Yang X, Fu X, Gong M, Naidoo D and Forbes A 2021 Creation and control of high-dimensional multi-partite classically entangled light *Light: Sci. Appl.* **10** 50

[3] He C, Shen Y and Forbes A 2022 Towards higher-dimensional structured light *Light: Sci. Appl.* **11** 205

[4] He H, Kong C, Chan K Y, So W L, Fok H K, Ren Y-X, Lai C S W, Tsia K K and Wong K K Y 2020 Resolution enhancement in an extended depth of field for volumetric two-photon microscopy *Opt. Lett.* **45** 3054–7

[5] Weiner A M 2000 Femtosecond pulse shaping using spatial light modulators *Rev. Sci. Instrum.* **71** 1929–60

[6] Kolner B H and Nazarathy M 1989 Temporal imaging with a time lens *Opt. Lett.* **14** 630–2

[7] Goda K, Solli D R, Tsia K K and Jalali B 2009 Theory of amplified dispersive Fourier transformation *Phys. Rev.* A **80** 043821

[8] Goda K, Tsia K K and Jalali B 2009 Serial time-encoded amplified imaging for real-time observation of fast dynamic phenomena *Nature* **458** 1145–9

[9] Hancock S W, Zahedpour S, Goffin A and Milchberg H M 2019 Free-space propagation of spatiotemporal optical vortices *Optica* **6** 1547–53

[10] Yessenov M, Free J, Chen Z, Johnson E G, Lavery M P J, Alonso M A and Abouraddy A F 2022 Space-time wave packets localized in all dimensions *Nat. Commun.* **13** 4573

[11] Jhajj N, Larkin I, Rosenthal E W, Zahedpour S, Wahlstrand J K and Milchberg H M 2016 Spatiotemporal optical vortices *Phys. Rev.* **6** 031037

[12] Kurtz F, Ropers C and Herink G 2020 Resonant excitation and all-optical switching of femtosecond soliton molecules *Nat. Photon.* **14** 9–13

[13] Chong A, Wan C, Chen J and Zhan Q 2020 Generation of spatiotemporal optical vortices with controllable transverse orbital angular momentum *Nat. Photon.* **14** 350

[14] Coullet P, Gil G and Rocca F 1989 Optical vortices *Opt. Commun.* **73** 403–8

[15] Peng J and Zeng H 2018 Build-up of dissipative optical soliton molecules via diverse soliton interactions *Laser Photonics Rev.* **12** 1800009

[16] Zhou Y, Ren Y-X, Shi J and Wong K K Y 2020 Breathing dissipative soliton explosions in a bidirectional ultrafast fiber laser *Photon. Res* **8** 1566–72

[17] Zhou Y, Ren Y-X, Shi J and Wong K K Y 2022 Dynamics of breathing dissipative soliton pairs in a bidirectional ultrafast fiber laser *Opt. Lett.* **47** 1968–71

[18] Afanasjev V V and Akhmediev N 1996 Soliton interaction in nonequilibrium dynamical systems *Phys. Rev.* E **53** 6471–5

[19] Malomed B A 1991 Bound solitons in the nonlinear Schrödinger–Ginzburg–Landau equation *Phys. Rev.* A **44** 6954–7

[20] Zhou Y, Ren Y-X, Shi J and Wong K K Y 2022 Breathing dissipative soliton molecule switching in a bidirectional mode-locked fiber laser *Adv. Photonics Res.* **3** 2100318

[21] Zhou Y, Shi J, Ren Y-X and Wong K K Y 2022 Reconfigurable dynamics of optical soliton molecular complexes in an ultrafast thulium fiber laser *Commun. Phys.* **5** 302

[22] Zhou Y, Ren Y-X, Shi J, Mao H and Wong K K Y 2020 Buildup and dissociation dynamics of dissipative optical soliton molecules *Optica* **7** 965–72

[23] Gecevičius M, Drevinskas R, Beresna M and Kazansky P G 2014 Single beam optical vortex tweezers with tunable orbital angular momentum *Appl. Phys. Lett.* **104** 231110

[24] Grier D G 2003 A revolution in optical manipulation *Nature* **424** 810–6

[25] Friese M E J, Enger J, Rubinsztein-Dunlop H and Heckenberg N R 1996 Optical angular-momentum transfer to trapped absorbing particles *Phys. Rev.* A **54** 1593

[26] Lavery M P J, Speirits F C, Barnett S M and Padgett M J 2013 Detection of a spinning object using light's orbital angular momentum *Science* **341** 537–40

[27] Lavery M P J, Barnett S M, Speirits F C and Padgett M J 2014 Observation of the rotational Doppler shift of a white-light, orbital-angular-momentum-carrying beam backscattered from a rotating body *Optica* **1** 1–4

[28] Dennis M R, O'Holleran K and Padgett M J 2009 Singular optics: optical vortices and polarization singularities *Progress in Optics* ed E Wolf (Amsterdam: Elsevier) pp 293–363

[29] Bliokh K Y 2021 Spatiotemporal vortex pulses: angular momenta and spin–orbit interaction *Phys. Rev. Lett.* **126** 243601

[30] Bliokh K Y, Rodríguez-Fortuño F J, Nori F and Zayats A V 2015 Spin–orbit interactions of light *Nat. Photon.* **9** 796–808

IOP Publishing

Optical Vortices
Fundamentals and applications
Yuanjie Yang, Yu-Xuan Ren and Carmelo Rosales-Guzmán

Chapter 7

Plasmonic vortices

Surface plasmon polaritons (SPPs) are a type of surface electromagnetic wave that propagate along metal-dielectric interfaces. They have proven to be extremely versatile and have found wide-ranging applications in integrated photonic devices, optical storage, optical sensing, and more. Recently, there has been a significant surge of interest in studying the fundamental principles and exploring the applications of SPPs carrying orbital angular momentum, commonly referred to as SPP vortices or plasmonic vortices. In this chapter, we provide a comprehensive overview of the fundamental concepts behind plasmonic vortices and delve into the latest advancements in their generation and wide-ranging applications. Our discussion encompasses the utilization of SPPs at lightwave frequencies and extends to the exciting realm of spoof SPPs at microwave and terahertz frequencies.

7.1 Vortices in surface plasmon polaritons: an introduction

Surface plasmon polaritons (SPPs) are surface electromagnetic excitations propagating along conductor-dielectric interfaces. These electromagnetic surface waves arise through the coupling of the electromagnetic fields to collective oscillations of the conductor's electrons, exhibiting unique propagation characteristics [1, 2]. By applying Maxwell's equations with specific boundary conditions to the flat interface between a conductor and a dielectric, the dispersion relation of SPP can be derived as [3]

$$k_{\text{spp}} = \frac{\omega}{c} \sqrt{\frac{\varepsilon_m \varepsilon_d}{\varepsilon_m + \varepsilon_d}} \tag{7.1}$$

where k_{spp} is the wavevector of the excited SPPs, ε_m and ε_d are permittivity of metal and dielectric, respectively. The wavelength of excited SPP wave is

doi:10.1088/978-0-7503-5844-6ch7

$$\lambda_{\text{spp}} = \frac{2\pi}{k_{\text{spp}}'} = \lambda_0 \, \text{Re}\left(\sqrt{\frac{\varepsilon_m + \varepsilon_d}{\varepsilon_m \varepsilon_d}}\right) \tag{7.2}$$

where λ_0 is the wavelength in free space, and k_{spp}' and k_{spp}'' are the real and imaginary part of k_{spp}, respectively. It shows that the wavelength of SPP is relatively shorter than that in free space, providing the ability to compress light into a subwavelength scale and make the field enhancement. Thus SPPs have triggered a wide range of applications, including directional guiding [4–6] and super-resolution focusing imaging [7], to mention but a few.

Moreover, owing to the absorption during the propagation process, the energy of evanescent waves decays exponentially away from the interface and is confined in a subwavelength scale. The propagation length is regarded as the distance where the intensity decreases to $1/e$ of its maximum intensity, given by

$$\delta_{\text{spp}} = \frac{1}{2k_{\text{spp}}''} = \frac{\lambda_0}{2} \, \text{Im}\left(\sqrt{\frac{\varepsilon_m + \varepsilon_d}{\varepsilon_m \varepsilon_d}}\right) \tag{7.3}$$

Typically, the propagation length is several micrometers in the visible region and the attenuation of excited SPP increases with a decreasing wavelength.

Importantly, the properties of SPPs can be altered by tailoring different structures of the surface, offering potential for the development of photonic devices such as on-chip discrimination of the topological charges of optical vortex beams [8–11]. Similarly, just as traditional vortex beams, angular momentum can be introduced into a two-dimensional surface in the form of plasmonic vortices. Such surface evanescent wave carrying OAM has been extensively studied [12], particularly in its generation and manipulation. A SPP vortex is able to trap and rotate particles along circular trajectory, especially small objects that traditional tweezers cannot control [13], which is advantageous in many aspects such as manipulation of metal particles [13].

SPPs are collective oscillations of electrons coupled with surface electromagnetic wave at the dielectric/metal interface, exhibiting unique propagation characteristics [1, 2]. have found numerous applications in integrated photonic devices, optical storage, and optical sensing, etc.

7.2 Generation of plasmonic vortices

Recent advances in metasurfaces and surface plasma theory expedite various methods for the generation of plasmonic vortices. In this section, we start by introducing the generation of SPP vortices based the use of plasmonic vortex lens and metasurfaces.

7.2.1 Generation of SPP vortices using PVLs

It was proposed in 2006 that evanescent Bessel beams can be generated on a silver film with etched Archimedean spiral slits under illumination of circularly polarized beam, the topological charge of an excited SPP vortex is determined by the chiralities of both the spiral structure and the incident beam [14]. One basic but

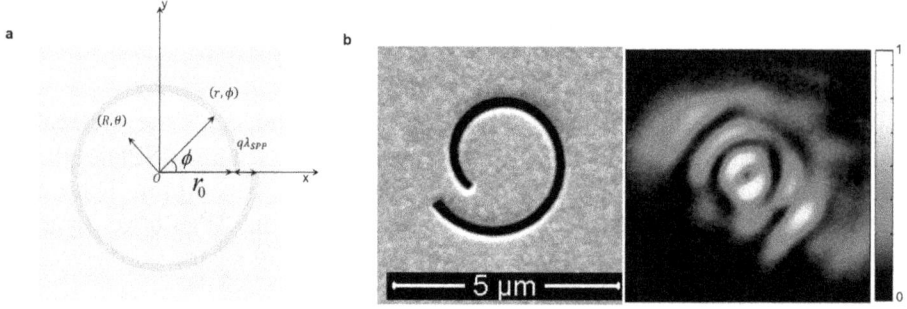

Figure 7.1. Generation of SPP vortices by the use of PVLs. (a) Diagram of a right-handed Archimedean spiral. (b) SEM image of a left-handed Archimedean spiral and the corresponding SNOM image of generated plasmonic vortices. a) Reproduced from [8]. © 2022 IOP Publishing Ltd. All rights reserved. b) Reprinted with permission from [16]. Copyright (2010) American Chemical Society.

effective method to generate plasmonic vortices is using the plasmonic vortex lenses (PVLs), a set of spiral protrusions or continuous slits in a metallic plate [15]. The most typical structure is the Archimedean spiral, as shown in figure 7.1(a). The distance from the origin to the slit can be expressed by

$$r = r_0 \pm q\lambda_{\text{spp}}\frac{\phi}{2\pi} \tag{7.4}$$

where ϕ is the azimuthal angle and r_0 is the initial radius in cylindrical coordinate system, $\pm$ denotes corresponding chirality and q is the ratio of the spiral pitch to λ_{spp}.

It is worth noting, it was analytically proven that PVL with Archimedean spiral structure can effectively generate SPP vortices with lateral size as small as $4\lambda_{\text{spp}}$ [17]. Later, as illustrated in figure 7.1(b), it was experimentally confirmed that SPP vortex can be generated by a PVL consisting of a single Archimedean spiral slot [16].

Noted that such a chiral structure is polarization-sensitive, SAM carried by the incident beams can be transformed to the OAM accordingly, thus beams with different circularly polarized state can excite plasmonic vortices with different topological charges. SPPs excited from each point of the slits interfere with each other and propagate to the center of the structure. The plasmonic field excited by a circularly polarized vortex beam $\vec{E}_{\text{in}} = e^{il\phi}e^{i\sigma\phi}(\vec{e}_z + i\sigma\vec{e}_\phi)$ at an observation point (R, θ) can be written as

$$\vec{E}_{\text{spp}}(R, \theta) = \vec{e}_z E_z e^{-k_z z} \int e^{il\phi}e^{i\sigma\phi} \exp(iR\vec{k}_r \cdot \vec{e}_R)\exp\left[-i(r_0 \pm q\phi\lambda_{\text{spp}}/2\pi)\vec{k}_r \cdot \vec{e}_r\right] \\ \times (r_0 \pm q\phi\lambda_{\text{spp}}/2\pi)\mathbf{d}\phi \tag{7.5}$$

where $\vec{k}_r$ and $\vec{k}_z$ are the transverse and radial wavevector of SPPs, respectively. Neglecting the loss of SPP, i.e. Im(k_r) $\simeq 0$, and using $\vec{k}_r = -k_r\vec{e}_r = -2*\pi/\lambda_{\text{spp}}\vec{e}_r$, we can simplified equation (7.5) into

$$\vec{E}_{\text{spp}}(R, \theta) \simeq \vec{e}_z E_0 e^{-k_z z}e^{ik_r r_0} \int r_0 e^{i(l+\sigma-1)\phi}e^{iRk_r \cos(\theta-\phi)}\mathbf{d}\phi \tag{7.6}$$

with the integral identity for the Bessel function,

$$J_n(x) = \frac{1}{2\pi} \int_0^{\pi} e^{i(nt - x\sin t)} \mathrm{d}t \tag{7.7}$$

we can derive the field distribution of the SPP vortices as followed:

$$\vec{E}_{\mathrm{spp}}(R,\ \theta) = 2\pi \vec{e_z} E_0 r_0 e^{-k_z z} e^{ik_r r_0} J_{l+\sigma-1}(k_r R) \tag{7.8}$$

Moreover, different propagation distances leads to the phase difference. Crucially, it is feasible to combine the radial distance with the phase difference through Archimedean spirals; periodic phase change of $2\pi q$ can be introduced by varying propagation distance of SPPs in the radial direction. Except for a single spiral slit, a PVL with multiple-turn Archimedean spirals was proposed to improve the coupling efficiency for the generation of plasmonic waves, and a smaller focal spot with higher intensity can be obtained [18]. Note that the azimuthal component of the circularly polarized incident beam doesn't contribute to the excitation of plasmonic waves owing to the fact that it is TE polarized with respect to the spirals. In order to further improve the coupling efficiency, a hybrid PVL was proposed consisting of an Archimedean spiral slit and alternating spiral triangle array, the latter being used to couple the azimuthal component into SPPs exactly. Compared to a pure spiral slit, the conversion efficiency can be significantly improved to 94.69% [19, 20].

Instead of using a whole circle of the spiral, segmented Archimedean spirals are used to design a novel PVL that provides another degree of freedom as the geometrical topological charge m, i.e., the number of windings. The geometry of such a PVL can be described as [21]:

$$r_m(\phi) = r_0 + \frac{\lambda_{\mathrm{spp}}}{\Delta}[\phi - (a-1)\Delta], \quad a = 1, 2, \dots, m \tag{7.9}$$

where $\Delta = 2\pi/m$ denotes the azimuthal range of each segment. Figure 7.2(a) shows the schematic representation of converging angular spectra of excited plasmonic waves inside the PVL with $m = 2$ by illuminating with a left-handed circularly polarized light. As such, owing to the geometric structure of Archimedean spirals, phase profiles of excited plasmonic vortices are $\exp(2i\theta)$ and $\exp(3i\theta)$ under the illumination of a radially and right-handed circularly polarized lights, respectively. As shown in figures 7.2(b)–(d), Kim et al numerically and experimentally verified the feasibility of generating SPP vortices with arbitrary topological charge by illuminating with a circularly polarized light ($\sigma = \pm 1$) [21]. Consequently, the topological charge of the plasmonic vortex is dependent on the geometrical topological charge m and the chirality σ of incident light.

Later, to challenge the paradigm in PVLs, Spektor et al proposed plasmonic vortex cavities with golden wall mirrors and ultraflat surfaces for the multiplication of OAM. The outward-propagating SPP vortex interacts with those boundaries that initially excited them. It is partially back-reflected to the center again and accumulates azimuthal phase, which is twice the geometric phase of the cavity, thus previously unobserved higher-order SPP vortices can be generated after reflections [22].

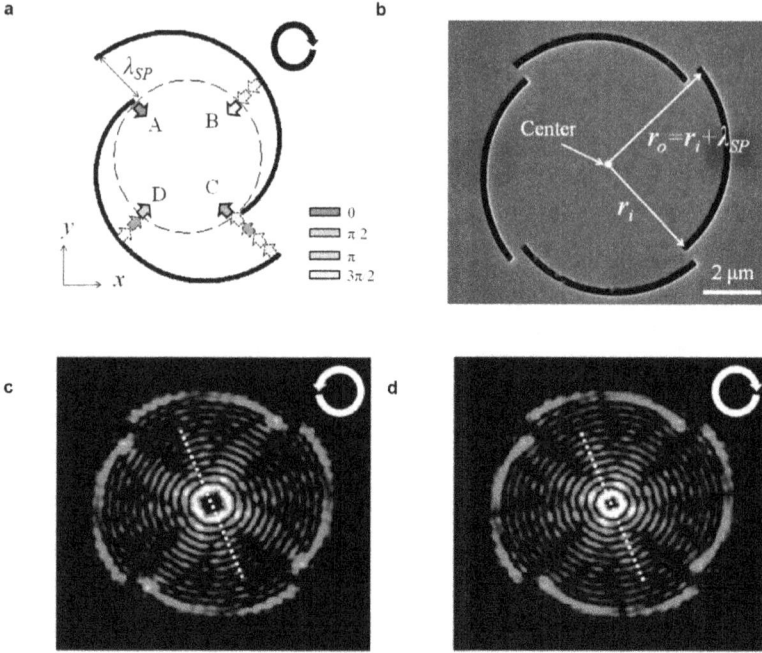

Figure 7.2. (a) Schematic of the converging angular spectrum inside the PVL under left-handed polarized illumination. (b) SEM image of the multi-segmented PVL. (c and d) A PVL with segmented Archimedean spirals and the corresponding near-field intensity distribution of the excited SPP vortices. Reprinted with permissions from [21]. Copyright (2010) American Chemical Society.

7.2.2 Spin-to-orbit coupling of SPP vortices

Analogous to free space spin–orbit coupling, Cho *et al* proposed that the SAM as well as the OAM can be coupled into SPP vortices, and numerically verified the aforementioned hypotheses by illuminating a circularly polarized vortex beam on the PVL [23]. The excited plasmonic field can be given by [21, 24]

$$E_{\mathrm{spp}}(R, \theta) \propto \sum_{a=1}^{m} \int_{(a-1)\Delta}^{a\Delta} \exp[i(l + \sigma + m)\phi + ik_r R \cos(\theta - \phi)]$$
$$\times \left\{ r_0 + \frac{\lambda_{\mathrm{spp}}}{\Delta}[\phi - (a - 1)\Delta] \right\} \mathrm{d}\phi. \tag{7.10}$$

Normally, the part $\lambda_{\mathrm{spp}}/\Delta[\phi - (a - 1)\Delta]$ is ignored because its value is much smaller than r_0, thus the above plasmonic field can be further simplified as

$$E_{\mathrm{spp}}(R, \theta) \propto 2\pi i^{l+\sigma+m} J_{l+\sigma+m}(k_r R)\exp[i(l + \sigma + m)\theta] \tag{7.11}$$

where the $J_n(\alpha)$ is a *n*-order Bessel function of the first kind. Since then, it has been generally accepted that the topological charge of a plasmonic vortex excited by such a PVL can be expressed as [23]

$$l_{\mathrm{spp}} = l + \sigma + m \tag{7.12}$$

i.e., the addition of geometrical topological charge of the structure, the SAM as well as the OAM of incident beam. Based on segmented Archimedean spirals, Spektor *et al* experimentally generated the plasmonic vortices and provided direct recording of SPP vortices at optical frequency [25]. The outward and inward counterpropagating plasmonic waves interfere with each other to form an azimuthally rotating vortex wave. By using time-resolved two-photon photoemission electron microscopy, the spatiotemporal dynamics of formation, revolution and decay of the SPP vortices is investigated, where the handedness of the wavefront in the decay process is reversed as compared to its formation, providing fundamental support for the understanding of OAM generation.

7.2.3 Deuterogenic plasmonic vortices

However, it has been found recently that there are some phenomena in experiments that are difficult to explain using equation (7.12). Namely, for a segmented PVL with illumination carrying SAM σ and OAM l, there are some unexpected SPP vortices besides the one, the topological charge is expected by $l_{spp} = l + \sigma + m$. Phenomenally speaking, as shown in figure 7.3(b), the generated SPP vortices is consisted with a big outer ring and a small inner ring, where the topological charge perfectly agrees with the expectation of equation (7.12) but the small inner ring carries an inconsistent topological charge significantly. Due to its relatively low intensity and being less recognizable, this unexpected small inner ring has never been reported by previous works.

Based on the theory of spin–orbit coupling, Yang *et al* numerically and experimentally revealed the existence of multiple plasmonic vortices. By taking the ignored part in equation (7.10) $\lambda_{spp}/\Delta[\phi - (a - 1)\Delta]$ into account, the plasmonic field can be rewritten as [26]

$$
\begin{aligned}
E_{spp}(R, \theta) \propto 2\pi \left(r_0 + \frac{1}{2}\lambda_{spp} \right) i^{l+\sigma+m} J_{l+\sigma+m}(k_r R)\exp[i(l + \sigma + m)\theta] \\
- i\lambda_{spp} \sum_{s=\pm 1}^{\pm\infty} \frac{1}{s} i^{l+\sigma+m-ms} J_{l+\sigma+m-ms}(k_r R)\exp[i(l + \sigma + m - ms)\theta]
\end{aligned}
\tag{7.13}
$$

where the first term shows the previous expected plasmonic vortex with topological charge $l + \sigma + m$, and the second term shows other deuterogenic SPP vortices, the topological charge of which can be given by [26]

$$
l_N = l + \sigma + Nm, \qquad N = 0, \pm 1, \pm 2, \ldots
\tag{7.14}
$$

More importantly, it can be observed that the weight of multiple plasmonic vortices is inversely proportional to the s, i.e., the dominant modes are $l_0 = l + \sigma$ and $l_1 = l + \sigma + m$, as illustrated in figures 7.3(a)–(d).

The deuterogenic OAM modes seem to break the angular momentum conservation law. However, the fact is that it reveals a general principle of spin-orbital coupling rather than breaks the conservation law. The previous theory is based on the assumption that there is only a single plasmonic vortex generated by a PVL, and

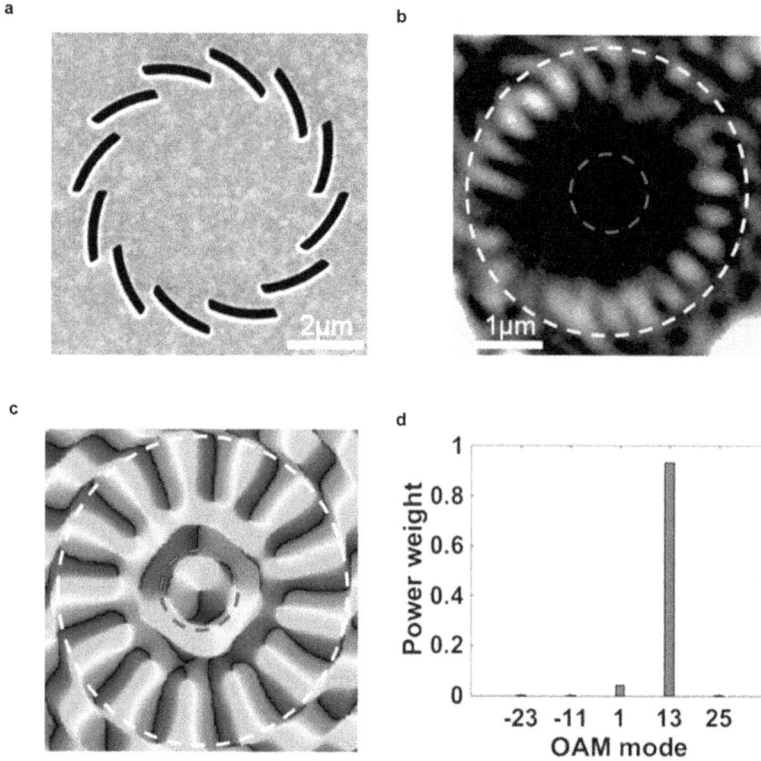

Figure 7.3. Nontrivial deuterogenic SPP vortices excited by the PVL. (a) SEM image of the PVL. (b) Zoomed-in view of the central part of the SPP vortices intensity pattern. (c) The corresponding simulated phase patterns of the deuterogenic SPP vortices. (d) The OAM spectrum of the generated SPP vortices. Reprinted with permission from [26]. Copyright (2020) American Chemical Society.

the OAM of such a vortex thus comes from two parts, namely, the angular momentum $(l + \sigma + m)$ is the coupling of the momentum of the incident beam and the geometric momentum of PVL. However, equation (7.14) shows that a PVL can simultaneously produce multiple SPP vortices instead of a single one. This is because constructive interference of the surface plasmons from the adjacent spirals of the PVL occurs whenever the path difference is the multiple of λ, namely, $n\lambda$. This means that the whole process of angular momentum conversion is a compound one. In other words, besides the angular momentum conservation during the process of generation of the PV with topological charge $l + \sigma + m$, the angular momentum is conserved during the generation of the central PV with topological charge $l + \sigma$ as well. The OAM of the central PV is converted from the SAM and the OAM of the incident vortex beam, and the PVL does not contribute to the angular momentum in this process. Therefore, the spin–orbit coupling process happens for each PV, and the angular momentum is conserved for each process.

Heretofore, a fundamental and complete theory has been formed for the generation of plasmonic vortices based on PVLs, which is promising for emerging vortex-based nanotechnology such as plasmonic tweezers and many other on-chip applications.

7.2.4 Generation of SPP vortices using plasmonic metasurfaces

As introduced above, traditional PVLs based on the Archimedean spirals produce plasmonic vortices by the interference of SPPs excited along the spirals with varying propagation phase. There exists another efficient approach to generate plasmonic vortices, known as metasurfaces, which is believed to possess a strong ability to dynamically control electromagnetic waves ranging from visible light to microwaves. Metasurfaces are planar optical devices composed of subwavelength meta-atoms that modulate the polarization state and amplitude as well as the phase of light waves [27]. Recently, metasurface technology has been widely applied as multifunctional metadevices in many aspects [28, 29]. Meanwhile, such prominent manipulation ability makes it a great candidate to generate vortex beams as well [30, 31]. Combined with the existing PVLs, specific phase distribution can be generated by plasmonic metasurfaces that further increase the degree of freedom to manipulate the excited plasmonic field.

Analogous to free-space light manipulation of metasurfaces, Lee *et al* proposed a metasurface consisting of multiple rectangular nanoslits, enabling generation of higher-order SPP vortices [34]. A traditional continuous spiral is discretized as a set of ring-distributed nanoslits, each of which can be regarded as a dipole source with the direction perpendicular to the longer axis. Different from conventional PVLs, additional OAM can be added from the tilted angle variance of nanoslits along the azimuthal direction other than radial position shift. The size of the meta-atom is of vital importance, which is related to the phase uniformity and transmittance of SPPs. It is shown that the topological charge of the plasmonic vortex excited by the double-ring metasurface is $2n - 1$, where n is the geometrical order denoting the rotation speed of the nanoslits. Generally speaking, the phase regulation of metasurfaces can be briefly divided into dynamic phase and geometric phase. Except for specific distribution to introduce the geometric phase, it is rational to use the dynamic phase induced by optical path difference to generate plasmonic vortices. Tan *et al* proposed a plasmonic metasurface that combines both dynamic and geometric phase through controlling the spatial position (arranged in Archimedean-spiral pattern) and the orientation angle of the nanoslits to achieve tunable OAM manipulation [32].

As illustrated in figure 7.4(a), spin-dependent SPP vortices can be generated by the metasurface composed of spatially varying nanoslits [32]. The orientation angle of the inner and outer rings can be written as $\psi_{\mathrm{in}}(\phi) = nm\phi$, $\psi_{out}(\phi) = nm\phi + \pi/2$, respectively. Rectangular nanoslits at the beginning and end of the spiral rotate $2n\pi$ during the process. The plasmonic field at the observation point (R, θ, z) can be further expressed as [33]

$$E_{\mathrm{spp}}(R, \theta, z) = A_0 e^{-k_z z} \sum_{a=1}^{m} \int_{(a-1)\Delta}^{a\Delta} (\cos[(nm - 1)\phi] - i\sigma \sin[(nm - 1)\phi] \exp(ik_{SPP}d)) \times \exp(i\sigma nm\phi)\exp[ik_{SPP}(r - R\cos(\theta - \phi))]\mathrm{d}\phi \qquad (7.15)$$

where A_0 is a constant and other parameters are the same as above.

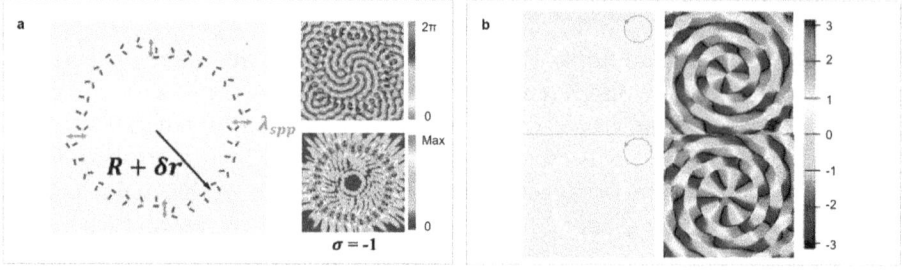

Figure 7.4. Generation of SPP vortices by using plasmonic metasurfaces based on the Archimedean spirals. (a) Metasurface combing the dynamic as well as geometric phase (b) Generation of plasmonic vortices carrying different OAM by changing the spiral pitch. Reproduced from [32] with permission from the Royal Society of Chemistry.

Moreover, the formation of SPP vortices carrying different OAM can be implemented with a fixed number of spirals by changing the spiral pitch q, as illustrated in figure 7.4(b) [33]. Also, instead of using two-layer rectangular slits with radius difference, SPP vortices can be excited in the center by a metasurface consisting of cross-shaped nanoslits distributed in an Archimedean spiral [35]. Using such design scheme, plasmonic vortices with optional topological charges can be generated only in the case of circularly polarized incidence.

In order to generate plasmonic waves carrying OAM under linearly polarized excitation, a metasurface composed of asymmetric cross-shaped slits was proposed that can achieve dynamic switching between subwavelength focusing and plasmonic vortex formation through different polarized directions [36]. It is noted that the order of the excited vortex can alter by 2 through changing the polarization state of incident light. Therefore, to control the rotation orientation of higher-order SPP vortices, unidirectional meta-slits and the heterochirality principle were combined [37].

For left circular polarized light, a counterclockwise rotational plasmonic vortex can be generated launched by the inner spirals, whereas SPPs excited at the outer structure propagate outward. In the case of right circular polarization, a clockwise rotating vortex is formed by the outer spirals. The rotation direction of the SPP vortex is flipped by merely altering the chirality of illumination, resulting in its OAM changes from 11 to -11.

As well as metasurfaces with Archimedean-spiral structures, innovated plasmonic metasurfaces were proposed to enable versatile functionalities. Later, a novel metasurface was proposed to produce SPP waves carrying arbitrary topological charge (fraction or integer) through the antisymmetric and symmetric phase [38]. More interestingly, under the illumination of spin-zero, i.e., the linearly polarized beam, an off-axis plasmonic vortex with breaking symmetry can be generated at the center. And the excited OAM wave can be shifted to any position deviating from the center by introducing a position shift $\vec{r}_\sigma$ in metasurface designing, which is potential in applications such as micromanipulation.

It shows that additional degree of freedom can be provided by combining traditional PVL structure with the concept of metasurface to manipulate plasmonic waves, whereas the SAM of incident beam influence the topological charge of excited SPPs. To eliminate the polarization dependence of traditional metasurfaces, specific structure with geometric phase was designed to compensate the spin-orbital interaction. As such, plasmonic waves carrying identical OAM can be excited by illuminating linearly, left- or right-circularly polarized incident beams, and even any arbitrary polarization state. The polarization-independent plasmonic vortex generator that considers a pair of nanoslits as a meta-atom, the difference in radius is $d = \lambda_{spp}/2$ was illustrated by Moon *et al* in 2019 [39].

Nevertheless, interleave independent metasurfaces are usually required for the generation of SPP vortices with varying topological charges, impairing the compactness of the optical device [40]. Recently, a phyllotaxis-inspired vortex nanosieve was proposed to excite multiple optical vortices with different topological charges, both on a chip and in free space, where one nanohole contributes to various vortices simultaneously. Multiple SPP waves were demonstrated by using ultrafast time-resolved photoemission electron microscopy [41]. Such a vortex generator would facilitate related applications such as high-capacity information metaphotonics.

7.3 Applications of SPP vortices and beyond

7.3.1 Applications of SPP vortices

Optical tweezers, as one of the most important techniques that offers precise control of small particles, has attracted appreciable interest due to its unique properties, such as non-contact manipulation [42–44]. In 2018, Ashkin was awarded the Nobel prize in Physics for the discovery of optical trapping that has been rapidly developed and applied in many fields, especially in biological systems. However, owing to the diffraction limit, the size of the focal spot is required to be on the same order as the wavelength of light, leading to the restriction of trapping precision [45]. Particularly, it is difficult to trap small metallic particles using traditional optical tweezers because of its strong scattering properties and absorption [46, 47]. As introduced above, SPPs are able to break the diffraction limit and generate subwavelength focusing spots, thus plasmonic configurations can be used to ameliorate those drawbacks. Generally, optical forces for manipulation can be classified into two categories; scattering and gradient forces. The former is related to the momentum interchange between particles and incident light on the interface due to discontinuities of the refraction index. The latter is associated with the gradient of the electromagnetic field that overcomes the scattering force to form a trap. In terms of plasmonic forces, focusing hotspots are usually compressed with higher gradient, making it possible to trap metallic particles. In recent decades, manipulation of micro- and even nano-meter-sized particles, such as semiconductors, metals and biological objects, have been reported by the use of plasmonic tweezers [48–50].

One of the most widely known demonstrations of the rotation of particles through SPP vortices was achieved by focusing a radially polarized beam with a certain OAM onto a metal surface [51]. Switching the radially polarized beam to an azimuthally polarized one, it shows that no force would exist and particles stop rotating due to the fact that SPPs cannot be excited by azimuthally polarized vortex beams. Except for the SAM, OAM carried by the incident beam can also be transferred into the excited SPP waves, providing an extra degree of freedom to manipulate particles in the evanescent field. Shen *et al* shows the rotational motion of a gold particle controlled by plasmonic vortices with $l = 2$ and $l = 5$, where the radius increases with an increasing topological charge. It shows that small objects are actuated along a circular trace in such a plasmonic vortex field, the rotational velocity of which is determined by the OAM of the input beam. Later, Tsai *et al* showed that a plasmonic field can be selectively excited as either an SPP vortex for microparticle rotation or a focal spot for trapping depending on the chirality of the incident beams [52].

As illustrated by Tsai *et al*, selective rotation or trapping of isotropic dielectric microparticles was achieved through a single Archimedean spiral. With a left-handed circularly polarized incident beam ($\sigma = -1$), a plasmonic wave with no OAM can be generated that provides stable optical potential to trap microparticles at the center, while with a right-handed circularly polarized illumination ($\sigma = 1$), a SPP vortex with topological charge $l = 2$ can be generated that is available to rotate microparticles effectively. However, only one-dimensional (1D) polarization-dependent forces were calculated, and analysis of 2D or 3D forces is of importance for the development of on-chip optical systems. Later in 2019, Hesselink *et al* further analyzed the corresponding forces in higher dimensions based on the same structure in Tsai's work, providing more insights into particle micromanipulation [53].

Iterative demonstration proposed an improved plasmonic vortex tweezer composed of a spiral grating with three arms and a nanotip, enabling dynamic manipulation of nano-objects in three dimensions [54]. Specifically, the gradient force induced by plasmonic waves with $l = 0$ can trap particles near the top of nanotip, whereas plasmonic vortex with higher-order topological charge can be decoupled to freely propagating light beam capable of releasing and pushing nanoparticles along particular trajectories.

Moreover, it is known that SPPs can be excited on a smooth surface without any structure. To avoid the complex process of fabricating nanostructures, plasmonic tweezers with a structureless metallic surface was subsequently proposed [55]. Such configurations can generate SPP waves at arbitrary positions by merely moving the excitation source, which is of convenience for dynamic manipulation. Zhang *et al* shows the optical system used for plasmonic vortex tweezers, in which a spiral phase plate is required to generate vortex beams. Meanwhile, the metal particles can be bounced out of the focus by the use of optical vortices unless it is located inside the focal doughnut initially, while plasmonic vortex tweezers can control the particles even from far away (circa 10λ) [56]. Such a prominent property helps remove the restrictions on the initial position of metal particles. Interesting outlooks including the rotational speed and trapping stability remain to be explored in the future.

7.3.2 Surface plasmonic vortices beyonds optical frequency

Due to the effective excitation of SPPs severely relying on the illumination of electromagnetic field into metals, plasmonic vortices are inherently limited to the optical regime. At low frequencies, metals such as gold and copper behave like perfect electric conductors, thus cannot induce SPP waves at the metal/dielectric interface. To tackle this issue, the concept of spoof SPPs, which are very similar to conventional SPPs excited at optical regime, has been brought forward to convert free-space electromagnetic waves at low frequency into surface waves. It is found that structured metals with subwavelength grooves or holes can support low-frequency electromagnetic modes, regarded as analogues of SPPs in the visible and near-infrared regime [57]. The unique properties of SPP vortices excited in the visible spectrum have aroused the interest to generate plasmonic vortices in microwave and terahertz range, which are called spoof SPP vortices.

7.3.2.1 Spoof SPP vortices in microwave regime

Similar to the conventional plasmonic vortices, owing to its geometrical superiority that connects the radial variation with the azimuthal variation, the spiral structure also plays an important role in the stimulation of microwave spoof SPP vortices. As shown in figure 7.5, Huidobro *et al* [58] demonstrated the generation of magnetic surface plasmons by a meta-particle structure, which comprises six spiral-shaped grooves wrapping around a small cylinder. With a linearly polarized wave as the excitation, an oscillated electric dipole or magnetic dipole form spoof SPPs can be generated. Later, the generation of spoof SPPs were used in further studies based on such meta-particle structures [59, 60].

In order to improve the efficiency of spoof SPP vortices generation, a structure consisting of the meta-particle and a comb-shaped spoof SPPs waveguide placed in

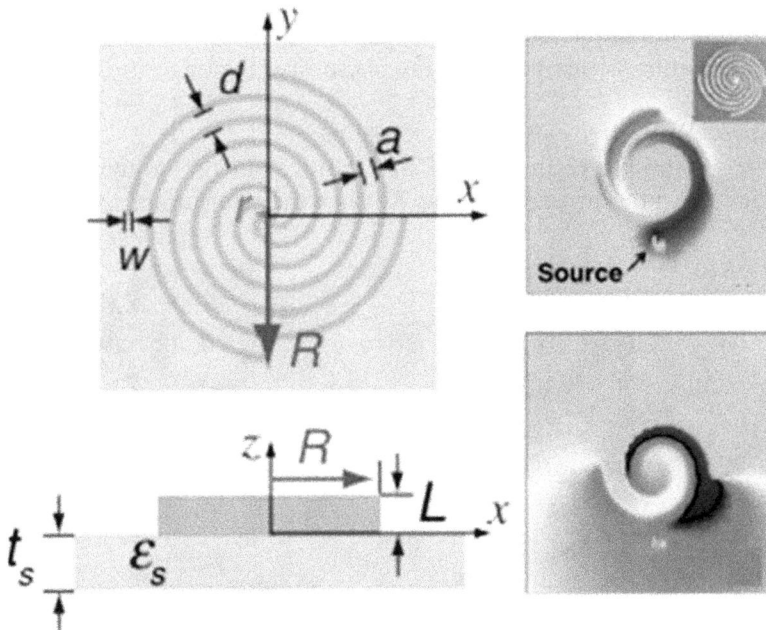

Figure 7.5. Generation of spoof SPP vortices in microwave regime with meta-particle that consist of spiral-shaped grooves. Reproduced from [58]. CC BY 3.0.

an optimized coupling distance g was proposed [61]. The comb-shaped waveguide can induce an asymmetric spatial distribution of the incident spoof SPP field, thus spoof SPP vortices can be induced more effectively by the meta-particle structure. It showed numerically and experimentally that a microwave spoof SPP vortex with topological charge $l = +1$ at 9.9 GHz located 0.5 mm above the system can be generated, and the topological charge can be switched to $l = -1$ by inversing the direction of the excitation. Meanwhile, the generation of spoof SPP vortices with topological charge $l = \pm 2$ were demonstrated at 10.13 GHz by changing the geometric structure of the meta-particle.

To generate a spoof SPP vortex with higher-order OAM, Su *et al* proposed another approach instead of the spiral-shaped meta-particles [62]. They show the meta-particle structure consists of rim-textured metallic cylinder corrugated with radial periodic grooves. For different excitation with frequency at 5.37, 7.82, 8.83, 9.30, and 9.57 GHz, spoof SPP vortices with OAM from $l = -1$ to $l = -5$ were generated.

More recently, based on the analogous structure, a double-layered microwave plasmonic resonator that can excite the spoof SPP vortices and measure the OAM dichroism capability of a chiral structure was designed [63]. Each layer of the resonator is composed of concentric rim-textured metallic patterns, which is rotated to the opposite direction to break the chiral symmetry of the whole structure and to induce OAM modes.

Furthermore, owing to the tremendous potential in a variety of applications, various methods have been proposed for the generation of vortex beams based on spoof SPPs. Recently, effective strategies such as a looped double-layer spoof SPP waveguide [64], circularly arranged dipole array [65] and a periodically modulated spoof SPP ring resonator [66] were designed to convert the near-field plasmonic waves into radiating vortex beams in the microwave region.

7.3.2.2 Spoof SPP vortices in terahertz regime

Besides the microwave regime, spoof SPP vortices in the terahertz regime were reported as well in recent years. Figure 7.6 shows a metasurface that combines the

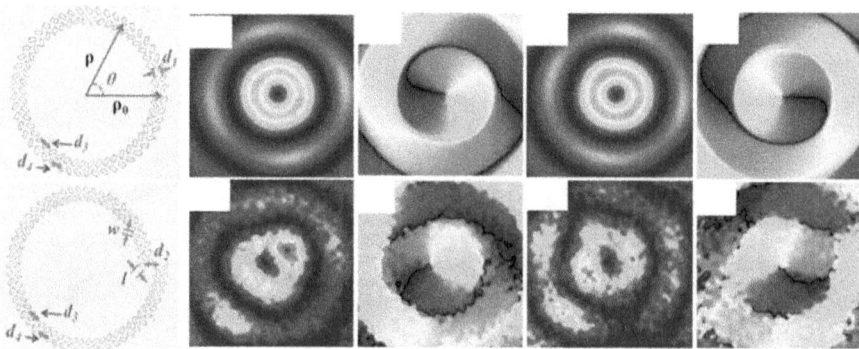

Figure 7.6. Generation of spoof SPP vortices in terahertz regime based on the two-layered plasmonic metasurface with circularly polarized illumination. [67] John Wiley & Sons. © 2018 WILEY-VCH Verlag GmbH & Co. KGaA, Weinheim.

Archimedean spiral and double-ring structure to generate and manipulate terahertz plasmonic vortices [67]. The phase distribution of the inner layer is $\Phi_0(\phi) = 3/4\pi + n\phi$ and the outer layer is set as $\Phi_1(\phi) = 1/4\pi + n\phi$. With circularly polarized illumination, the terahertz plasmonic vortices carrying OAM $l = \pm 1, \pm 2, \pm 3$ were excited at 0.33 THz by adjusting the slit-rotation factor n to 1 or 1.5. Later in 2020, the extension of such structure was reported to enable the multiplexing of THz plasmonic vortices [68]. It is well known that a spoof SPP vortex can be excited by a two-layered geometric metasurface, and an extra space shift can be performed by adding a displacement in the design of metasurface. Then, multiple spoof SPP vortices can be generated by accumulating the layers.

Most of those works transformed the incident circularly polarized waves to SPP vortices with different topological charges dependent on its SAM, which can be explained by the photonic spin-Hall effect [69]. Nevertheless, it is shown that the conversion from optical SAM to plasmonic OAM can be selective through coupled paired resonators, realizing the conversion for one spin state while well suppressed for the other. It is proposed a metasurface, of which the basic cell consisting of two split-ring-shaped slits basic resonators oriented along orthogonal directions. Generally, for meta-atoms with fourfold rotation symmetry, related spin-dependent phenomena can be rather weak because it is isotropic for the incident beam and the circular polarization state in momentum space is degenerated. In that case, meta-atoms with twofold rotation symmetry was designed, providing the feasibility for spin-to-orbital conversion. Moreover, spoof plasmonic vortices in THz regime is excited by left circularly polarized illumination, however, the plasmonic field is well suppressed for right-circularly polarized incidence. In addition, meta-particles with rim-texture cylinder structure composed of circularly periodic grooves was also reported as an effective method for generating spoof SPP vortices in terahertz regime [70].

Furthermore, a novel method with helically grooved metal wires was proposed to generate the spoof SPP vortices and is able to support the propagation of chiral spoof SPPs along the grooves [71]. Later, Rüting et al further studied the relation between the structure of helically grooved metal wire and the OAM of generated chiral spoof SPPs [72]. Then, based on the previous researches, it was proven that the chirality of helical grooves can break the symmetry of the cylindrical metal waveguide, and decompose the degeneracy of spoof SPP HE mode into the chiral spoof SPP propagating on the wire. Their simulations show that, under illumination with frequency $f = 1.3$ THz, the chiral spoof SPPs is able to propagate in the near-field region, circa 10 μ m away from the end of metal wire [73].

7.4 Conclusion

The peculiar characteristics of vortex beams, related to doughnut-shaped intensity profiles and phase singularities, enable versatile functionalities such as optical communications with high-density data transmission. Albeit traditional optical devices composed of bulky components have been traditionally used to generate vortex beams, on-chip devices are still desired for some specific requirement. In order to miniaturize vortex generators and improve the controllability of vortex beams, as introduced in this review, compact nanophotonic devices including PVLs and plasmonic metasurfaces have been extensively investigated. Under illumination of circularly polarized beams, plasmonic vortices can be generated by use of PVLs based on the Archimedean spiral. Moreover, PVLs with multi-turn and segmented Archimedean spirals were proposed to improve the coupling efficiency, which are polarization-dependent due to the symmetry-broken geometry of the Archimedean spiral. Thus, novel structures are proposed to achieve miscellaneous functions such as off-centered and multiple-focus SPP vortices. Moreover, generation methods of spoof SPP vortices in microwave and Terahertz regime have been reviewed in detail. It shows that plasmonic vortices can be used in micromanipulation, especially for these particles that traditional tweezers cannot control. The discovery of plasmonic vortices has led to some applications and more advances are expected.

Heretofore, to further satisfy ever-growing needs of vortex beams, the concept of OAM has been researched not only in real space [74], but also in spatiotemporal domain [75, 76] and momentum space [77, 78]. Besides optical vortices, various proposals have been put forth for the extension of vortex beams carrying intrinsic OAM, including acoustic [79, 80] and electron vortices [81–84], and more recently vortex beams of nonelementary particles [85]. It is worth mentioning that plasmonic vortex lattices can be generated by a hexagon-shaped device under illumination of circularly polarized light [86, 87], providing potential for the research of optical micromanipulation. We envision that phase singularities in nanophotonics, plasmonic vortices for instance, can further promote related applications in both classical and quantum realms.

References

[1] Bertolotti M, Sibilia C and Guzman A M 2017 *Evanescent Waves in Optics: An Introduction to Plasmon vol 206 of Springer Series in Optical Sciences* (Berlin: Springer) https://doi.org/10.1007/978-3-319-61261-4

[2] Zayats A V, Smolyaninov I I and Maradudin A A 2005 Nano-optics of surface plasmon polaritons *Phys. Rep.* **408** 131–314

[3] Barnes W L, Dereux A and Ebbesen T W 2003 Surface plasmon subwavelength optics *Nature* **424** 824–30

[4] Min C, Liu J, Lei T, Si G, Xie Z, Lin J, Du L and Yuan X 2016 Plasmonic nano-slits assisted polarization selective detour phase meta-hologram *Laser Photon. Rev.* **10** 978–85

[5] Yin L, Vlasko-Vlasov V K, Pearson J, Hiller J M, Hua J, Welp U, Brown D E and Kimball C W 2005 Subwavelength focusing and guiding of surface plasmons *Nano Lett.* **5** 1399–402

[6] Lin J, Balthasar Mueller J P, Wang Q, Yuan G, Antoniou N, Yuan X-C and Capasso F 2013 Polarization-controlled tunable directional coupling of surface plasmon polaritons *Science* **340** 331–4

[7] David A, Gjonaj B, Blau Y, Dolev S and Bartal G 2015 Nanoscale shaping and focusing of visible light in planar metal-oxide-silicon waveguides *Optica* **2** 1045–8

[8] Bai Y, Yan J, Lv H and Yang Y 2022 Plasmonic vortices: a review *J. Opt.* **24** 084004

[9] Chen J, Chen X, Li T and Zhu S 2018 On-chip detection of orbital angular momentum beam by plasmonic nanogratings *Laser Photon. Rev.* **12** 1700331

[10] Feng F, Si G, Min C, Yuan X and Somekh M 2020 On-chip plasmonic spin-hall nanograting for simultaneously detecting phase and polarization singularities *Light: Sci. Appl.* **9** 95

[11] Mei S *et al* 2016 On-chip discrimination of orbital angular momentum of light with plasmonic nanoslits *Nanoscale* **8** 2227–33

[12] Ni J, Huang C, Zhou L M, Gu M, Song Q, Kivshar Y and Qiu C W 2021 Multidimensional phase singularities in nanophotonics *Science* **374** eabj0039

[13] Zhang Y, Min C, Dou X, Wang X, Urbach H P, Somekh M G and Yuan X 2021 Plasmonic tweezers: for nanoscale optical trapping and beyond *Light Sci. Appl.* **10** 59

[14] Ohno T and Miyanishi S 2006 Study of surface plasmon chirality induced by Archimedes' spiral grooves *Opt. Express* **14** 6285–90

[15] Gorodetski Y, Niv A, Kleiner V and Hasman E 2008 Observation of the spin-based plasmonic effect in nanoscale structures *Phys. Rev. Lett.* **101** 043903

[16] Chen W, Abeysinghe D C, Nelson R L and Zhan Q 2010 Experimental confirmation of miniature spiral plasmonic lens as a circular polarization analyzer *Nano Lett.* **10** 2075–9

[17] Yang S, Chen W, Nelson R L and Zhan Q 2009 Miniature circular polarization analyzer with spiral plasmonic lens *Opt. Lett.* **34** 3047–9

[18] Miao J, Wang Y, Guo C, Tian Y, Guo S, Liu Q and Zhou Z 2011 Plasmonic lens with multiple-turn spiral nano-structures *Plasmonics* **6** 235–9

[19] Chen W, Nelson R L and Zhan Q 2012 Efficient miniature circular polarization analyzer design using hybrid spiral plasmonic lens *Opt. Lett.* **37** 1442–4

[20] Chen W, Rui G, Abeysinghe D C, Nelson R L and Zhan Q 2012 Hybrid spiral plasmonic lens: towards an efficient miniature circular polarization analyzer *Opt. Express* **20** 26299–307

[21] Kim H, Park J, Cho S W, Lee S Y, Kang M and Lee B 2010 Synthesis and dynamic switching of surface plasmon vortices with plasmonic vortex lens *Nano Lett.* **10** 529–36

[22] Spektor G, Prinz E, Hartelt M, Mahro A K, Martin A and Orenstein M 2021 Orbital angular momentum multiplication in plasmonic vortex cavities *Sci. Adv.* **7** eabg5571

[23] Cho S W, Park J, Lee S Y, Kim H and Lee B 2012 Coupling of spin and angular momentum of light in plasmonic vortex *Opt. Express* **20** 10083–94

[24] Rui G, Zhan Q and Cui Y 2015 Tailoring optical complex field with spiral blade plasmonic vortex lens *Sci. Rep.* **5** 13732

[25] Spektor G *et al* 2017 Revealing the subfemtosecond dynamics of orbital angular momentum in nanoplasmonic vortices *Science* **355** 1187–91

[26] Yang Y, Wu L, Liu Y, Xie D, Jin Z, Li J, Hu G and Qiu C W 2020 Deuterogenic plasmonic vortices *Nano Lett.* **20** 6774–9

[27] Wang X, Nie Z, Liang Y, Wang J, Li T and Jia B 2018 Recent advances on optical vortex generation *Nanophotonics* **7** 1533–6

[28] Zheng G, Mühlenbernd H, Kenney M, Li G, Zentgraf T and Zhang S 2015 Metasurface holograms reaching **80**

[29] Mei S, Mehmood M Q, Hussain S, Huang K, Ling X, Siew S Y, Liu H, Teng J, Danner A and Qiu C-W C-W 2016 Flat helical nanosieves *Adv. Funct. Mater.* **26** 5255–62

[30] Karimi E, Schulz S A, De Leon I, Qassim H, Upham J and Boyd R W 2014 Generating optical orbital angular momentum at visible wavelengths using a plasmonic metasurface *Light Sci. Appl.* **3** e167

[31] Zhao A, Pham A and Drezet A 2021 Plasmonic fork-shaped hologram for vortex-beam generation and separation *Opt. Lett.* **46** 689–92

[32] Tan Q, Guo Q, Liu H, Huang X and Zhang S 2017 Controlling the plasmonic orbital angular momentum by combining the geometric and dynamic phases *Nanoscale* **9** 4944–9

[33] Wu L-X, Li X and Yang Y-J 2019 Generation of surface plasmon vortices based on double-layer archimedes spirals *Acta Phys. Sin.* **68** 234201

[34] Lee S-Y, Kim S-J, Kwon H and Lee B 2015 Spin-direction control of high-order plasmonic vortex with double-ring distributed nanoslits *IEEE Photon. Technol. Lett.* **27** 705–8

[35] Yang H, Chen Z, Liu Q, Hu Y and Duan H 2019 Near-field orbital angular momentum generation and detection based on spin-orbit interaction in gold metasurfaces *Adv. Theory Simul.* **2** 1900133

[36] Chen C-F, Ku C-T, Tai Y-H, Wei P-K, Lin H-N and Huang C-B 2015 Creating optical near-field orbital angular momentum in a gold metasurface *Nano Lett.* **15** 2746–50

[37] Prinz E, Spektor G, Hartelt M, Mahro A-K, Aeschlimann M and Orenstein M 2021 Functional meta lenses for compound plasmonic vortex field generation and control *Nano Lett.* **21** 3941–6

[38] Tang B, Zhang B and Ding J 2019 Generating a plasmonic vortex field with arbitrary topological charges and positions by meta-nanoslits *Appl. Opt.* **58** 833–40

[39] Moon S-W, Jeong H-D, Lee S, Lee B, Ryu Y-S and Lee S-Y 2019 Compensation of spin-orbit interaction using the geometric phase of distributed nanoslits for polarization-independent plasmonic vortex generation *Opt. Express* **27** 19119–29

[40] Zhou H, Dong J, Zhou Y, Zhang J, Liu M and Zhang X 2015 Designing appointed and multiple focuses with plasmonic vortex lenses *IEEE Photon. J.* **7** 1–7

[41] Jin Z, Janoschka D, Deng J *et al* 2021 Phyllotaxis-inspired nanosieves with multiplexed orbital angular momentum *eLight* **1** 5

[42] Paterson L, MacDonald M P, Arlt J, Sibbett W, Bryant P E and Dholakia K 2001 Controlled rotation of optically trapped microscopic particles *Science* **292** 912–4

[43] Grier D G 2003 A revolution in optical manipulation *Nature* **424** 810–6

[44] Yuanjie Y, Yuxuan R, Mingzhou C, Yoshihiko A and Carmelo R-G 2021 Optical trapping with structured light: a review *Adv. Photon.* **3** 034001

[45] Marago O M, Jones P H, Gucciardi P G, Volpe G and Ferrari A C 2013 Optical trapping and manipulation of nanostructures *Nat. Nanotechnol.* **8** 807–19

[46] Bosanac L, Aabo T, Bendix P M and Oddershede L B 2008 Efficient optical trapping and visualization of silver nanoparticles *Nano Lett.* **8** 1486–91

[47] Zhan Q 2004 Trapping metallic Rayleigh particles with radial polarization *Opt. Express* **12** 3377–82

[48] Righini M, Zelenina A S, Girard C and Quidant R 2007 Parallel and selective trapping in a patterned plasmonic landscape *Nat. Phys.* **3** 477–80

[49] Chen K-Y, Lee A-T, Hung C-C, Huang J-S and Yang Y-T 2013 Transport and trapping in two-dimensional nanoscale plasmonic optical lattice *Nano Lett.* **13** 4118–22

[50] Berthelot J, Aćimović S S, Juan M L, Kreuzer M P, Renger J and Quidant R 2014 Three-dimensional manipulation with scanning near-field optical nanotweezers *Nat. Nanotechnol.* **9** 295–9

[51] Shen Z, Hu Z J, Yuan G H, Min C J, Fang H and Yuan X C 2012 Visualizing orbital angular momentum of plasmonic vortices *Opt. Lett.* **37** 4627–9

[52] Tsai W-Y, Huang J-S and Huang C-B 2014 Selective trapping or rotation of isotropic dielectric microparticles by optical near field in a plasmonic archimedes spiral *Nano Lett.* **14** 547–52

[53] Zaman M A, Padhy P and Hesselink L 2019 Solenoidal optical forces from a plasmonic archimedean spiral *Phys. Rev.* A **100** 013857

[54] Liu K, Maccaferri N, Shen Y, Li X, Zaccaria R P, Zhang X, Gorodetski Y and Garoli D 2020 Particle trapping and beaming using a 3D nanotip excited with a plasmonic vortex *Opt. Lett.* **45** 823–6

[55] Min C, Shen Z, Shen J, Zhang Y, Fang H, Yuan G, Du L, Zhu S, Lei T and Yuan X 2013 Focused plasmonic trapping of metallic particles *Nat. Commun.* **4** 2891

[56] Zhang Y, Shi W, Shen Z, Man Z, Min C, Shen J, Zhu S, Urbach H P and Yuan X 2015 A plasmonic spanner for metal particle manipulation *Sci. Rep.* **5** 15446

[57] Garcia-Vidal F J, Fernández-Domínguez A I, Martin-Moreno L, Zhang H C, Tang W, Peng R and Cui T J 2022 Spoof surface plasmon photonics *Rev. Mod. Phys.* **94** 025004

[58] Huidobro P A, Shen X, Cuerda J, Moreno E, Martin-Moreno L, Garcia-Vidal F J, Cui T J and Pendry J B 2014 Magnetic localized surface plasmons *Phys. Rev.* X **4** 021003

[59] Gao Z, Gao F, Zhang Y and Zhang B 2015 Complementary structure for designer localized surface plasmons *Appl. Phys. Lett.* **107** 191103

[60] Gao Z, Gao F, Zhang Y, Shi X, Yang Z and Zhang B 2015 Experimental demonstration of high-order magnetic localized spoof surface plasmons *Appl. Phys. Lett.* **107** 041118

[61] Su H, Shen X, Su G, Li L, Ding J, Liu F, Zhan , Liu Y and Wang Z 2018 Efficient generation of microwave plasmonic vortices via a single deep-subwavelength meta-particle *Laser Photon. Rev.* **12** 1800010

[62] Su G, Su H, Hu L, Qin Z, Shen X, Ding J, Liu F, Lu M, Zhan P and Liu Y 2021 Demonstration of microwave plasmonic-like vortices with tunable topological charges by a single metaparticle *Appl. Phys. Lett.* **118** 241106

[63] Zhang X and Cui T J 2020 Single-particle dichroism using orbital angular momentum in a microwave plasmonic resonator *ACS Photon.* **7** 3291–7

[64] Yin J Y, Ren J, Zhang L, Li H and Cui T J 2018 Microwave vortex-beam emitter based on spoof surface plasmon polaritons *Laser Photon. Rev.* **12** 1600316

[65] Zhang Y, Zhang Q, Chan C H, Li E, Jin J and Wang H 2019 Emission of orbital angular momentum based on spoof localized surface plasmons *Opt. Lett.* **44** 5735–8

[66] Liao Z, Zhou J N, Luo G Q, Wang M, Sun S, Zhou T, Ma H F, Cui T J and Liu Y 2020 Microwave-vortex-beam generation based on spoof-plasmon ring resonators *Phys. Rev. Appl.* **13** 054013

[67] Zang X *et al* 2019 Manipulating terahertz plasmonic vortex based on geometric and dynamic phase *Adv. Opt. Mater.* **7** 1801328

[68] Zang X *et al* 2020 Geometric metasurface for multiplexing terahertz plasmonic vortices *Appl. Phys. Lett.* **117** 171106

[69] Xu Q *et al* 2019 Coupling-mediated selective spin-to-plasmonic-orbital angular momentum conversion *Adv. Opt. Mater.* **7** 1900713

[70] Arikawa T, Hiraoka T, Morimoto S, Blanchard F, Tani S, Tanaka T, Sakai K, Kitajima H, Sasaki K and Tanaka K 2020 Transfer of orbital angular momentum of light to plasmonic excitations in metamaterials *Sci. Adv.* **6** eaay1977

[71] Fernández-Domínguez A I, Williams C R, García-Vidal F J, Martín-Moreno L, Andrews S R and Maier S A 2008 Terahertz surface plasmon polaritons on a helically grooved wire *Appl. Phys. Lett.* **93** 141109

[72] Fernández-Domínguez F R A I, Martín-Moreno L and García-Vidal F J 2012 Subwavelength chiral surface plasmons that carry tuneable orbital angular momentum *Phys. Rev.* B **86** 075437

[73] Yao H Z and Zhong S 2015 Wideband circularly polarized vortex surface modes on helically grooved metal wires *IEEE Photon. J.* **7** 1–7

[74] Sroor H, Huang Y-W, Sephton B, Naidoo D, Vallés A, Ginis V, Qiu C-W, Ambrosio A, Capasso F and Forbes A 2020 High-purity orbital angular momentum states from a visible metasurface laser *Nat. Photon.* **14** 498–503

[75] Dai Y, Zhou Z, Ghosh A, Mong R S K, Kubo A, Huang C-B and Petek H 2020 Plasmonic topological quasiparticle on the nanometre and femtosecond scales *Nature* **588** 616–9

[76] Chong A, Wan C, Chen J and Zhan Q 2020 Generation of spatiotemporal optical vortices with controllable transverse orbital angular momentum *Nat. Photon.* **14** 350–4

[77] Wang B, Liu W, Zhao M, Wang J, Zhang Y, Chen A, Guan F, Liu X, Shi L and Zi J 2020 Generating optical vortex beams by momentum-space polarization vortices centred at bound states in the continuum *Nat. Photon.* **14** 623–8

[78] Yang Z-Q, Shao Z-K, Chen H-Z, Mao X-R and Ma R M 2020 Spin-momentum-locked edge mode for topological vortex lasing *Phys. Rev. Lett.* **125** 013903

[79] Fu Y, Shen C, Zhu X, Li J, Liu Y, Cummer Steven A and Xu Y 2020 Sound vortex diffraction via topological charge in phase gradient metagratings *Sci. Adv.* **6** eaba9876

[80] Hong Z, Zhang J and Drinkwater B W 2015 Observation of orbital angular momentum transfer from bessel-shaped acoustic vortices to diphasic liquid-microparticle mixtures *Phys. Rev. Lett.* **114** 214301

[81] Uchida M and Tonomura A 2010 Generation of electron beams carrying orbital angular momentum *Nature* **464** 737–9

[82] Verbeeck J, Tian H and Schattschneider P 2010 Production and application of electron vortex beams *Nature* **467** 301–4

[83] Lloyd S, Babiker M and Yuan J 2012 Quantized orbital angular momentum transfer and magnetic dichroism in the interaction of electron vortices with matter *Phys. Rev. Lett.* **108** 074802.1–4

[84] Vanacore G M *et al* 2019 Ultrafast generation and control of an electron vortex beam via chiral plasmonic near fields *Nat. Mater.* **18** 573–9

[85] Luski A, Segev Y, David R, Bitton O, Nadler H, Ronny Barnea A, Gorlach A, Cheshnovsky O, Kaminer I and Narevicius E 2021 Vortex beams of atoms and molecules *Science* **373** 1105–9

[86] Wang Y, Xu Y, Feng X, Zhao P, Liu F, Cui K, Zhang W and Huang Y 2016 Optical lattice induced by angular momentum and polygonal plasmonic mode *Opt. Lett.* **41** 1478–81

[87] Srivastava S, Jain P and Maiti T 2019 Design of scalable optical decoder based on hexagonal plasmonic modes induced on topological insulator surface states *Sci. Rep.* **9** 9190

IOP Publishing

Optical Vortices
Fundamentals and applications
Yuanjie Yang, Yu-Xuan Ren and Carmelo Rosales-Guzmán

Chapter 8

Tailoring structured light with spatial light modulators

In this chapter we will provide a more detailed description of the use of liquid crystal spatial light modulators for the generation of structured light beams. We will begin our description by providing a short overview of the liquid crystal technology, which is the basis of understanding how light can be tailored using this device. We will continue our description by explaining the generalities about light beam shaping using greyscale images. We will end this chapter by describing in detail two of the main techniques for light beam shaping: Phase-only and complex-amplitude modulation.

8.1 The liquid crystal spatial light modulators

A liquid crystal spatial light modulator is a pixelated device consisting of millions of cells filled with liquid crystal rod-like molecules. Three of the main ways in which the long axis of the liquid crystal molecules within each pixel can be aligned are: parallel aligned nematic (PAN), vertically aligned nematic (VAN), or twisted nematic (TN). Regarding the first two, here the LC molecules are sandwiched between a silicon substrate and a transparent electrode, whereby their orientation is controlled through a predefined voltage V. In this way, when no voltage is applied the molecules remain static in a preferred orientation, but as the voltage increases, the molecules rotate by an angle $\Theta(V)$, as a function of the strength of the voltage V. As a consequence of such rotations, the effective extraordinary index of refraction $n_e(\Theta)$ of the liquid crystal decreases. As a result, an input light beam acquires at every pixel a local phase shift, which is also a function of the applied voltage and is given by [1],

$$\phi(V) = \frac{2\pi}{\lambda}[n_e(V) - n_o]d. \tag{8.1}$$

doi:10.1088/978-0-7503-5844-6ch8
8-1

Here λ represents the wavelength of the input light beam, d is the thickness of the pixel, n_o is the ordinary refractive index and $n_e(V)$ is the extraordinary refractive index, which is a function of the applied voltage.

Typical commercially available liquid crystal displays are electro-optical, in which the orientation of the molecules (the effective refraction index) is controlled via an electric field. Furthermore, they can modulate the properties of light upon transmission or upon reflection. The former is based on transparent liquid crystal displays, whereas the second is based on reflective LCoS displays. In addition, the specific configurations of the molecular alignment within each pixel completely determines how the incident light beam gets modulated, that is, phase-only, amplitude-only, or both, amplitude and phase modulation. For example, in the PAV and VAN configurations, the liquid crystal molecules are sandwiched between a silicon and a transparent electrode. This allows one to control the orientation of the liquid crystal molecules as a function of the magnitude of the applied voltage on both ends. In this way, when no voltage is applied, the molecules rest in a preferred orientation, but as the voltage increases, the molecules tend to rotate by ako,k, specific angle. In this way, modulation of the voltage applied to each pixel induces a rotation of liquid crystal molecules within substrates, which in turn modifies the effective refraction index a light of beam experiences upon its transmission. An schematic representation of the constituting components of an LCoS-SLM display is illustrated in figure 8.1. Here, for illustrative purposes we only show the configuration corresponding to four pixels, each of which has different orientations of the liquid crystal molecules. The layer containing the liquid crystal molecules is sandwiched between two transparent alignment films. This layer is then sandwiched between a transparent electrode layer (top) and a silicon substrate (bottom). Finally, everything is covered with a flat glass substrate and connected to pixelated metal electrodes, which control the orientation of the liquid crystal molecules at each pixel. Crucially, in commercially available SLMs the applied voltage is controlled through greyscale images, typically 256 grey levels in such a way that each grey level is

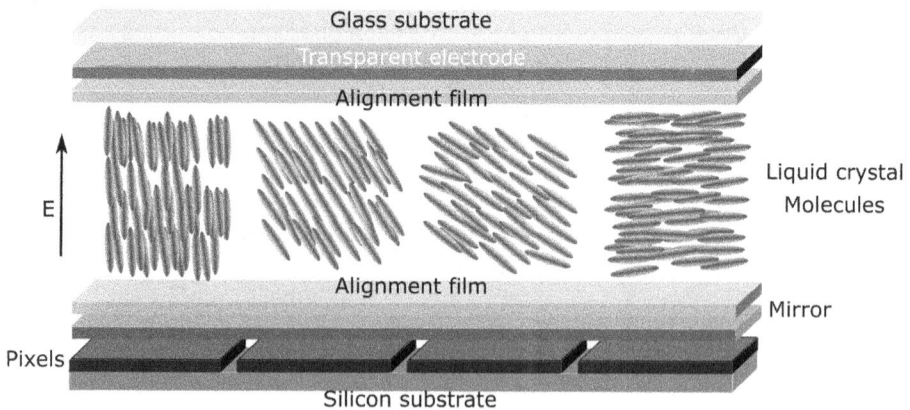

Figure 8.1. Schematic representation of a liquid crystal on silicon spatial light modulator display. Here, E represents the applied electric field.

associated with a unique and discrete increment of the voltage and therefore of the phase. For example, black is associated with 0 phase shift and white with π, along with a linear relation for intermediate values, as schematically illustrated in figure 8.2. Hence, we only need to display on the SLM's screen an 8-bit (256 grey levels) pixelated image of the same spatial resolution as the SLM (typically 1920×1080) to induce phase increments of $2\pi/256$ at each pixel on the LC display. The images can be easily generated with any software and saved if the formats are supported by each specific device, and are typically known as computer-generated holograms and sometimes also as digital holograms, even though this term might not be correct. By way of example, figure 8.3 shows four examples of the greyscale images required to generate optical vortices, in these examples we show the fork-like holograms corresponding to topological charges $\ell = 1, \ell = 3, \ell = 6$ and $\ell = 7$. Note the typical fork-like pattern characteristic of these beams. For completeness, the corresponding experimentally generated beams are shown on the bottom rows in correspondence to the greyscale image.

It is worth mentioning that the liquid crystal is birefringent; it possesses two different indices of refraction, as evidenced from equation (8.1). This implies that only the polarization component of light parallel to the extraordinary axis will be modulated, typically this will be the horizontal polarization component. This has been used as an advantage rather than as a drawback in the generation of vector beams, as briefly discussed before. As a final comment, while most SLMs are phase-only devices, meaning they can only modulate the phase of a complex-amplitude optical field. Nonetheless, it is also possible to modulate their amplitude through appropriate modulation schemes, such as complex-amplitude modulation, as will be discussed in section 8.2.2.

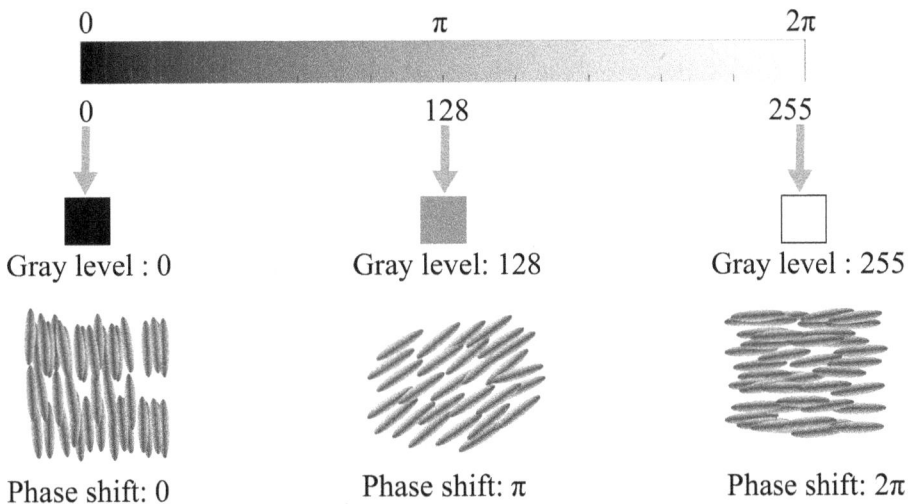

Figure 8.2. Schematic representation of the phase retardation as a function of the applied voltage, which is controlled through greyscale images.

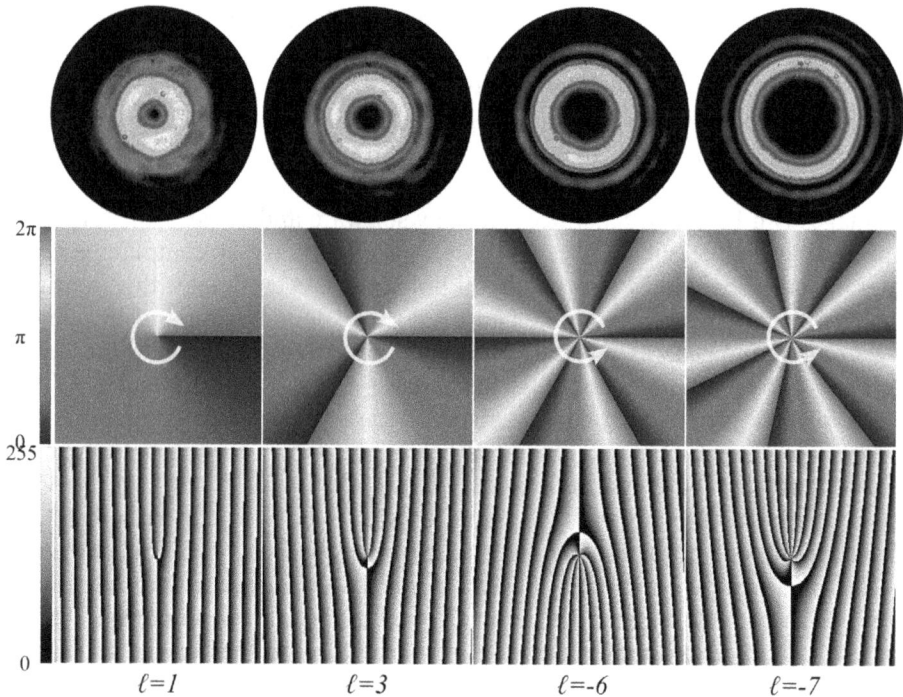

Figure 8.3. Example of typical greyscale images (top) to generate optical vortices embedded with orbital angular momentum.

8.2 Tailoring light with spatial light modulators

As explained earlier, an SLM is made typically using more than 2 million pixels (deepening on the specific model), liquid crystal cells that are controlled independently via an applied voltage, which has been previously calibrated to allow a rotation of the same as a function of the voltage. As a result, each pixel behaves as a birefringent material, which induces a local phase shift for a particular polarization state (typically horizontal) of an incoming light. Such phase shifts can be anywhere between 0 and 2π with most common devices, even though modern versions can achieve up to 8π phase shifts, depending also on the specific wavelength of the input beam. The specific phase shift applied by each pixel is controlled by grey tones, typically 255 tones, where black (0) is associated to no phase shift and white (255) to a 2π phase shift. It is also worth emphasising that the pixelation nature of SLMs forms a two-dimensional grating, causing diffraction in all directions of the light that falls between the pixels. As a result, the conversion efficiency of light is limited to about only 90%, also depending on the specific model. In addition, a portion of the input light does not interact with the computer-generated hologram, due, for example, to imperfect calibration, reflection off the electrodes, amongst others, and therefore it does not acquire any modulation. This portion of spurious light, even though small (typically between 5% and 20%), has to be separated from the modulated light to avoid interference effects which degrades the quality of the

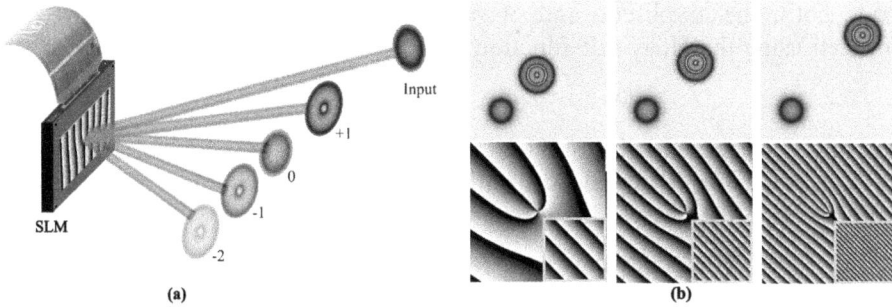

Figure 8.4. (a) A diffraction grating added to the computer-generated holograms separates modulated from unmodulated light. (b) The separation of the diffraction orders is proportional to the period of the grating.

modulated light beam. Hence, to isolate the modulated light from unmodulated, a grating is typically added to the desired hologram, which can be linear or sinusoidal. This is schematically shown in figure 8.4(a) for the specific case of an optical vortex and a linear diffraction order. As illustrated, the first diffraction order is the most bright, but there is also a portion of modulated and unmodulated light in the others. Further, the diffraction orders propagate from the SLM at an angle that is proportional to the period of the grating, as illustrated in figure 8.4(b) for a linear grating oriented diagonally, which separates the diffraction orders in the diagonal direction, here for the sake of clarity we only show the zero and first orders. A more detailed description of the SLM principle as well as additional applications can be found in [1–3] and references therein.

8.2.1 Phase modulation

Up to now we have only explained that in order to modulate light using a liquid crystal SLM, a greyscale image is displayed on the liquid crystal screen and we have provided a few examples of typical fork-like holograms to create light beams endowed with orbital angular momentum. However, SLMs allow us to create more complex light patterns, being the solutions to the wave equation, either in its exact or paraxial form, of more interest which in general are given by complex function possessing an amplitude and a phase term. As such, shaping of common light fields requires in general a computer-generated hologram generated from a transfer function capable of modulating both the amplitude and phase of the complex light field. Since most SLMs allow only the direct control of the phase, several techniques have being proposed to indirectly modulate the amplitude using phase-only holograms [4–11].

In a first approximation, a phase-only computer-generated hologram capable to reproduce the amplitude changes in complex light fields must be able to encode the changes of sign of the amplitude term, as well as the phase variation onto a single transfer function. To explain this, let us consider, for example, the complex light field, we will further consider that the amplitude function $A(x, y)$ represents the amplitude and $\exp i\phi[(x, y)]$ the phase term. A transmittance function capable to encode both, the amplitude and phase term as a phase-only hologram can be computed using the Heaviside function as [5],

$$U(x, y) = A(x, y)\exp i\phi[(x, y)], \tag{8.2}$$

where $A(x, y)$ represents the amplitude and $\exp i\phi[(x, y)]$ the phase term. A phase-only function capable to encode both, the amplitude and phase term can be computed using the Heaviside function as [5],

$$t(x, y) = \exp\{i[\exp i\phi[(x, y)] + \pi\Theta(x, y)]\}, \tag{8.3}$$

where $\Theta(x, y)$ is the Heaviside function that takes into account the sign variations of the amplitude given by

$$\Theta(x, y) = \Theta[A(x, y)]. \tag{8.4}$$

This transmittance function in combination with a linear diffraction grating will take the form

$$t(x, y) = \exp\{i[\exp i\phi[(x, y)] + \pi\Theta(x, y)] + (ux + vy)\}, \tag{8.5}$$

where u and v are the spatial frequencies of the grating along the horizontal and vertical direction, respectively. By way of example, we will show how this technique applied to the well-known Laguerre–Gauss beams, natural solutions of the paraxial wave equation in cylindrical coordinates (ρ, φ), which are described mathematically at $z = 0$ as

$$LG_p^\ell(\rho, \varphi) = \frac{1}{\omega_0}\sqrt{\frac{2p!}{\pi(\ell + p)!}}\left(\frac{\sqrt{2}\rho}{\omega_0}\right)^\ell L_p^\ell\left[\frac{2\rho^2}{\omega_0^2}\right]\exp\left(-\frac{\rho^2}{\omega_0^2}\right)\exp(i\ell\varphi). \tag{8.6}$$

Here, the functions $L_p^\ell(\cdot)$ represent the generalized Laguerre polynomials, ℓ is associated to the orbital angular momentum of the beams, p is a positive integer that accounts for the number of maxima $(p + 1)$ along the transverse direction and the parameter ω_0 is the beamwaist of the mode. For this specific case, the amplitude term is given by

$$A(\rho) = \frac{1}{\omega_0}\sqrt{\frac{2p!}{\pi(\ell + p)!}}\left(\frac{\sqrt{2}\rho}{\omega_0}\right)^\ell L_p^\ell\left[\frac{2\rho^2}{\omega_0^2}\right]\exp\left(-\frac{\rho^2}{\omega_0^2}\right), \tag{8.7}$$

whereas the phase term is given by

$$\phi(\varphi) = \exp(i\ell\varphi). \tag{8.8}$$

Hence, for this specific example, the phase-only hologram, including a linear grating, will be given by

$$H(\rho, \varphi) = \mathrm{mod}\{[\ell\varphi + \pi\Theta[A(\rho, \varphi)]] + (ux + vy), 2\pi\}, \tag{8.9}$$

where $\mathrm{mod}\{\cdot, 2\pi\}$ represents the modulus function. Two specific examples of the amplitude (left panels), phase (middle panels) and holograms computed using this

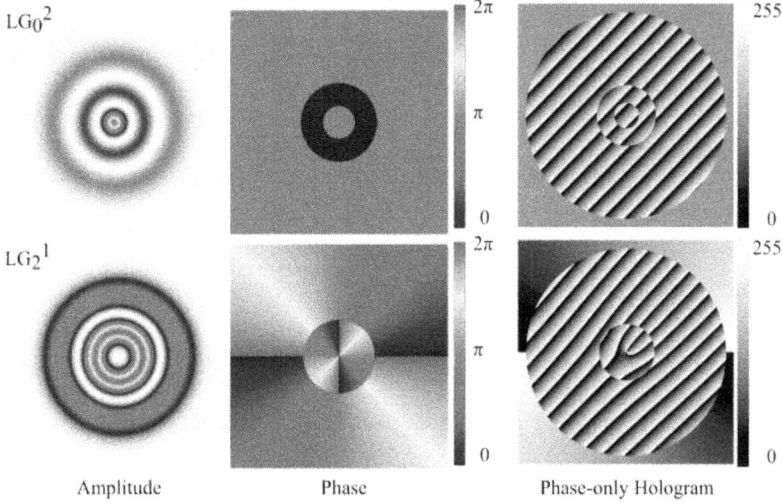

$LG_0{}^2$

$LG_2{}^1$

Amplitude   Phase   Phase-only Hologram

Figure 8.5. Amplitude (left panels) and phase (middle panels) profile of the Laguerre–Gauss modes LG_0^2 and LG_2^1 and their corresponding phase-only holograms.

technique (right panels) are shown in figure 8.5 for the Laguerre–Gaussian modes LG_0^2 (top panels) and LG_2^1 (bottom panels).

8.2.2 Complex amplitude modulation

An alternative method to achieve amplitude and phase modulation is Complex Amplitude Modulation (CAM), which can be performed in various ways; for example, in a naive way by multiplying the amplitude and phase terms [9]. A more appropriate way to implement CAM was proposed by Arrizon *et al* [6, 10], which produces light fields with relatively high quality when compared to others [4, 9] and will be described next. To this end, we will use again the optical field as described by equation (8.2), which can be written in general as

$$U(\mathbf{r}) = A(\mathbf{r})\exp[i\phi(\mathbf{r})] \tag{8.10}$$

where now $A(\mathbf{r}) \in [0, 1]$ and $\phi(\mathbf{r}) \in [-\pi, \pi]$. In essence, complex-amplitude modulation consist then on modulating the amplitude indirectly by converting it into a phase term. In other words, it consists of finding a function $h(\mathbf{r})$ of the form,

$$h(\mathbf{r}) = \exp\{i\Phi[A(\mathbf{r}), \phi(\mathbf{r})]\} \tag{8.11}$$

where $\Phi[A(\mathbf{r}), \phi(\mathbf{r})]$ is a phase function to be determined, which contains information of both the amplitude and the phase.

Such a phase function can be found by expanding equation (8.11) as a Fourier series as

$$h(\mathbf{r}) = \sum_{q=-\infty}^{\infty} c_q^A \exp[iq\phi(\mathbf{r})], \tag{8.12}$$

where the expansion coefficients c_q^A are given by

$$c_q^A = \frac{1}{2\pi} \int_{-\pi}^{\pi} \exp\{i\Phi[A(\mathbf{r}),\ \phi(\mathbf{r})]\}\exp[-iq\phi(\mathbf{r})]d\phi. \qquad (8.13)$$

For the specific case $q = 1$, the above expression takes the form

$$c_1^A = \frac{1}{2\pi} \int_{-\pi}^{\pi} \exp\{i\Phi[A(\mathbf{r}),\ \phi(\mathbf{r})]\}\exp[-i\phi(\mathbf{r})]d\phi, \qquad (8.14)$$

If we now write $h(\mathbf{r})$ for this term we obtain the expression

$$h(\mathbf{r}) = c_1^A \exp[i\phi(\mathbf{r})], \qquad (8.15)$$

which is similar to equation (8.10) if we let $c_1^A = aA(\mathbf{r})$, with a a positive constant in the interval [0, 1]. Hence, the problem has reduced to solve the equation

$$\frac{1}{2\pi} \int_{-\pi}^{\pi} \exp\{i\Phi[A(\mathbf{r}),\ \phi(\mathbf{r})]\}\exp[-i\phi(\mathbf{r})]d\phi = aA(\mathbf{r}), \qquad (8.16)$$

which is equivalent to

$$\frac{1}{2\pi} \int_{-\pi}^{\pi} \cos\{i\Phi[A(\mathbf{r}),\ \phi(\mathbf{r})]\}d\phi + \frac{i}{2\pi} \int_{-\pi}^{\pi} \sin\{i\Phi[A(\mathbf{r}),\ \phi(\mathbf{r})]\} = aA(\mathbf{r}). \quad (8.17)$$

Equation (8.14) can be fulfilled if

$$\int_{-\pi}^{\pi} \sin\{i\Phi[A(\mathbf{r}),\ \phi(\mathbf{r})]\}d\phi = 0, \qquad (8.18)$$

$$\int_{-\pi}^{\pi} \cos\{i\Phi[A(\mathbf{r}),\ \phi(\mathbf{r})]\}d\phi = 2\pi A. \qquad (8.19)$$

A possible solution to equations (8.18) and (8.19) can be

$$\Phi[A(\mathbf{r}),\ \phi(\mathbf{r})] = f[A(\mathbf{r})]\sin[\phi(\mathbf{r})], \qquad (8.20)$$

which Arrizon refers to as a type 3 hologram [6] and allows the generation of light beams with high quality.

All that remains now is to find the function $f[A(\mathbf{r})]$, which only depends on the amplitude of the optical field. A possible way to find it is by expanding $h(\mathbf{r}) = \exp\{if[A(\mathbf{r})]\sin[\phi(\mathbf{r})]\}$ using the well-known Jacobi–Anger identity

$$h(\mathbf{r}) = \exp\{if[A(\mathbf{r})]\sin[\phi(\mathbf{r})]\} = \sum_{s=-\infty}^{\infty} J_s\{f[A(\mathbf{r})]\}\exp[is\phi(\mathbf{r})], \qquad (8.21)$$

where J_s is the Bessel function of order s.

By comparing the previous equation with equation (8.12) we immediately obtain the relation

$$c_q^A(\mathbf{r}) = J_q\{f[A(\mathbf{r})]\}. \qquad (8.22)$$

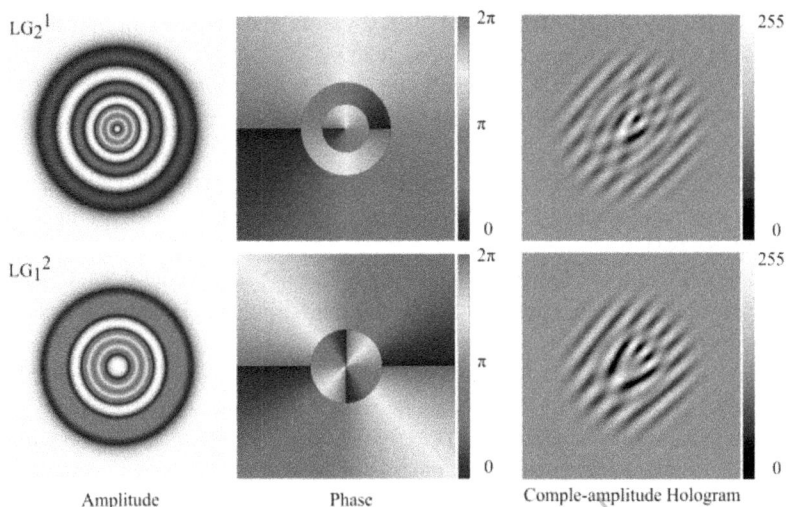

Figure 8.6. Amplitude (left panels) and phase (middle panels) profile of the Laguerre–Gauss modes LG_2^1 and LG_1^2 and their corresponding complex-amplitude modulation holograms.

Again, by taking $q = 1$ we obtain the relation

$$c_1^A(\mathbf{r}) = J_1\{f[A(\mathbf{r})]\} = A(\mathbf{r})a, \qquad (8.23)$$

where $J_1(x)$ is the first order Bessel function. Hence, $f[A(\mathbf{r})]$ can be obtained by inverting the first order Bessel function, which unfortunately cannot be done and therefore a numerical inversion is required. As such, the function $f[A(\mathbf{r})]$ takes the form

$$f[A(\mathbf{r})] = J_1^{-1}[A(\mathbf{r})a]. \qquad (8.24)$$

A numerical inversion of $J_1(x)$ is limited to the interval $[0, x_m]$, with $x_m = 1.84$ is the value where $J_1(x)$ reaches its first maximum $J_1(x_m) = 0.5819$. This restriction limits the values of $\phi[A(\mathbf{r}), \phi(\mathbf{r})]$ to the reduced domain $[-0.586\pi, 0.586\pi]$, which implies that the maximum phase shift that can be obtained with this method is $\Delta\phi = 1.17\pi$. By way of example, the computer-generated hologram for Laguerre–Gauss beams will have the form

$$H(\rho, \varphi) = J_1^{-1}[aA(\rho)]\sin(\ell\varphi + ux + vy), \qquad (8.25)$$

where $A(\rho)$ is given by equation (8.7) and J_1^{-1} represent the numerical inversion of the first order Bessel function. As an example, figure 8.6 shows the amplitude (left panels) and phase (middle panels) profile of the Laguerre–Gauss modes LG_1^2 and LG_2^1, as well as the corresponding complex-amplitude modulation hologram.

References

[1] Rosales-Guzmán C and Forbes A 2017 *How to Shape Light with Spatial Light Modulators* (Bellingham, WA: SPIE Press) https://doi.org/10.1117/3.2281295

[2] Forbes A, Dudley A and McLaren M 2016 Creation and detection of optical modes with spatial light modulators *Adv. Opt. Photon.* **8** 200–27

[3] Lazarev G, Hermerschmidt A, Krüger S and Osten S 2012 *Optical Imaging and Metrology: Advanced Technologies* ed W Osten and N Reingand (New York: Wiley) 1–29 https://doi.org/10.1002/9783527648443

[4] Ohtake Y, Ando T, Fukuchi N, Matsumoto N, Ito H and Hara T 2007 Universal generation of higher-order multiringed Laguerre–Gaussian beams by using a spatial light modulator *Opt. Lett.* **32** 1411–3

[5] Matsumoto N, Ando T, Inoue T, Ohtake Y, Fukuchi N and Hara T 2008 Generation of high-quality higher-order Laguerre-Gaussian beams using liquid-crystal-on-silicon spatial light modulators *J. Opt. Soc. Am.* A **25** 1642–51

[6] Arrizón V, Ruiz U, Carrada R and González L 2007 Pixelated phase computer holograms for the accurate encoding of scalar complex fields *J. Opt. Soc. Am.* A **24** 3500–7

[7] Davis J A, Valadéz K O and Cottrell D M 2003 Encoding amplitude and phase information onto a binary phase-only spatial light modulator *Appl. Opt.* **42** 2003–8

[8] van Putten E G, Vellekoop I M and Mosk A P 2008 Spatial amplitude and phase modulation using commercial twisted nematic lcds *Appl. Opt.* **47** 2076–81

[9] Clark T W, Offer R F, Franke-Arnold S, Arnold A S and Radwell N 2016 Comparison of beam generation techniques using a phase only spatial light modulator *Opt. Express* **24** 6249–64

[10] Ando T, Ohtake Y, Matsumoto N, Inoue T and Fukuchi N 2009 Mode purities of Laguerre–Gaussian beams generated via complex-amplitude modulation using phase-only spatial light modulators *Opt. Lett.* **34** 34–6

[11] Aguirre-Olivas D, Mellado-Villasenor G, de-la Llave D S and Arrizón V 2015 Efficient generation of Hermite–Gauss and Ince–Gauss beams through kinoform phase elements *Appl. Opt.* **54** 8444–52

IOP Publishing

Optical Vortices
Fundamentals and applications
Yuanjie Yang, Yu-Xuan Ren and Carmelo Rosales-Guzmán

Chapter 9

Tailoring optical vortex with digital micromirror device

9.1 Introduction of amplitude DMD

The digital micromirror device (DMD) is a micromechanical device that has millions of tiny mirror units of dimension on the order of a few microns. Each mirror unit can be individually addressed through the driving circuit. As the mirror unit is switched on, the portion of the tiny light beam can be directed to the downstream optics, while the mirror unit at the off state directs the tiny beam to an optical sink. Although the DMD switches the light in a binary module, it can modulate both the phase and amplitude of the light wavefront. Spatial modulation can be achieved by projecting the greyscale images on the device. The device enjoys a much faster speed once the binary patterns are used.

The DMD is potentially more advantageous than a traditional liquid-crystal spatial light modulator in terms of spectrum sensitivity, speed, and polarization modulation. For instance, the DMD refreshes with a 20 kHz frame rate, and has a high fill factor of over 90 %. The micromirror units are arranged in an array of 1024 × 768 pixels, and are individually addressed electronically [1]. Each micromirror with width 13.68 μm could be individually addressed to situate at either +12° or −12° with respect to the surface normal in response to the addressing electric signal '1' and '0' [2, 3]. These two states represent the 'on state' and 'off state', respectively (figure 9.1). The on-state micromirror directs the micro beam to the downstream optical path, while the off-state micromirror reflects the micro beam onto a beam dump. As a result, the two states of the micromirrors in the DMD are controlled by the digital signals '0' and '1' (figure 9.1). The amplitude type of the spatial light modulator has been corroborated to be useful in laser studies [4].

Since 1987, the DMD has become one of the most important devices in projection display [5–8]. Although the DMD has been successfully applied in display technology to develop minature projectors, recent advances suggest that the

doi:10.1088/978-0-7503-5844-6ch9   9-1

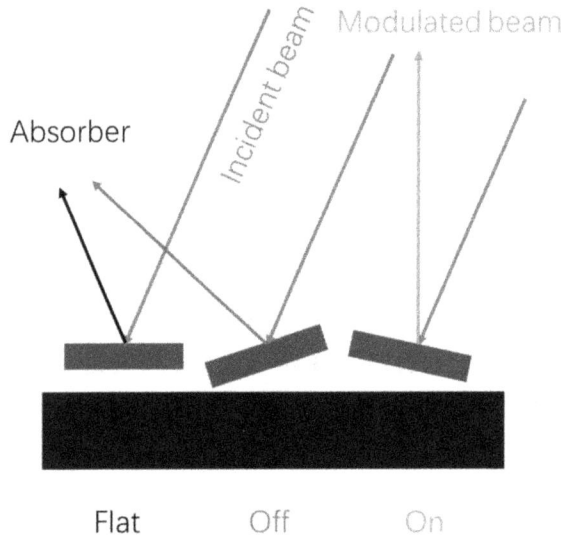

Figure 9.1. Schematics of a DMD.

DMD can be used for optical beam shaping [9]. In contrast to the liquid-crystal SLM, the DMD performs better in terms of the fill factor, refresh rate, polarization insensitivity, and broadband response. Most lasers have an output beam with a Gaussian spatial distribution associated with the geometric shape of the laser resonator.

The binary-amplitude DMD creates arbitrary spatial optical modes with high speed, efficiency and fidelity. With the DMD, several optical modes were success-fully shaped, including the Laguerre–Gaussian [2], Bessel modes [10], Ince–Gaussian modes [11], and flat-top modes [21]. Compared with common LC-SLM, the DMD has a higher fill factor rate of around 90% (LC counterparts less than 70%), faster refreshing frequency of 5.2 kHz (LC around 60HZ) and sensitivity over different wavelength [12, 13].

Laser beam shaping includes the precise control of the amplitude and phase of the output beam [14, 15]. In this chapter, we briefly introduce the major algorithms used to calculate the binary hologram for DMD, and some applications in shaping fundamental spatial modes. The transversal appearance of perfect vortex beam looks like an annular ring with radius determined by the topological charge [16].

9.2 Algorithms to generate amplitude hologram for DMD

The hologram can usually be created through interference between the target field and a plane wave. The optical phase singularity is related with orbital angular momentum (OAM). The net phase change of the circuit encompassing the nodal point is quantized as [12]

$$l = \frac{1}{2\pi} \oint_C dr \cdot \nabla \chi = \frac{1}{2\pi} \oint_C d\chi, \qquad (9.1)$$

where the integer l is the topological charge. DMD projected with an off-axis hologram can effectively tailor the beam to the target field. For instance, the interferogram for the helical phase $\exp(il\varphi)$ and the unit plane wave $\exp(-ikr)$ is

$$I = |\exp(il\varphi) + \exp(-ikr)|^2 = 2 + 2\cos(l\varphi + kr), \quad (9.2)$$

where φ is the azimuthal angle. Equation (9.2) provides a means to create the interferogram for beams with OAM. By replacing the helical phase term in equation (9.2) with an arbitrary target complex amplitude, this would produce interferograms for nearly arbitrary fields.

Alternatively, the spatial phase $\phi(x, y)$ can be encoded through the Lee method with a carrier frequency ν_0 as follows:

$$h(x, y) = 1/2[1 + \cos(2\pi(x - y)\nu_0 - \phi(x, y))], \quad (9.3)$$

The above function can also be written in the exponential form

$$h(x, y) = \frac{1}{2} + \frac{1}{4}e^{i2\pi(x-y)\nu_0}e^{-i\phi(x, y)} + \frac{1}{4}e^{i2\pi(y-x)\nu_0}e^{i\phi(x, y)}. \quad (9.4)$$

The spatial light modulators are adopted for spatial phase modulation. A binary-amplitude hologram encodes the complex field through amplitude DMD. The binary Lee method simultaneously encodes the amplitude and phase of the target field [12], with the hologram binarized in the following:

$$H(x, y) = \frac{1}{2} + \frac{1}{2}\,\text{sgn}\left[\cos\left(\frac{2\pi x}{x_0} + \pi p(x, y)\right) - \cos(\pi\omega(x, y))\right], \quad (9.5)$$

where sgn is the sign function, $\omega(x, y)$ and $p(x, y)$ are slowly varying terms, and x_0 is the carrier periodicity [17]. The slowly varying terms are related with the target optical field $A(x, y)e^{i\phi(x, y)}$ through

$$\omega(x, y) = \frac{1}{\pi}\arcsin[A(x, y)], . \quad (9.6)$$

$$p(x, y) = \frac{1}{\pi}\phi(x, y), . \quad (9.7)$$

Equations (9.4) and (9.5) provide a means to numerically calculate the slowly varying terms for arbitrary target optical fields.

Although the Lee method is sufficient to calculate binary holograms for shaping arbitrary transverse filed with DMD, full control over the spatial phase and amplitude can also be achieved using a super-pixel encoding method [11, 18]. In 2003, Arrizón proposed a complex modulation approach using dual-pixel encoding. In 2014, Goorden *et al* developed a superpixel method for the complex modulation [18]. The DMD is divided into super-pixels for further encoding of the whole target field. $n \times n$ micromirrors are grouped into a single superpixel, the synthesized phase and amplitude regulated by the superpixel can be evaluated by selectively switching

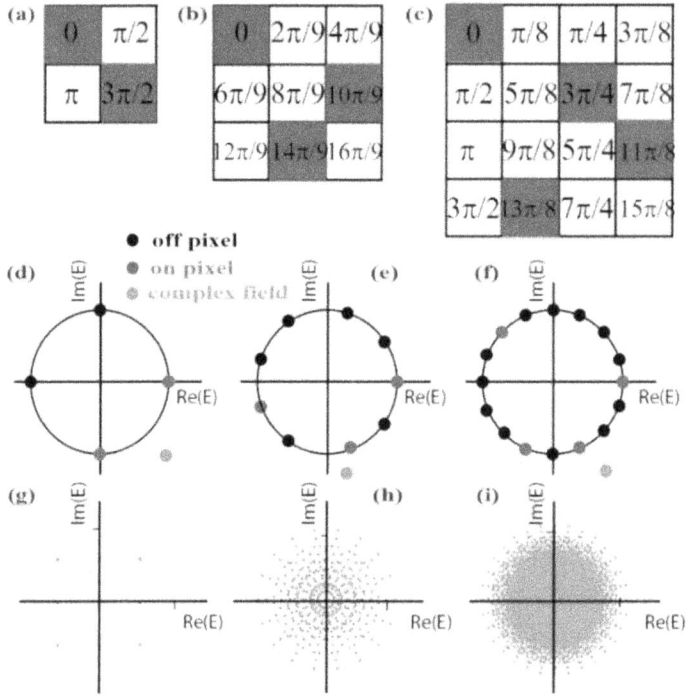

Figure 9.2. Demonstration of superpixel encoding. (a) 2 × 2, (b) 3 × 3, and (c) 4 × 4 pixels. The contribution from each pixel uniformly populated on a unit circle (d–f). Synthetic contribution from selected pixels produces the complex field (g–i). Reproduced with permission from [11].

on the mirrors in the giant pixel. The phase response of each micromirror in the superpixel depends on the pixel locations. The response in the target plane of a superpixel is the sum over all the individual active pixel responses.

Figures 9.2(a)–(c) demonstrate the superpixel of 2 × 2, 3 × 3, 4 × 4 pixels. Each micromirror in the superpixel has a phase difference with respect to the top-left pixel. The values of phase difference are written in each square in figures 9.2(a)–(c). By selectively switching the pixels, e.g., the shadowed square pixels in figures 9.2(a)–(c), the resultant complex fields are expressed in dark green dots through a vector sum over all the dark purple dots.

For $n \times n$ square pixels, the synthesized complex field will increase dramatically with n. The number of possible fields is 9 for $n = 2$ (figure 9.2(g)). This position ensures the target phase responses of neighbouring pixels inside a superpixel are $2\pi/n^2$ out of phase in the x-direction and $2\pi/n$ out of phase along the y-direction. The values increase to 343 and 6561 for $n = 3$ and $n = 4$, respectively (see figures 9.2 (h) and (i)). The phase resolution for $n = 3$ case could be estimated to be $2\pi/343 = 0.018$ rad (1.05°). As a comparison, commercial LC-SLM could provide 8-bit modulation with a phase resolution of $2\pi/2^8 = 0.025$ rad (1.41°).

Various algorithms can be used to determine the holograms, e.g., the division of desired fields by input modes and the error-diffusion technique. We here introduce another version of the hologram based on image dithering of the off-axis hologram to improve the temporal stability of the shaped light field.

The binary value of DMD(x, y) can be determined by comparing the desired amplitude reflectance $r(x, y)$ modified by the propagated errors from nearby pixels. The error function is

$$e(x, y) = r_1(x, y) - \text{DMD}(x, y) \tag{9.8}$$

and the reflectance function is replaced by

$$r_1(x + a, y + b) = r_1(x + a, y + b) + c(a, b) \times e(x, y), \tag{9.9}$$

where a and b are row and column coordinate shifts of the nearest neighbor pixels yet to be processed. The weighting coefficients for the four nearest neighbors are given by [19]

$$c(1, -1) = -3/16, \quad c(1, 0) = -5/16,$$

$$c(1, 1) = -1/16, \quad \text{and} \quad c(0, 1) = -7/16. \tag{9.10}$$

After replacement, the pixel value is binarized and the error for such a pixel is calculated to propagate to nearby pixels. An iterative algorithm was further applied to refine the reflectance pattern.

The performance of the algorithms depends on the practical applications, and proper choice of the algorithm includes the measure of the calculation speed, diffraction efficiency, beam fidelity, etc. In the following sections, we will demonstrate the shaping of some typical laser modes with various algorithms.

9.3 Tailoring super-Gaussian beam with DMD

The first example of beam sculpturing with DMD is the super-Gaussian beam. The cross-sectional intensity distribution of monochromatic coherent Gaussian beam can be described by the following formula:

$$I_G(r, z) = I_0 \left(\frac{\omega_0}{\omega(z)} \right)^2 \exp \left(-\frac{2r^2}{\omega^2(z)} \right), \tag{9.11}$$

where $I_G(r, z)$ is the intensity of the beam, r and z are the radial and propagation distances, $I_0 = I(0, 0)$ is the intensity at the center of beam waist, and $\omega(z)$ is the z-dependent beam waist.

Consider a super-Gaussian with target profile of the form [20],

$$I_{\text{SG}}(r, z = 0) = I_{\text{SG0}} \exp \left(-2 \left(\frac{r}{\omega_b} \right)^{2N} \right), \tag{9.12}$$

where $I_{\text{SG}}(r, z)$ is the intensity of the super-Gaussian beam, I_{SG0} is the on-axis intensity, N $(N \geqslant 2)$ is the super-Gaussian order, and ω_b is the waist at $1/e^2$.

Amplitude-type deflection patterns are designed to shape the super-Gaussian beam through dividing the super-Gaussian distribution by the Gaussian profile. The reflectance function is generated through $R(x, y) = \mathrm{SL}(x, y)/G(x, y)$. The desired amplitude reflectance, $r_1 = \sqrt{R(x, y)}$, is further optimized by the error-diffusion algorithm.

The DMD modulates the light intensity through sequential modulation of the width of the electric addressing signal. The micromirrors switch promptly to project the continuous holograms and provide a quasi-continuous pattern. Therefore, the output power suggests a nonlinear relationship with the greyscale of the pattern, i.e., the gamma curve. Linear correction is required to correct for the projected hologram and minimize the distortion, such that the output intensity displays a linear relationship with the gray values of the modulation pattern. The display feature of the DMD varies with the correction factor g as experimentally verified from 1.0 to 2.8.

The gamma curve changes from the quasi-exponent profile to a nearly linear shape (figure 9.3). The adjustment of the correction factor does improve the whole gamma curve, and the improvement still shows a good linear shape for the three linear segments: low, middle, and high greyscale regions.

Figure 9.3(b) features the middle section of the nearly linear gamma response when the greyscale takes the optimal value of 2.8 (filled red circle). By contrast, the middle section of the gamma curve with greyscale value of 2.8 and contrast 0.05 is plotted together in figure 9.3(b). The dynamic range of the greyscale value is augmented by fine-tuning the contrast. The full gamma curve is plotted and linearly

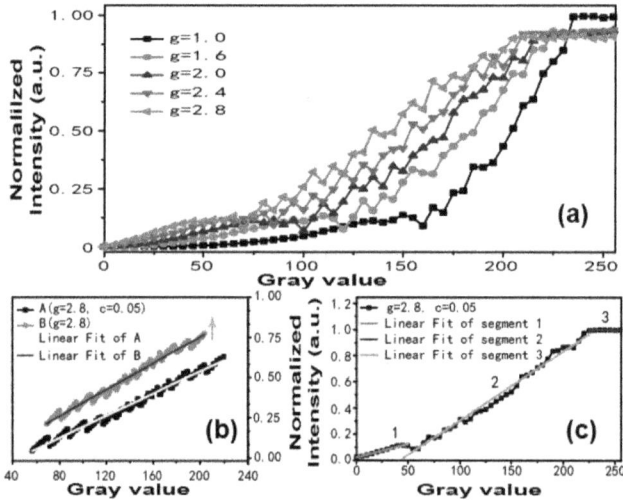

Figure 9.3. The gamma curve correction. (a) Gamma curve evolves from an exponential (black square) to linear (pink left triangle) profile when the greyscale factor g increases from 1.0 to 2.8. (b) The linear fit to the middle part. (c) The gamma curve can be regulated to an empirically good shape, with greyscale and contrast of 2.8 and 0.05, respectively. Adapted by permission from Springer Nature Customer Service Centre GmbH: (Springer) (Science China Physics, Mechanics & Astronomy) [22] Copyright (2014).

fitted with three segments (figure 9.3(c)). The relationship between the light intensity and gray value of the projected images reads

$$y = \begin{cases} 0.02x + 0.25 & 0 \leqslant x \leqslant 50; \\ 0.06x - 2.31 & 51 \leqslant x \leqslant 225; \\ 0.006x + 10.27 & 226 \leqslant x \leqslant 255; \end{cases} \tag{9.13}$$

where y stands for the light intensity and x is the greyscale value.

The DMD modulates the laser light spatially through binary-amplitude modulation. The irradiance of the laser beam flexibly redistributed, and a super-Gaussian beam has been successfully shaped from the Gaussian beam with the use of DMD. A flat-top beam features a uniform intensity in the center with sharp edges. Analytically, it can be modeled as a super-Gaussian shape equation (9.12).

The reflection pattern is first calculated for DMD by dividing the target super-Gaussian distribution with the input intensity profile. With this, a super-Gaussian beam is shaped successfully. Figure 9.4(a) shows the sectional images of the super-Gaussian beam with the intensity profiles along the dashed lines overlaid. Figure 9.4 (b) demonstrates the 1D fit to the intensity profile with the beam radius and order of $\omega(0) = 1.6$ mm and $n = 2$. The 3D pseudo-color images for figure 9.4(a) are provided in figure 9.4(c). The beam profiles at different propagation locations shown in figure 9.4(d) suggests the diffraction against propagation of the tailored super-Gaussian beam in free space. The pixel grid effect causes the fluctuations in figures 9.4(b), (e), and (h) and can be further eliminated.

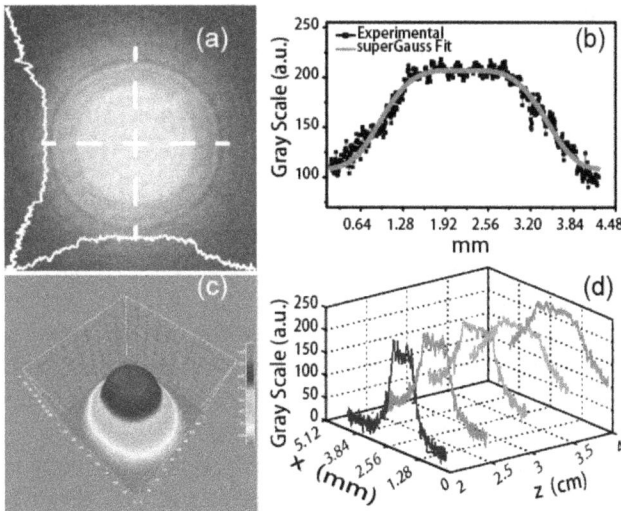

Figure 9.4. The super-Gaussian beam. (a) Cross-sectional image of the beam at a propagation distance of 3.5 cm. The white curves represent the beam profile along the orthogonal directions marked by the dashed lines. (b) The super-Gaussian fit ($n = 2$) to the beam profile (squares). (c) 3D pseudo-color intensity distribution. (d) Linear profile along the propagation direction. Adapted by permission from Springer Nature Customer Service Centre GmbH: (Springer) (Science China Physics, Mechanics & Astronomy) [22] Copyright (2014).

9.4 Tailoring perfect vortex beam with DMD

The optical vortex suggests a ring-shaped intensity profile, and the diameter of the vortex highly depends on the topological charge. In various applications, e.g., optical trapping, one needs larger topological charge but does not want the beam size to depend on the topological charge. The perfect vortex would satisfy both requirements, for which the complex amplitude is [22]

$$E_l(\rho, \theta) = \delta(\rho - \rho_0) \exp(il\theta), \tag{9.14}$$

where (ρ, θ) are the polar coordinates, δ demonstrates the Dirac delta function, ρ_0 is the radius of the main annulus, and l is the topological charge (TC). The electric field of the perfect vortex beam can be expanded using Bessel series [22]:

$$E_l(\rho, \theta) = \exp(il\theta)\text{circ}(\frac{\rho}{\rho_0})\sum_{n=1}^{+\infty}\frac{J_l\left(\alpha_{l,n}\frac{\rho_0}{a}\right)}{[J_{l+1}(\alpha_{l,n})]^2}J_l\left(\alpha_{l,n}\frac{\rho}{a}\right), \tag{9.15}$$

$$\text{circ(r)} = \begin{cases} 1, & r \leqslant 1 \\ 0, & r > 1 \end{cases}, \tag{9.16}$$

where a is the upper boundary of the radial coordinate ρ, and $\alpha_{l,n}$ are the zero points of the Bessel functions. Practically, finite terms in equation (9.15) approximate to the perfect vortex. However, distortion may appear due to strong dependence on the topological charge l, although the amplitude of the perfect vortex should not be related to topological charge (equation (9.14)). Approximating the Dirac function with a narrow-width Gaussian function, the perfect vortex is approximated by a finite width annulus ring with transverse amplitude profile [23],

$$E_l(\rho, \theta) = \exp\left(-\frac{(\rho - \rho_0)^2}{\Delta\rho^2}\right)\exp(il\theta), \tag{9.17}$$

where ρ_0 and $\Delta\rho$ respectively denotes the radius and width of the annulus.

The generation of the vortex beam depends on the modulation of light with a vortex phase. However, the amplitude and phase of the Bessel vortices strongly depend on the topological charge. In many applications, especially the micro-manipulation, a narrow dark hollow beam with a large topological charge is strongly needed to increase the rotation speed of microparticles [22, 23]. The transverse intensity of a perfect vortex has a single ring with its radius independent on the topological charge [5].

The full amplitude and phase of the Gaussian apodized perfect vortex are created through both the Lee method and the super-pixel methods. Figure 9.5 shows the theoretical and experimental outcome using Lee and super-pixel methods for topological charge $l = 20$. The resultant intensity and phase distribution for a binary-amplitude hologram with the Lee method are demonstrated in figures 9.5(c) and (d), respectively. Figure 9.5(e) shows the profile of the experimental perfect vortex beam. Figures 9.5(f), (g) and (h) show the intensity, phase, and intensity

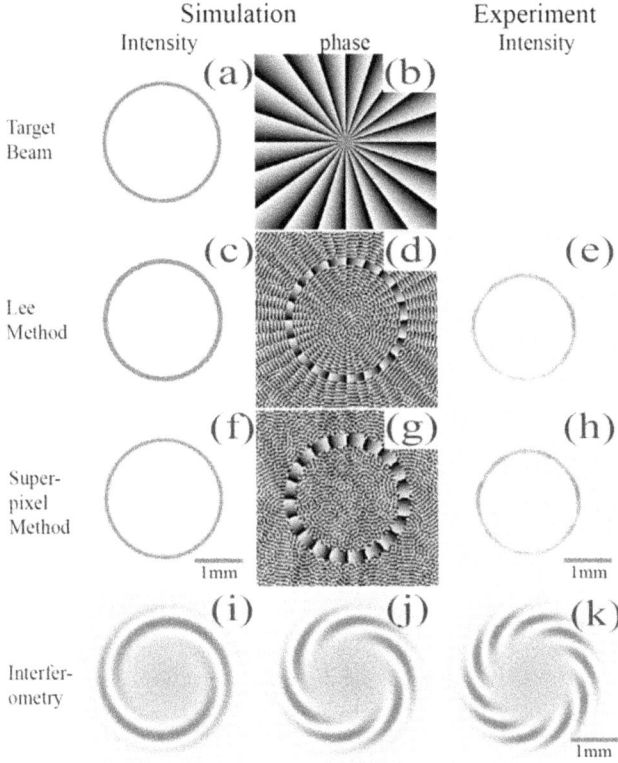

Simulation   Experiment

Figure 9.5. Theoretical and experimental transverse profiles of the perfect vortex with topological charge $l = 20$, $\rho_0 = 1$ mm, $\Delta\rho = 0.05$ mm. Transverse (a) intensity and (b) phase of the perfect vortex. Numerically reconstructed (c) intensity, (d) phase, and (e) the measurement of the perfect vortex through the Lee method. Numerical (f) intensity, (g) phase, and (h) measurement with the super-pixel method. (i–k) The interference pattern of Gaussian beam and perfect vortices with $l = 2, 6, 10$, respectively. Adapted with permission from [24] © 2015 Optical Society of America.

profiles with the super-pixel method. Both the Lee and super-pixel methods could be employed to create the perfect vortex with higher fidelity and efficiency for topological charges as large as 90.

The interferograms for a perfect vortex and Gaussian beam are shown in figures 9.5(i)–(k). The spiral fringe pattern confirms the presence of OAM in the perfect vortex. The number of spirals signifies the order or topological charge of the beam, and the rotation direction is determined by the sign of TC [12].

The angular spectrum for the perfect vortex is

$$G(f_x, f_y, 0) = \iint_{-\infty}^{+\infty} E_l(x, y, 0) \exp\left[-i2\pi(f_x x + f_y y)\right] \mathrm{d}x\mathrm{d}y \qquad (9.18)$$

where $G(f_x, f_y, 0)$ represent the angular spectrum at $z = 0$, f_x, f_y) is the spatial angular frequency. As the beam propagates along the z axis, the angular spectrum at the z-plane is

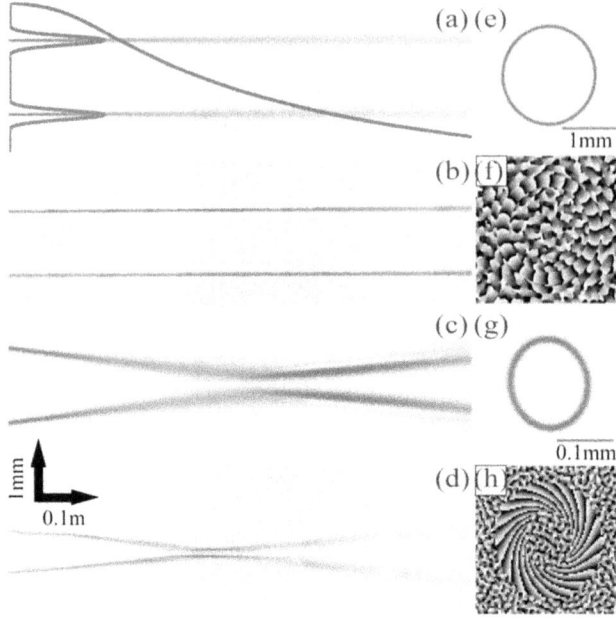

Figure 9.6. Longitudinal intensity profile for perfect vortex with $l = 20$, $\rho_0 = 1$ mm, $\Delta\rho = 0.05$ mm. (a) theoretical and (b) experimental intensity profile for the perfect vortex, and the transverse (e) intensity and (f) phase of the beam at exit plane with $z = 1$ m, scale bar 1 mm. The (c) theoretical and (d) experimental longitudinal profile near the focus. The transverse (g) intensity and (h) phase of the beam at the focus, scale bar 0.1 mm. Adapted with permission from [24] © 2015 Optical Society of America.

$$G(f_x, f_y, z) = G(f_x, f_y, 0) \exp\left(i2\pi z\sqrt{f^2 - f_x^2 - f_y^2}\right), \tag{9.19}$$

The transverse Electric field at z plane is

$$E_l(x, y, z) = \iint_{-\infty}^{+\infty} G(f_x, f_y, z) \exp\left[i2\pi(f_x x + f_y y)\right] \mathrm{d}f_x\, \mathrm{d}f_y. \tag{9.20}$$

The propagating field for a perfect vortex with topological charge $l = 20$ is shown in figure 9.6(a). The experimental measurement is corroborated by simulation (figure 9.6(b)). The geometric parameters are $\rho_0 = 1$ mm, $\Delta\rho = 0.05$ mm, $z = 1$ m. The transverse intensity and phase profiles at the exit plane (figures 9.5(e) and (f)) suggests a good spiral phase and amplitude within a thin ring for a very long distance.

The complex amplitude of the perfect vortex passing a focusing lens reads,

$$E_l(\rho, \theta, 0) = \exp\left(-\frac{(\rho - \rho_0)^2}{\Delta\rho^2}\right) \exp\left(il\theta\right) \exp\left(-ik\frac{x^2 + y^2}{2f}\right), \tag{9.21}$$

where f represents the focal length of the lens. By concentrating the perfect vortex beam with a lens ($f = 50$ mm), the focusing property is characterized by measuring the propagation field near the focus. Both the simulation (figure 9.6(c)) and experiment (figure 9.6(d)) suggest that the focusing field of the perfect vortex exhibits a narrow hollow channel. The intensity (figure 9.6(g)) and phase (figure 9.6(h)) maps at the focal plane corroborate the ring-like intensity profile and the topological phase structure.

The perfect vortex is characterized by its dark hollow and stable radius independent on its topological charge. The stable transmission and focusing property of the beam are investigated both experimentally and numerically. The perfect vortex keeps a considerably high fidelity for a topological charge as large as 90. Large topological charge means that the photons in the perfect vortex beam possess high a magnitude of OAM.

9.5 Tailoring the vortex Hermite–Gaussian beam

Mathematically, the laser modes are solutions to the scalar Helmholtz equation [25], $(\nabla^2 + k^2)E = 0$. Due to symmetry of the laser cavity, the scalar Helmholtz equation has various solutions, e.g., the Hermite–Gaussian (HG) modes in Cartesian coordinates, Laguerre–Gaussian (LG) modes in cylindrical coordinates, Mathieu modes and Ince–Gaussian (IG) modes in elliptic coordinates, and Webber modes in parabolic coordinates. Those modes form complete sets of orthogonal bases, and the arbitrary spatial modes in respective coordinates could be decomposed by series of the fundamental modes. Hermite–Gaussian (HG) mode, as a fundamental transverse electromagnetic mode, can be generated either inside the cavity or outside the laser cavity through beam shaping. With the fast switch ability, the DMD can also shape the HG beam using binary encoding of the beam pattern.

The HG modes are eigenfunctions in orthogonal coordinates. The time-varying paraxial solution in Cartesian coordinates for the electric field has the form $E(x, y, z, t) = A(x, y, z)\exp[i(kz - \omega t)]$. By applying the slowly varying envelope approximation, $\frac{\partial^2 A}{\partial z^2} \ll k^2 A$, $\frac{\partial^2 A}{\partial z^2} \ll k\frac{\partial A}{\partial z}$, the Hermite–Gauss HG_{mn} modes can be solved by separation of variables in x and y coordinates. Mathematical form for these solutions are

$$A(x, y, z) = A_0 H_m\left(\frac{\sqrt{2}x}{\omega(z)}\right) H_n\left(\frac{\sqrt{2}y}{\omega(z)}\right) \frac{\omega_0}{\omega(z)} \exp\left[-i\varphi_{mn}(z)\right] \exp\left[\frac{ik}{2q(z)}r^2\right], \quad (9.22)$$

where A_0 is a constant electric field amplitude, ω_0 is the beam waist radius, $\omega(z)$ is the beam radius at position z, $z_R = \pi\omega_0^2/\lambda$ is the Rayleigh range, $q(z) = z - iz_R$ is the complex beam parameter, $\varphi_{mn}(z) = (m + n + 1)\tan^{-1}(z/z_R)$ is the Gouy phase shift, and $H_m(x)$ denotes the Hermite polynomials that satisfy the differential equation

$$\frac{d^2 H_m}{dx^2} - 2x\frac{dH_m}{dx} + 2mH_m = 0. \quad (9.23)$$

The solutions for the differential equation (9.23) are

$$H_m(x) = (-1)^m e^{x^2} \frac{d^m}{dx^m} e^{-x^2}. \tag{9.24}$$

The first four Hermite polynomials are $H_0(x) = 1$, $H_1(x) = 2x$, $H_2(x) = 4x^2 - 2$, and $H_3(x) = 8x^3 - 12x$. Higher-order Hermite polynomials are determined by the recursion relation $H_{m+1}(x) = 2xH_m(x) - H'_m(x)$.

When $|x - u| \ll z$, the Fresnel transform describes the beam propagation along the optical axis z

$$A(x, z) = \sqrt{\frac{-ik}{2\pi z}} \int_{-\infty}^{+\infty} A(u, 0) \exp\left(\frac{ik}{2z}(u - x)^2\right) du, \tag{9.25}$$

where z is the separation between the two planes, u and x are the transverse coordinates in the initial and final planes, respectively, A is the complex amplitude and k is the wavenumber. Assume the beam waist is at the origin of the z coordinate, the complex amplitude of the 1D HG beam at origin ($z = 0$) is,

$$A(u, 0) = H_n\left(\frac{\sqrt{2}u}{\omega_x}\right) \exp\left[-\left(\frac{u}{\omega_x}\right)^2\right], \tag{9.26}$$

where a is a real number. Inserting equation (9.26) into equation (9.25) yields

$$A(x, z) = \sqrt{-\frac{ik}{2\pi z}} \int_{-\infty}^{+\infty} H_n\left(\frac{\sqrt{2}u}{\omega_x}\right) \exp\left[\frac{ik}{2z}(u - x)^2 - \frac{u^2}{\omega_x^2}\right] du. \tag{9.27}$$

Applying the reference integral (integral 7.374.8 in the reference [27]), $\int_{-\infty}^{+\infty} \exp(-(x - y)^2) H_n(\alpha x) dx = \sqrt{\pi}(1 - \alpha^2)^{n/2} H_n(\alpha y \sqrt{1 - \alpha^2})$, the diffraction integral could be simplified as

$$\begin{aligned} E_n(x, z) = i^n \frac{\omega_x}{\omega_x(z)} H_n\left[\frac{\sqrt{2}x}{\omega_x(z)}\right] \exp\left[-\frac{x^2}{\omega_x^2(z)}\right] \\ \times \exp\left[\frac{ikx^2}{2R(z)} - i(n + \frac{1}{2})\arctan\left(\frac{z}{z_R}\right)\right], \end{aligned} \tag{9.28}$$

where $z_R = k\omega_x^2/2$, $\omega_x(z) = \omega_x[1 + (z/z_R)^2]^{1/2}$, $R(z) = z[1 + (z_R/z)^2]$. Equation (9.28) analytically describes the propagation behavior of the 1D HG mode. Figure 9.7(j) shows the experimental setup to map the propagation of HG mode. Figures 9.7(k) and (l) display the propagating field reconstructed from a series of sectional images with an axial step of 1 mm in the x–z and y–z planes, respectively. The fitted beam radii are plotted with round dots in figure 9.7(m), and the insets show the sectional beam profiles at the respective coordinates. The Rayleigh range and the beam waist radius were determined through the fitting as $z_R = 13.75$ mm and $b = 46$ μm, respectively.

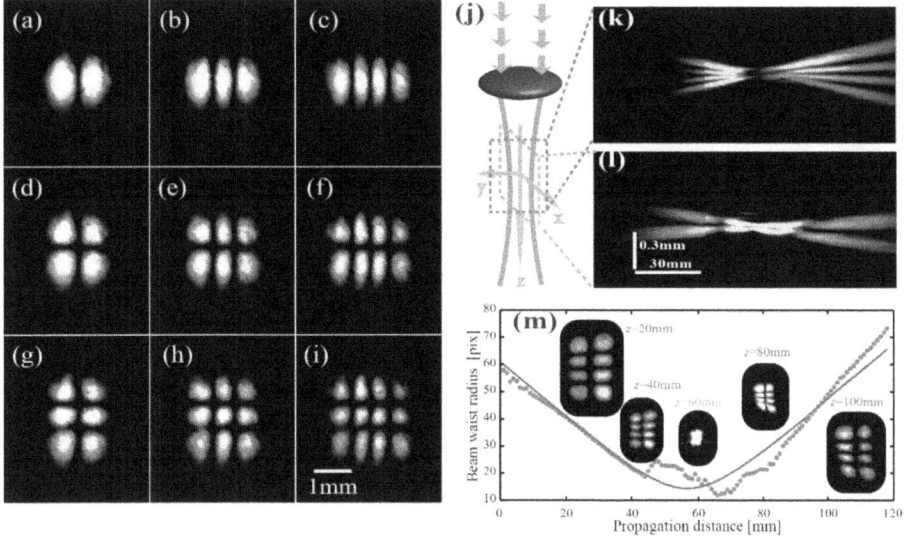

Figure 9.7. Sculpturing HG beam with DMD. Transverse profiles for (a) $HG_{1,0}$, (b) $HG_{2,0}$, (c) $HG_{3,0}$, (d) $HG_{1,1}$, (e) $HG_{2,1}$, (f) $HG_{3,1}$, (g) $HG_{1,2}$, (h) $HG_{2,2}$, (i) $HG_{3,2}$ modes. Scale bars for (a–i) are shown in (i). (j) Schematic of the setup for 3D intensity mapping. (k) Sideview profile for in the plane covering the left four major lobes. (l) Sideview profile across the centre of the two major lobes in the second line on the transverse pattern. Scale bars are identical for (k) and (l). (m) The beam radius as function of propagation distance. Insets show the transverse profile at the respective locations. Reproduced from [26]. © 2015 IOP Publishing Ltd.

Apart from the spatial modes, the photons have two types of angular momenta: the spin angular momentum (SAM) and the orbital angular momentum (OAM). The HG mode encoded with OAM suggests a quasi-continuously deformed shape.

We construct the elliptical vortex HG (vHG) mode through the combination of HG modes with the following form [11]:

$$A_n(x, y, 0) = B \exp\left(-\frac{x^2}{\omega_x^2} - \frac{y^2}{\omega_y^2}\right)(1 + a^2)^{-\frac{n}{2}}$$

$$\times \sum_{p=0}^{n} \frac{n!(ia)^p}{p!(n-p)!} H_p\left(\frac{\sqrt{2}x}{\omega_x}\right) H_{n-p}\left(\frac{\sqrt{2}y}{\omega_y}\right), \quad (9.29)$$

where a is a real parameter. By applying the reference series, $\sum_{p=0}^{n} \frac{n!t^p}{p!(n-p)!} H_p(\xi) H_{n-p}(\eta) = (1 + t^2)^{\frac{n}{2}} H_n(\frac{t\xi + \eta}{\sqrt{1+t^2}})$, the complex amplitude of the vHG mode (equation (9.29)) is simplified as

$$A_n(x, y, 0) = B \exp\left(-\frac{x^2}{\omega_x^2} - \frac{y^2}{\omega_y^2}\right) \times \left(\frac{1 - a^2}{1 + a^2}\right)^{\frac{n}{2}} H_n\left(\sqrt{2}\frac{ia\omega_y x + \omega_x y}{\omega_x \omega_y \sqrt{1 - a^2}}\right). \quad (9.30)$$

The projection of orbital angular momentum on the optical axis is derived through

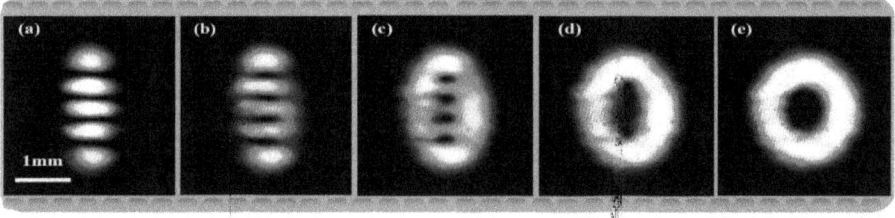

Figure 9.8. Smooth variation of the OAM. The orbital angular momenta J_z/I are (a) 0, (b) −1.88, (c) −3.20, (d) −3.84, (e) −3.99, respectively. The topological index n adopts 4 in the experiment. Reproduced from [26]. © 2015 IOP Publishing Ltd.

$$J_z = -\mathrm{i} \int\int_R A_n^* \left(x\frac{\partial A_n}{\partial y} - y\frac{\partial A_n}{\partial x} \right) \mathrm{d}x\mathrm{d}y. \tag{9.31}$$

Substituting equation (9.30) into equation (9.31) yields $J_z = -\pi B^2 2^{n-1} n! na (1+a^2)^{-1}(\omega_x^2 + \omega_y^2)$. The power of the beam is $I = \int\int_{R^2} A_n^* A_n \mathrm{d}x\mathrm{d}y = \pi B^2 2^{n-1} n! \omega_x \omega_y$. The normalized OAM of the vHG beam with respect to the laser power is

$$\frac{J_z}{I} = \left(\frac{-na}{1+a^2} \right)\left(\frac{\omega_x^2 + \omega_y^2}{\omega_x \omega_y} \right). \tag{9.32}$$

This relationship describes OAM independence of the laser power. When $\omega_x = \omega_y = \omega_0$, the elliptic vHG beam reduces to circular vHG, and

$$\frac{J_z}{I} = \frac{-2na}{1+a^2}. \tag{9.33}$$

The vortex HG mode can be produced through binary encoding. As demonstration, we show vortex HG mode with modal index $n = 4$. It is feasible to regulate the vortex HG shape, and control the projection of OAM on the optics axis.

Due to the fast switching rate and flexibility for refreshing the pattern, the rotation of the spatial modes could be realized with a frame rate over 22 kHz. We further demonstrated the dynamic control over the shape through creating the vortex HG (vHG) modes with increased OAM. To understand the optical vortex singularity, the dynamic deformation from a traditional HG beam to a vortex HG beam provides a deep insight to benefit across singular optics and optical manipulation. The deformed spatial vHG mode creates a beam with arbitrary real magnitude of OAM. Figures 9.8(a)–(e) show typical frames extracted from the movie. The topological index n adopts 4, and the orbital angular momenta J_z/I are (a) 0, (b) −1.88, (c) −3.20, (d) −3.84, (e) −3.99, respectively.

Although the z-projection of OAM varies with a, the total OAM conserves during the deformation. The shape of the beam becomes flatter in the vertical direction for large a, when the value of a approaches 1. Moreover, all the n isolated nulls are situated at the beam center (figure 9.8(e)), and such a mode resembles the LG mode with order $(0, n)$. Similarly, the vHG mode is reduced to HG mode with order $(0, n)$ when $a = 0$.

9.6 Tailoring vortex Ince–Gaussian beam with DMD

The paraxial wave equation (PWE), $\nabla_\perp^2 \Psi + 2ik\partial_z\Psi = 0$, is an approximate to the Helmholtz equation for traveling waves. In elliptic coordinates, the solution for this PWE could be constructed in a modulated version of the Gaussian beam,

$$\text{IG}(\boldsymbol{r}) = E(\xi)N(\eta)\exp{(iZ(z))}\Psi_G(\boldsymbol{r}), \qquad (9.34)$$

where E, N, Z are real functions of the space coordinates, $\Psi_G(\boldsymbol{r})$ is the complex amplitude of a Gaussian beam in Cartesian coordinates, and IG represents Ince–Gaussian mode.

The elliptic coordinates (ξ, η) relate with the Cartesian coordinates [28], $x = f(z)\cos h(\xi)\cos(\eta)$, $y = f(z)\sin h(\xi)\sin(\eta)$, where $\xi \in [0, +\infty)$ and $\eta \in [0,2\pi)$ are the radial and angular elliptic coordinates. $f(z) = f_0\omega(z)/\omega_0$ defines the semi-focal separation. In elliptic coordinate, the curves with constant ξ are confocal ellipses, and curves with constant η are confocal hyperbolas.

By inserting the trial solution in the PWE, the functions $E(\xi)$ and $N(\eta)$ satisfy a unified Ince equation, which is a special case of the Hill equation [29, 30],

$$\frac{\text{d}^2X}{\text{d}\eta^2} + \varepsilon\sin{(2\eta)}\frac{\text{d}X}{\text{d}\eta} + (a - p\varepsilon\cos{(2\eta)})X = 0. \qquad (9.35)$$

Solutions for equation (9.35) are trigonometric series with unknown coefficients, which are determined by recurrence relations [29]. For simplicity, the solutions are expressed in even and odd Ince polynomials $C_p^m(\eta, \varepsilon)$ and $S_p^m(\eta, \varepsilon)$, respectively. The order p, degree m, and ellipticity parameter ε determine the transverse beam profile. The Ince polynomials are normalized through the trigonometric functions [30], i.e.,

$$\int_{-\pi}^{\pi}\left[C_p^m(\eta, \varepsilon)\right]^2 d\eta = \int_{-\pi}^{\pi}\left[S_p^m(\eta, \varepsilon)\right]^2 d\eta = \pi. \qquad (9.36)$$

For even Ince polynomials, $0 \leqslant m \leqslant p$, while for odd Ince polynomials, $1 \leqslant m \leqslant p$, the indices m and p have the same parity, e.g., $(-1)^{m-p} = 1$. We define another parity parameter δ, $\delta = 0$ represents even Ince–Gaussian beam; while for odd Ince–Gaussian beam, $\delta = 1$.

The even and odd Ince–Gaussian modes with mode numbers p, m, and ellipticity parameter ε could be respectively reconstructed as

$$\text{IG}_{p, m}^e(\xi, \eta, z; \varepsilon) = \frac{C\omega_0}{\omega(z)}C_p^m(i\xi, \varepsilon)C_p^m(\eta, \varepsilon) \times \exp\left[-\frac{ikr^2}{\omega(z)} - (p + 1)\Psi_{GP}(z)\right], \quad (9.37)$$

$$\text{IG}_{p, m}^o(\xi, \eta, z; \varepsilon) = \frac{S\omega_0}{\omega(z)}S_p^m(i\xi, \varepsilon)S_p^m(\eta, \varepsilon) \times \exp\left[-\frac{ikr^2}{\omega(z)} - (p + 1)\Psi_{GP}(z)\right], \quad (9.38)$$

where $\Psi_{GP}(z) = \arctan(z/z_R)$ is the Guoy phase shift, C and S are normalization constants, and $z_R = \pi\omega_0^2/\lambda$ is the Rayleigh length of the beam with waist radius ω_0.

The ellipticity ε, waist radius ω_0, and semifocal separation f_0 are associated through $\varepsilon = 2f_0^2/\omega_0^2$. The Ince–Gaussian mode is jointly determined by the mode order and the ellipticity. The Laguerre–Gaussian and Hermite–Gaussian beams are two complete sets of basis. Ince–Gaussian modes (IGMs) constitute another complete set of transversal eigenmodes for paraxial Helmholtz equation, and provide a transition mode set for the basis in Cartesian and cylindrical coordinates. When $\varepsilon = \infty$ and $\varepsilon = 0$, the IGMs will be reduced to the Hermite–Gaussian and Laguerre–Gaussian modes, respectively.

The first row in figure 9.9 demonstrates the experimental profiles for even IGM $IG_{7,5}^e$, and the second row shows patterns for the odd IGM $IG_{8,6}^o$. The ellipticities are (a,d) $\varepsilon = 0.8$, (b,e) $\varepsilon = 10$, and (c,f) $\varepsilon = 1000$. The sectional images shown in figures 9.9(a)–(d) resemble on the $LG_{p=1}^{l=5}$ and $LG_{p=1}^{l=6}$ modes, respectively, while those shown in figures 9.9(c) and (f) are similar to the $HG_{m=5}^{n=2}$ and $HG_{m=5}^{n=3}$ modes in sequence.

In elliptic coordinates, laser beams with OAM could be synthesized through superposition of the IGMs. For instance, the helical Ince–Gaussian modes (HIGM) are constructed through the superposition of even and odd IGMs,

$$\text{HIG}_{p,m}^{\pm}(\xi, \eta; \varepsilon) = \text{IG}_{p,m}^e(\xi, \eta; \varepsilon) \pm i\text{IG}_{p,m}^o(\xi, \eta; \varepsilon), \qquad (9.39)$$

whose phase rotates elliptically around the long axis of the ellipse. The rotation direction is defined by the sign in equation (9.39). Figure 9.10 shows the transverse profiles of the $IG_{10,6}(\xi, \eta; \varepsilon)$ at the beam waist. The ellipticities in each row are $\varepsilon = 2, 10, 100$, respectively. Each column corresponds to the even, odd, and helical $IG_{10,6}$ in sequence. For helical modes, the central vortex splits into m vortices. As ε increases, new vortices emerge in the outer rings of the pattern (figures 9.10(c), (f), and (i)). All the vortices on the outer rings have a topological charge of 1.

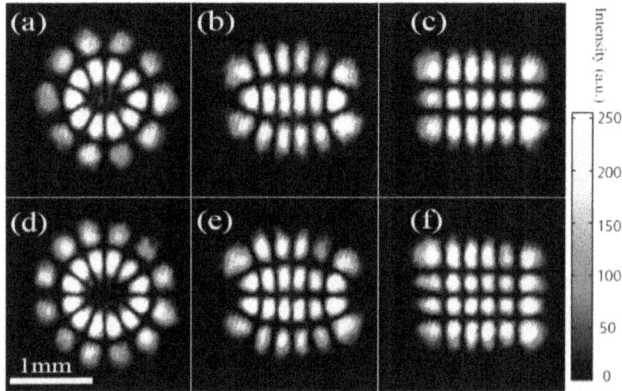

Figure 9.9. The intensity profiles for the even (first row, $IG_{7,5}^e$) and odd (second row, $IG_{8,6}^o$) IGMs with (a, d) $\varepsilon = 0.8$, (b, e) $\varepsilon = 10$, and (c, f) $\varepsilon = 1000$, respectively. Reproduced with permission from [11].

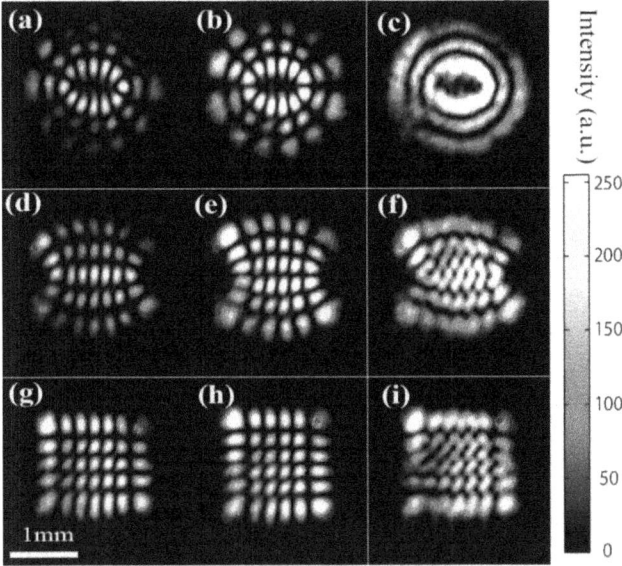

Figure 9.10. Transverse profiles for even (a, d, g), odd (b, e, h) and helical (c, f, i) IGMs with $p = 10$ and $m = 6$. The ellipticities in each row are (a, b, c) $\varepsilon = 2$, (d, e, f) $\varepsilon = 10$, and (g, h, i) $\varepsilon = 100$ respectively. Reproduced with permission from [11].

Similarly, a Mathieu Gauss beam constitutes another set of vector modes obtained from the Helmholtz equation in elliptical cylindrical coordinates [31, 32]. The solution can be expressed as even and odd Mathieu beams. The Mathieu beams are also high-order nondiffracting beams that excite interesting applications in cold atoms. The creation of helical IGMs demonstrates the ability of using DMD to tailor the light field to possess orbital angular momentum. This provides a means to study the local angular momentum states in elliptic coordinates and even the HG beams in Cartesian coordinates.

9.7 Tailoring the vortex Airy beam with DMD

Berry and Balazs predicted a nondiffracting Airy-like wave packet from the Schrödinger equation [33, 34]. The experimentally generated Airy-like beam was realized in optics by modulating the cubic phase in the spectrum space with a liquid-crystal spatial light modulator (LC-SLM) [35, 36]. The Airy wave packet was initially predicted by Berry and Balazs through deducing a special solution for the Schrödinger equation. By truncating the wave packet, the finite-energy Airy beam was first experimentally realized in the form of optics. The normalized paraxial equation of diffraction is [35]

$$i\frac{\partial u}{\partial \xi} + \frac{1}{2}\frac{\partial^2 u}{\partial s^2} = 0, \tag{9.40}$$

where u denotes the electric field envelope, $s = x/x_0$ is a dimensionless transverse coordinate, x_0 represents an arbitrary transverse scale, $\xi = z/kx_0^2$ is a normalized distance in the direction of propagation, and $k = 2\pi n/\lambda_0$ is the wavenumber in a vacuum. Experimental Airy beams are with finite power, and this is fulfilled by adding an exponentially decaying aperture function [35]. Such a truncated Airy beam in the transverse plane is described as follows [28]:

$$u(s, \xi = 0) = Ai(s)\exp{(as)}, \tag{9.41}$$

where a is the decaying factor to truncate the tail of the infinite Airy beam, and $Ai(s)$ represents the Airy function, whose integral form is $Ai(\eta) = 1/\pi \int_0^\infty \cos(t^3/3 + \eta t)dt$. The corresponding Fourier spectrum in the spectral space is expressed as

$$U_0(k) = \exp{(-ak^2)}\exp\left(\frac{i}{3}(k^3 - 3a^2k - ia^3)\right), \tag{9.42}$$

where k stands for a vector in the normalized k-space. The above equation demonstrates that the finite-energy Airy beam can be produced by modulating a Gaussian beam with a cubic phase. This implies that the Fourier spectrum of the finite Airy beam involves a Gaussian amplitude envelope and a cubic phase. The 1D Airy beam is experimentally obtained by modulating the incident Gaussian beam with a cubic phase spectrum, e.g., $\varphi_0(k) = \exp{(\frac{i}{3}k^3)}$. The counterpart for the 2D Airy beam is with the form of $\varphi_0(k_x, k_y) = \exp{(\frac{i}{3}(k_x^3 + k_y^3))}$. Therefore, the truncated Airy beam can be experimentally generated through modulating a Gaussian beam with a cubic phase [35, 37]. However, by replacing the spectral coordinates with the absolute value in the spectral cubic phase term, i.e., $\varphi(k_x,k_y) = \exp{(i(|k_x|^3 + |k_y|^3)/3)}$, we can produce a 2D SAB with its intensity distribution exhibiting four major lobes situated on corners of a square [38].

The cubic phase mask for creating the 1D Airy beam and the intensity of a theoretically predicted 1D Airy beam are respectively demonstrated in figures 9.11 (a) and (b). 2D versions of the cubic phase mask as well as the theoretical intensity distributions are shown in figures 9.11(c) and (d) in sequence. By replacing the spatial coordinates with the absolute value, the cubic phase mask exhibits rectangular symmetry (figure 9.11(e)). Such a symmetric cubic phase mask shapes the coherent collimated laser beam into SAB. The beam initially goes through a process of autofocus, and then its transversal profile spreads in the propagation space with a shape of four main lobes situated on corners of a square (figure 9.11(f)). A binary version of the computer-generated Lee hologram is demonstrated in figure 9.11(g).

The SAB is a caustic beam with a 'cusp structure', and the dynamics of such a cusp structure is governed by the three-ray superposition within the caustic region. The caustic classification arises from the phase symmetry rather than from the phase power, thus breaking the commonly accepted concept that fold and cusp caustics are related to the Airy and Pearcey functions, respectively [39]. Despite the qualitative similarities between HGs and SABs in the far field, the former has no autofocusing

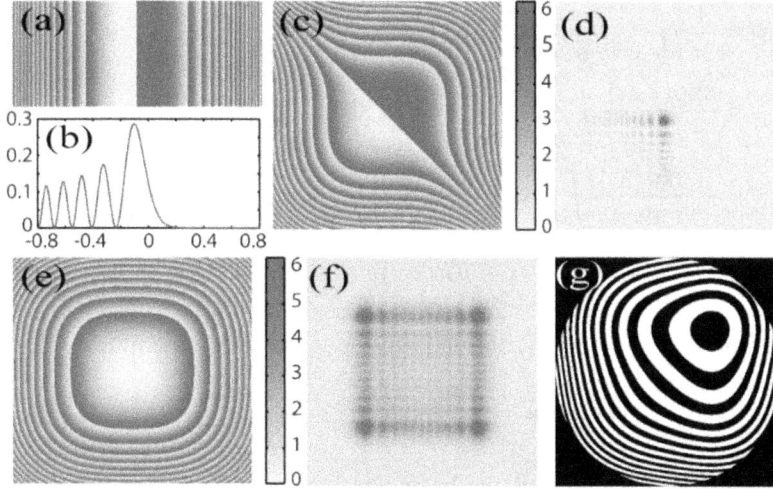

Figure 9.11. The cubic phase and beam profiles for SAB. The 1D (a) cubic phase and (b) Airy beam. The 2D (c) cubic phase and (d) Airy beam. (e) Cubic phase, and (f) SAB are encoded in a (g) binary Lee hologram. Adapted with permissions from [41] © 2021 Optica Publishing Group.

feature, which is a key characteristic of a cusp catastrophic structure due to the superposition of three rays. The optical needle beam is such a kind of focus featuring the narrow electromagnetic filed distribution and the elongated longitudinal focal line [40].

The autofocusing SAB was tailored in the spatial spectrum domain through even cubic phase and displays a single on-axis lobe at the focus. The SAB splits into four off-axis identical major lobes with outward acceleration. The resultant beam is given by,

$$u(x, y, z) = \frac{1}{2\pi} \iint_{-\infty}^{+\infty} \exp\left(-a\left(k_x^2 + k_y^2\right)\right) \exp\left(\frac{i}{3}(|k_x|^3 + |k_y|^3)\right)$$
$$\exp\left(iz\sqrt{k^2 - k_x^2 - k_y^2}\right) \exp\left(i(k_x x + k_y y)\right) dk_x dk_y, \tag{9.43}$$

The creation of complex amplitude relies on the encoding of the field with a hologram. Such technique is usually based on off-axis holography and computer algorithms [42, 43].

Even symmetric phase in spectral space induces the angular spectrum of 1D SAB [25],

$$U_1(k) = \exp\left(-ak^2\right) \exp\left(\frac{i}{3}(|k|^3 - 3a^2|k| - ia^3)\right). \tag{9.44}$$

The inverse Fourier transform of equation (9.44) determines the complex amplitude of 1D SAB [44],

$$u_1(s) = \frac{1}{2\pi} \int_{-\infty}^{+\infty} U_1(k) \exp{(\mathrm{i}ks)}\mathrm{d}k$$

$$= \frac{\exp(-as)}{2\pi} \int_0^\infty \exp\left(\frac{\mathrm{i}k^3}{3} - ks\right)\mathrm{d}k$$

$$+ \frac{\exp{(as)}}{2\pi} \int_\infty \exp\left(\frac{\mathrm{i}k^3}{3} + ks\right)\mathrm{d}k \qquad (9.45)$$

$$= \frac{\exp(-as)}{2}[Ai(-s) + iGi(-s)] + \frac{\exp{(as)}}{2}[Ai(s) + iGi(s)],$$

where Gi represents a Scorer function [39], the integral representation of which is

$$Gi(\eta) = \frac{1}{\pi} \int_0^\infty \sin\left(\frac{1}{3}t^3 + \eta t\right)\mathrm{d}t. \qquad (9.46)$$

The two-dimensional (2D) SAB can be constructed as $u_2(x, y, z = 0) = u_1(s_x)u_1(s_y)$. The electric field of the 2D SAB superimposed by spiral phase, vSAB, reads:

$$u(x, y) = u_2(x, y) \prod_{j=1}^N [(x - x_j) + i(y - y_j)]^l [(x - x_j)^2 + (y - y_j)^2]^{\frac{l}{2}}, \quad (9.47)$$

where (x_j, y_j) are the coordinates of the jth vortex dislocation, l represents the respective topological charge, and N is the total number of vortices.

The spiral phase superimposed into the autofocusing region enables the SAB beam to form a hollow optical channel. The propagation dynamics of the on-axis VSABs with $l = 1$ and 2 are experimentally verified by tomographically mapping the 3D beam profile (figure 9.12). The on-axis VSAB autofocuses to a bottle-like focal

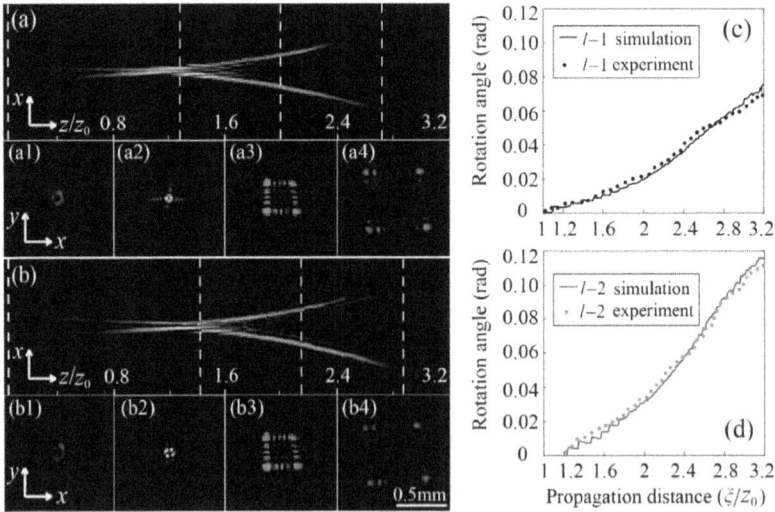

Figure 9.12. On-axis vSAB. (a) Sideview profile for on-axis vSAB with $l = 1$. (a1–a4) The transverse profiles at locations marked with dashed line in (a). (b) The same vSAB as (a) but with topological charge of $l = 2$. The rotation angle as function of propagation distance for (c) $l = 1$, and (d) $l = 2$. Adapted with permission from [44] © 2018 Optical Society of America.

line along propagation. This is corroborated by the beam propagation for an on-axis VSAB with $l = 1$ (figure 9.12). The transverse patterns (figures 9.12(a1)–(a4)) show the beam profiles at the respective locations. Figure 9.12(b) shows the similar results for on-axis SAB with topological charge of $l = 2$. The on-axis VSAB inherits autofocusing property from SAB, but focuses into a doughnut rather than concentrates into a needle (figures 9.12(a) and (b)). The non-zero topological charge shifts the autofocusing plane of the VSAB presenting a non-identical focal position (second dashed lines in figures 9.12(a) and (b)). In general, the autofocusing plane shifts more forward for larger topological charge. The ratios of autofocusing positions between VSAB and SAB are 1.1:1 and 1.25:1 for charges 1 and 2, respectively, as measured in the experiment (figure 9.12).

The interaction between the SAB and a spiral phase induces the self-rotation of the beam pattern (figures 9.12(c) and (d)). The on-axis VSABs rotate anticlockwise more significantly after the focal plane. As the topological charge increases, the rotation effect is more apparent. The rotation direction will be flipped for negative-charge vortices as evidenced by clockwise rotation in the SAB pattern with negative charge for the embedded on-axis vortex (figure 9.13).

The vSAB share the autofocusing and vortex phase structure. The on-axis vortex in SAB creates a hollow focus channel owing to the interplay of topological phase and autofocusing. The transverse pattern near the autofocusing channel rotates during propagation, and the rotation direction can be reversed by changing the sign of the topological charge.

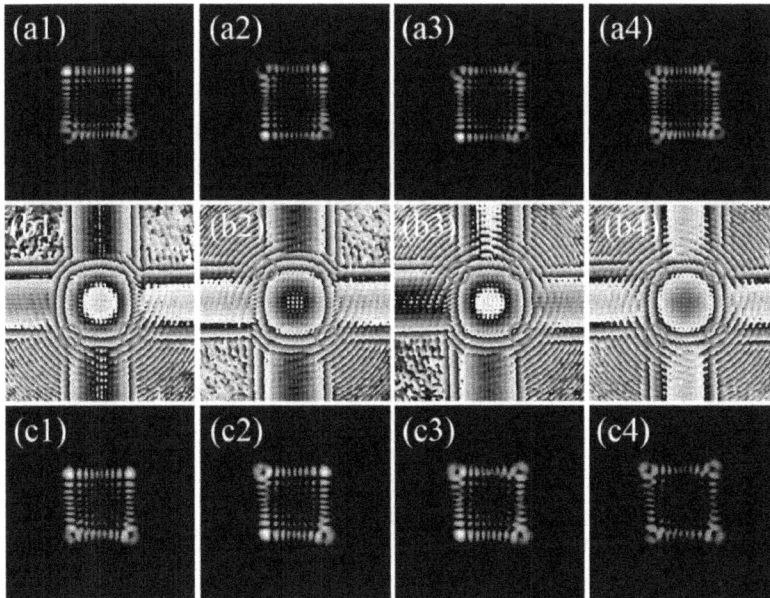

Figure 9.13. Intensity (a), phase (b), and measurement (c) for SAB with multiple vortices. Each column represents the vSAB with (1) lateral vortex pair, (2) diagonal vortex pair, (3) triple vortices, and (4) quadruple vortices. Red circles in (b) mark the phase singularities. Adapted with permission from [44] © 2018 Optical Society of America.

For a single vortex, the off-axis VSAB shows the projection of the vortex into one of the major lobes of the SAB, and one hollow channel generates in the corresponding major lobe. Off-axis VSAB with vortices imbedded on multiple major lobes reveals that the interplay between topological phase and the self-acceleration does suggest a doughnut in the corresponding major lobe. Once the vortices are embedded in the major lobes, they are protected from each other and show no apparent interaction between the vortices among different major lobes.

9.8 Summary

In this chapter, we generally introduce the basic principle of wavefront shaping using DMD, and the algorithms used for shaping some typical spatial modes. The proposed method is not limited to Ince–Gaussian beams, but can be applied to beams with other shapes, such as self-accelerating Mathieu beams and Webber accelerating beams, and the nondiffracting beams for various biomedical engineering applications [45–48]. With the recent advances in metasurface design [49] and fabrication of optical fibers, it is possible to integrate the nondiffracting beam with optical fibers for easy delivery of the nondiffracting beam [50].

The DMD has features such as fast refreshing rate, polarization insensitivity, and binary-amplitude modulation. These features have been demonstrated in laser wavefront shaping to produce high-order laser modes and nondiffracting beams. The DMD offers possibilities to structure light at a considerably fast speed. The traditional vortex and the vortex embedded in high-order spatial modes offers more varieties for diverse interdisciplinary applications.

References

[1] Chang C-M and Shieh H-P D 2000 Design of illumination and projection optics for projectors with single digital micromirror devices *Appl. Opt.* **39** 3202–8
[2] Ren Y-X, Li M, Huang K, Wu J-G, Gao H-F, Wang Z-Q and Li Y-M 2010 Experimental generation of Laguerre–Gaussian beam using digital micromirror device *Appl. Opt.* **49** 1838–44
[3] Ding X-Y, Ren Y-X, Gong L, Fang Z-X and Lu R-D 2014 Microscopic lithography with pixelate diffraction of a digital micro-mirror device for micro-lens fabrication *Appl. Opt.* **53** 5307–11
[4] Sun P, Liu D, Zhang Y, Li X, Zhang Y and Zhu J 2011 Evolution of low-frequency noise passing through a spatial filter in a high power laser system *Sci. China Phys. Mech. Astron.* **54** 411–5
[5] Hornbeck L J 1998 Current status and future applications for DMD-based projection displays *Proc. 5th Int. Display Workshop IDW'98* pp 713–6
[6] Hornbeck L J 1998 From cathode rays to digital micromirrors-a history of electronic projection display technology *Texas Instrum. Tech. J.* **15** 7–46
[7] Van Kessel P F, Hornbeck L J, Meier R E and Douglass M R 1998 A MEMS-based projection display *Proc. IEEE* **86** 1687–704
[8] Bansal V and Saggau P 2013 Digital micromirror devices: principles and applications in imaging *Cold Spring Harb. Protoc.* **2013** 404–11

[9] Perumal L and Forbes A 2023 Broadband structured light using digital micro-mirror devices (DMDs): a tutorial *J. Opt.* **25** 074003

[10] Gong L, Qiu X-Z, Ren Y-X, Zhu H-Q, Liu W-W, Zhou J-H, Zhong M-C, Chu X-X and Li Y-M 2014 Observation of the asymmetric Bessel beams with arbitrary orientation using a digital micromirror device *Opt. Express* **22** 26763–76

[11] Ren Y-X, Fang Z-X, Gong L, Huang K, Chen Y and Lu R-D 2015 Dynamic generation of Ince–Gaussian modes with a digital micromirror device *J. Appl. Phys.* **117** 133106

[12] Dennis M R, O'Holleran K and Padgett M J 2009 Singular optics: optical vortices and polarization singularities *Progress in Optics* ed E Wolf (Amsterdam: Elsevier) pp 293–363

[13] Vaity P and Rusch L 2015 Perfect vortex beam: fourier transformation of a Bessel beam *Opt. Lett.* **40** 597–600

[14] Dickey F M and Holswade S C 1996 Gaussian laser beam profile shaping *Opt. Eng* **35** pp. 3285–95

[15] Dickey F M 2003 Laser beam shaping *Opt. Photon. News* **14** 30–5

[16] Curtis J E and Grier D G 2003 Structure of optical vortices *Phys. Rev. Lett.* **90** 133901

[17] Mirhosseini M, Magaña-Loaiza O S, Chen C, Rodenburg B, Malik M and Boyd R W 2013 Rapid generation of light beams carrying orbital angular momentum *Opt. Express* **21** 30196–203

[18] Goorden S A, Bertolotti J and Mosk A P 2014 Superpixel-based spatial amplitude and phase modulation using a digital micromirror device *Opt. Express* **22** 17999–8009

[19] Liang J, Kohn J R N, Becker M F and Heinzen D J 2009 1.5% root-mean-square flat-intensity laser beam formed using a binary-amplitude spatial light modulator *Appl. Opt.* **48** 1955–62

[20] Mercier B, Rousseau J P, Jullien A and Antonucci L 2010 Nonlinear beam shaper for femtosecond laser pulses, from Gaussian to flat-top profile *Opt. Commun.* **283** 2900–7

[21] Ding X, Ren Y-X and Lu R 2015 Shaping super-Gaussian beam through digital micro-mirror device *Sci. China Phys. Mech. Astron.* **58** 034202

[22] Ostrovsky A S, Rickenstorff-Parrao C and Arrizón V 2013 Generation of the 'perfect' optical vortex using a liquid-crystal spatial light modulator *Opt. Lett.* **38** 534–6

[23] Chen M, Mazilu M, Arita Y, Wright E M and Dholakia K 2013 Dynamics of microparticles trapped in a perfect vortex beam *Opt. Lett.* **38** 4919–22

[24] Chen Y, Fang Z-X, Ren Y-X, Gong L and Lu R-D 2015 Generation and characterization of a perfect vortex beam with a large topological charge through a digital micromirror device *Appl. Opt.* **54** 8030–5

[25] Born M and Wolf E 1997 Electromagnetic theory of propagation, interference and diffraction of light *Principles of Optics* 6th edn (London: Cambridge University Press)

[26] Ren Y-X, Fang Z-X, Gong L, Huang K, Chen Y and Lu R-D 2015 Digital generation and control of Hermite–Gaussian modes with an amplitude digital micromirror device *J. Opt.* **17** 125604

[27] Gradshteyn I M R I S 2007 *Table of Integrals, Series, and Products* (London: Academic)

[28] Schwarz U T, Bandres M A and Gutiérrez-Vega J C 2004 Observation of Ince–Gaussian modes in stable resonators *Opt. Lett.* **29** 1870–2

[29] Ince E L 1925 A linear differential equation with periodic coefficients *Proc. Lond. Math. Soc.* **s2–23** 56–74

[30] Arscott F M 1967 XXI—the Whittaker-Hill equation and the wave equation in paraboloidal co-ordinates *Proc. R. Soc. Edinburgh, Sec.* A **67** 265–76

[31] Gutiérrez-Vega J C, Iturbe-Castillo M D and Chávez-Cerda S 2000 Alternative formulation for invariant optical fields: Mathieu beams *Opt. Lett.* **25** 1493–5

[32] Yang Y, Ren Y-X, Chen M, Arita Y and Rosales-Guzmán C 2021 Optical trapping with structured light: a review *Adv. Photonics* **3** 034001

[33] Berry M V and Balazs N L 1979 Nonspreading wave packets *Am. J. Phys.* **47** 264–7

[34] Unnikrishnan K 1999 Short- and long-time decay laws and the energy distribution of a decaying state *Phys. Rev.* A **60** 41–4

[35] Siviloglou G A, Broky J, Dogariu A and Christodoulides D N 2007 Observation of accelerating Airy beams *Phys. Rev. Lett.* **99** 213901

[36] Siviloglou G A and Christodoulides D N 2007 Accelerating finite energy Airy beams *Opt. Lett.* **32** 979–81

[37] Siviloglou G A, Broky J, Dogariu A and Christodoulides D N 2008 Ballistic dynamics of Airy beams *Opt. Lett.* **33** 207–9

[38] Vaveliuk P, Lencina A, Rodrigo J A and Martinez Matos O 2014 Symmetric Airy beams *Opt. Lett.* **39** 2370–3

[39] Vaveliuk P, Lencina A, Rodrigo J A and Matos O M 2015 Caustics, catastrophes, and symmetries in curved beams *Phys. Rev.* A **92** 033850

[40] Wang H, Shi L, Lukyanchuk B, Sheppard C and Chong C T 2008 Creation of a needle of longitudinally polarized light in vacuum using binary optics *Nat. Photon.* **2** 501–5

[41] Fang Z-X, Ren Y-X, Gong L, Vaveliuk P, Chen Y and Lu R-D 2015 Shaping symmetric Airy beam through binary amplitude modulation for ultralong needle focus *J. Appl. Phys.* **118** 203102

[42] Lee W-H 1979 Binary computer-generated holograms *Appl. Opt.* **18** 3661–9

[43] Vasara A, Turunen J and Friberg A T 1989 Realization of general nondiffracting beams with computer-generated holograms *J. Opt. Soc. Am.* A **6** 1748–54

[44] Fang Z-X, Chen Y, Ren Y-X, Gong L, Lu R-D, Zhang A-Q, Zhao H-Z and Wang P 2018 Interplay between topological phase and self-acceleration in a vortex symmetric Airy beam *Opt. Express* **26** 7324–35

[45] Kaminer I, Bekenstein R, Nemirovsky J and Segev M 2012 Nondiffracting accelerating wave packets of Maxwell's equations *Phys. Rev. Lett.* **108** 163901

[46] Zhang P, Hu Y, Cannan D, Salandrino A, Li T, Morandotti R, Zhang X and Chen Z 2012 Generation of linear and nonlinear nonparaxial accelerating beams *Opt. Lett.* **37** 2820–2

[47] Bandres M A and Rodríguez-Lara B M 2013 Nondiffracting accelerating waves: weber waves and parabolic momentum *New J. Phys.* **15** 013054

[48] Aleahmad P, Miri M-A, Mills M S, Kaminer I, Segev M and Christodoulides D N 2012 Fully vectorial accelerating diffraction-free Helmholtz beams *Phys. Rev. Lett.* **109** 203902

[49] Mao H, Ren Y-X, Yu Y, Yu Z, Sun X, Zhang S and Wong K K Y 2021 Broadband meta-converters for multiple Laguerre–Gaussian modes *Photon. Res.* **9** 1689–98

[50] Plidschun M, Ren H, Kim J, Förster R, Maier S A and Schmidt M A 2021 Ultrahigh numerical aperture meta-fiber for flexible optical trapping *Light: Sci. Appl.* **10** 57

IOP Publishing

Optical Vortices

Fundamentals and applications

Yuanjie Yang, Yu-Xuan Ren and Carmelo Rosales-Guzmán

Chapter 10

Structured light for optical communication

One of the applications of structured light beams which have gained significant popularity in recent time is in optical communications, at both the classical and quantum levels, either in free space or optical fibers. As such, in this chapter we provide an overview about how structured light beams are used in optical communications. The first part is devoted to the use of structured light beams in classical communications, this section is treated in three subsections where we discuss separately the cases of free space, optical fibers underwater optical channels. In the second section we delve into the use of structured light in quantum optics communications, in particular for the development of novel protocols for secure communications.

10.1 Classical communications with structured light

The 2009 Nobel Prize for Physics was awarded to Charles Kao for his pioneering ideas in telecommunications, suggesting the silica-based optical fiber as the technology of the future, setting the basis for a long tradition in optical communications [1]. This seminal work ignited a new era of research that made fiber-based technologies a reality, completely revolutionising the era of telecommunications. Improvements came from all directions, new types of lasers or optical fibers, but also new types of coding schemes based on the various properties of light, such as wavelength, time, phase, amplitude or polarization, as illustrated in figure 10.1. In particular, wavelength division multiplexing (WDM) allows for over 100 independent data-carrying channels in commercial systems, see figure 10.1(a). In addition to WDM, polarization division multiplexing also allows one to double the number of channels by codifying information in two orthogonal polarization states, see figure 10.1(d).

It is nowadays hard to conceive a world without Internet, an essential tool for many of our daily-life activities. As such, the data traffic on the Internet is growing incredibly fast, at around 40% every year, a trend which is far from reaching its

doi:10.1088/978-0-7503-5844-6ch10
10-1

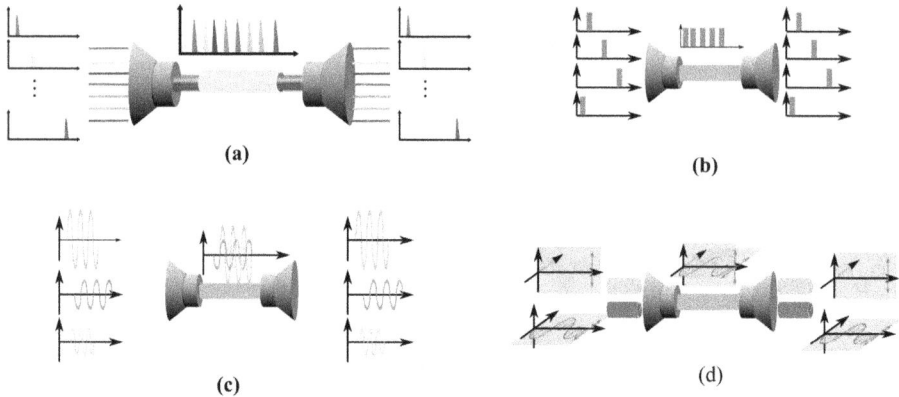

Figure 10.1. Information codification schemes. (a) Wavelength division multiplexing, (b) time division multiplexing, (c) amplitude and phase division Multiplexing and (d) polarization division multiplexing.

maximum. Amongst the main causes responsible for this increasing demand of bandwidth are social networking, as well as the video on-demand platform services, such as Netflix or Youtube, which rely on the premise is more apropriate of high-definition videos. it is worth noting that the bandwidth for data transmission over fire-based systems, which until recently was considered to be infinite, has reached its limit. As a result, a 'capacity crunch,' of our communications systems is within sight, which implies a severe limitation in the Internet capacity growth [2]. This limit is not technological, but fundamental, which implies a significant innovation in the ways we encode/decode and transmit the information, for both optical fibers and free space. Space-division multiplexing (SDM) over free space or optical fiber has emerged as a potential technology to give our current communications systems a new momentum. Here, as the name indicates, multiple independent data channels occupying different spaces are transmitted simultaneously, therefore multiplying the bandwidth capacity by the number of channels. Of particular interest is mode-division multiplexing (MDM), which utilises the different sets of spatial modes as the channels to transmit the information [3–5]. In what follows we will describe in more detail some of the techniques that have been proposed, in optical fibers, in free-space and finally underwater.

10.1.1 Optical communications in optical fibers

Most of our current communications systems are based on optical fibers transmitting information in the form of a beam of laser light. However, given the increasing demand for higher transmission capacities, the idea of using space-division multiplexing (SDM) has gained popularity, even though it was posted almost simultaneously with the beginnings of optical fiber communication. Along this line, perhaps the most obvious is the multicore fiber approach, which consist on the fabrication of fibers containing multiple cores, as illustrated in figure 10.2(a). Yet, there is an alternative approach, which consist on using orthogonal modes of light, each defining a spatially distinct channel, travelling along multimode optical

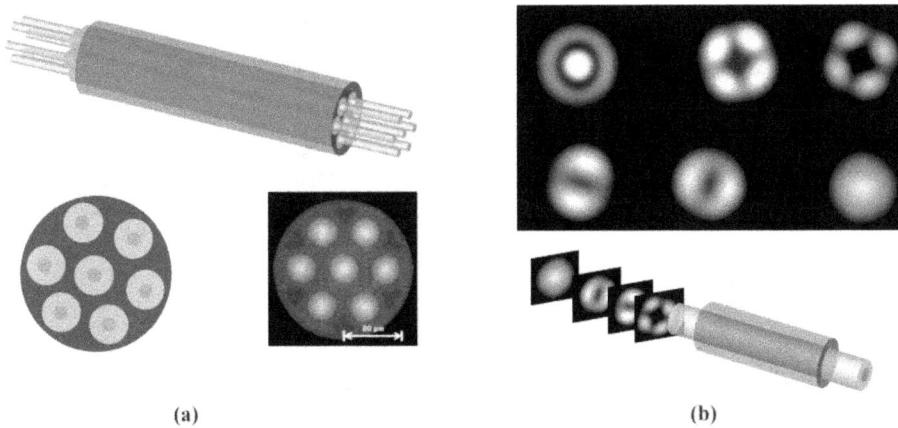

Figure 10.2. (a) In space-division multiplexing multiple independent data channels occupying different spaces are transmitted simultaneously. (b) In mode-division multiplexing orthogonal modes of light define distinct channels, which can propagate simultaneously occupying the same space.

fibers or free space, known as mode-division multiplexing (MDM), as illustrated in figure 10.2(b). Here we are only concerned with MDM, but readers interested in the general idea of SDM are suggested to take a look at the several review papers on the topic [6].

To begin with, it is worth mentioning that the transmission of spatial modes over multimode fibers still represents a challenge, since upon propagation the modes will show a significant spatial overlap and as a consequence, the signals are prone to random coupling. In addition, there would be differential group mode delay (DMGD), as well as differential modal loss or gain amongst the various modes. As a result, the energy of the data stream carried by a give mode will spread out into neighbouring modes, causing information crosstalk that mitigates the successful transfer of information. Along this line, multimode fibers with core/cladding diameters of 50/125 μm and 62.5/125 μm have been engineered to support more that 100 modes, nonetheless, DMGD is still an issue, since the spatial modes can accumulate very large delays. Hence, even though such fibers are still unsuitable for long-haul transmission, digital signal processing, in particular compensation electronics, might hold the solution for this technology to take off. An alternative solution might lie on the use of few-mode fibers (FMFs), where recent advances have demonstrated that optical fibers supporting a small number of modes possess a low DMGD. Along this line, the modes supported by such optical fibers are the vector modes HE_{11}, TM_{01}, TE_{01} and HE_{21}, see figure 10.3(a). Nevertheless, the linearly polarized (LP) modes LP_{01} and LP_{11}, constructed from linear combinations of the previous modes, see figure 10.3(a), are preferred as they are easier to excite and detect, giving a total of six structured light modes.

Recent work has also demonstrated the simultaneous transmission of such modes over a few-mode fiber, each representing an independent channel carrying 40 Gb s^{-1} over 96 km with a low DMGD. In this demonstration, the data stream transmitted on each channel was recovered using digital signal processing based on coherent

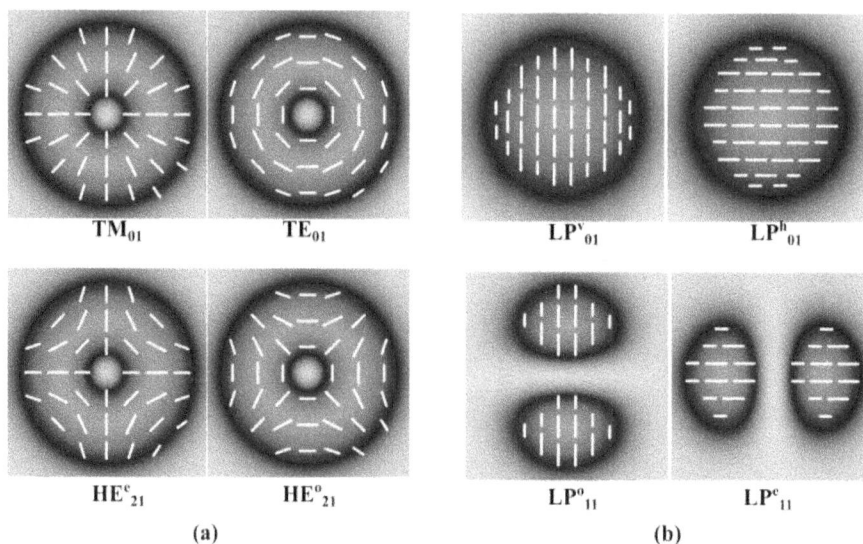

Figure 10.3. The principal modes of multimode optical fibers. (a) Vector modes. (b) Linearly polarized modes.

detection and multiple-input multiple-output processing. The use of more complex modal basis sets, including modes that carry orbital angular momentum, is also being explored, with the goal of reducing mode coupling [3]. This pioneering demonstration showed the viability of using spatial modes of light endowed with OAM as information channels capable to carry independent data streams over a single optical fiber and over a long distance. More specifically, four angular momentum modes at a single wavelength were transmitted over a distance of 1 km and at a data transmission speed of 400 gigabits per second. In addition, transmission speeds of 1.6 terabits per second were demonstrated with two OAM modes and 10 wavelengths. Of relevance is also the recent demonstration of a fiber-based system with the capability to transmit data streams over 2 km with vector beams directly generated from photonic integrated devices [7, 8]. More precisely, radially and azimuthally polarized vector beams, generated from silicon microring resonators etched with angular gratings, were multiplexed and used as data-carrying channels and transmitted over a 2 km large core optical fiber with low-level mode crosstalk.

10.1.2 Optical communications in free space

Mode division multiplexing (SDM) consists of storing and transmitting the information in orthogonal spatial modes of light, whereby each mode can carry an independent data stream, increasing in this way the link capacity in proportion to the number of orthogonal modes. Importantly, SDM can be combined with all existing modulation techniques (amplitude, polarization, frequency), allowing for an even higher bandwidth capacity. Nonetheless, many of the experimental demonstrations have been carried out indoors and under controlled conditions, using not only light beams endowed with OAM, but also other orthogonal bases [9, 10].

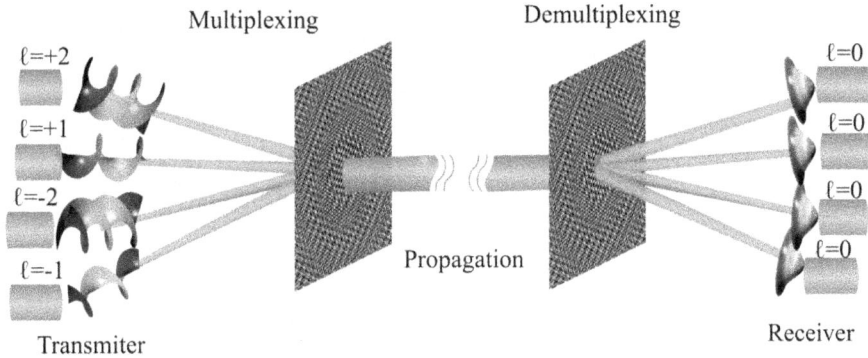

Figure 10.4. Mode division multiplexing in free space. Here four orthogonal spatial modes, each carrying an independent stream of data are multiplexed and transmitted occupying the same space. The orthogonality of the modes guarantees the separation of information at the receiver for its recovery.

Figure 10.4 shows schematically the implementation of SDM with structured light beams endowed with OAM. Here a set of four OAM modes with topological charges $\ell = 2$, $\ell = 1$, $\ell = -1$ and $\ell = -2$, each carrying an independent data stream, are multiplexed to propagate along a common optical path. In theory, their orthogonality property allows the information to travel without mixing, nonetheless, optical perturbations, such as those introduced by the atmospheric turbulence, produce a degradation of the information. At the receiving end, the information is decoded back using an inverse process to the generation one.

Nonetheless, several considerations have to be taken into account for designing a free-space optical link, some related to the modes themselves, such as beam divergence or crosstalk between neighbouring modes, and some others to the system itself, such as the aperture size of the receiver, or a careful alignment of both, the transmitter and the receiver. One of the pioneering demonstrations of data transmission using the orbital angular momentum degree of freedom of light, over a 15 m free-space optical link, was performed in 2004 [11]. In a first attempt to achieve more realistic optical links and over longer distances, a 3 km optical link was demonstrated with strong turbulence over the city of Vienna in 2014. In this experimental demonstration, a greyscale image was transmitted pixel by pixel using 16 different OAM superposition modes, featuring a petal-like intensity distribution, each representing a particular greyscale value. In a more promising attempt, a proof-of-concept experiment was carried out also using OAM superposition modes to encode a message and transmit it over a distance of 143 km between two Canary Islands, La Palma and Tenerife. The receiver comprised the use of artificial intelligence to distinguish between different OAM superpositions. In a major attempt to increase the number of available channels and therefore the bandwidth capacity of optical links, it was proposed to use the full set of Laguerre–Gaussian modes [9]. More precisely, the set of LG beams are defined by two parameters: an azimuthal index denoted by ℓ and a radial index denoted by p, the former responsible for the OAM. In this way, each OAM mode, characterized by its ℓ value, defines an infinite set of modes with different p values, as shown in figure 10.5(a),

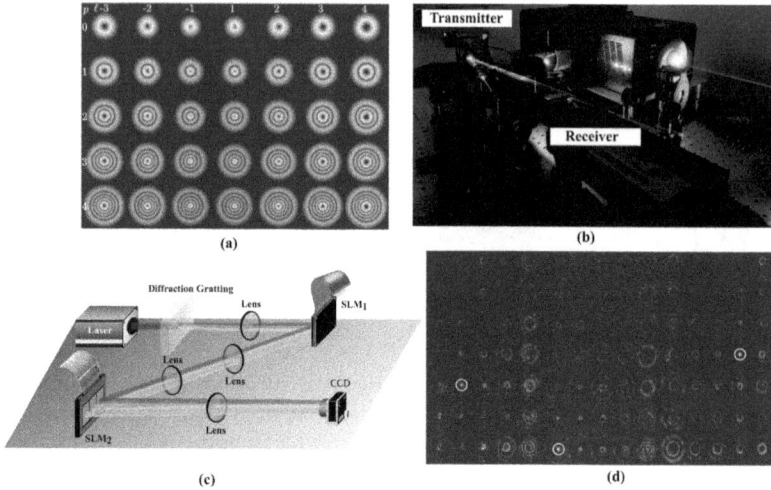

Figure 10.5. (a) The 35 Laguerre-Gauss modes in the experiment. (b) Laboratory picture where the transmiter and receiver are visible. (c) Schematic representation of the experimental setup. (d) Recovered intensity on a charge-coupled- device camera (CCD) of the modal decomposition, showing three detected modes.

increasing therefore the bandwidth capacity in proportion to p. In the experiment, 100 spatial modes were multiplexed/demultiplexed using a single digital hologram displayed on a spatial light modulator. Further, in this proof-of-principle experiment, a hybrid system consisting of wavelength-mode-division multiplexing was demonstrated. For this, over 100 spatial modes were encoded on three different wavelengths, as shown in figure 10.5(c). To demonstrate the feasibility of the proposal, greyscale and colour images were encoded pixel-by-pixel, assigning the different grey levels to a unique LG mode, transmitted over free space and decoded on the receiver side using also a multiplexed hologram displayed on a second SLM. For the sake of clarity, a schematic representation of the experimental setup shown in figure 10.5(c).

In the experiment, a continuous argon ion laser emitting in three different wavelengths (457 nm, 488 nm, 514 nm) were directed to an SLM, whose screen was separated into three independent sections one for each beam, each controlled independently, where the holograms corresponding to a set of 35 Laguerre–Gauss modes were displayed. To demonstrate the concept, colour images were transmitted pixel by pixel, by first decomposing each pixel into its three colour components (red, blue and green). In this way, each colour was assigned to a particular wavelength and its saturation level to one of the 35 different spatial modes, reaching in this way 35 levels of saturation. The colour information of each pixel was then transmitted in the form of spatial modes, which was decoded at the receiving end using a second SLM (SLM_2), also divided into three independent sections, employing modal decomposition. On each section a multiplexed hologram consisting of the complex conjugation of all 35 modes, encoded with different spatial carrier frequencies was displayed. Furthermore, each mode was identified by measuring the on-axis intensity of the projection in the far field using a lens and a CCD camera in a 2f configuration where all 105 modes appear spatially separated, as shown in figure 10.5(d). By way of example, here the three detected modes are encircled.

More recently, the use of vector beams as information channels, which also form infinite and orthogonal states of light, has gained popularity [12–14]. One of the main reasons is the recent demonstration that they are more resilient to atmospheric perturbations [15]. In one of the first experiments, Barreiro *et al* proposed a dense-coding scheme based on the use of vector vortex beams [16]. In this pioneering experiment it was shown that photon pairs entangled in a vector beam were able carry two bits of information. A more recent experiment demonstrated a free-space optical link consisting of four orthogonal vector beams, each transmitting 20 Gbits^{-1}. For this indoor proof-of-principle experiment, the vector beams were generated with q-plates, multiplexed and transmitted as one single light beam and demultiplexed at the receiver with q-plates, showing very little crosstalk [13]. Vector beams can also be used to construct higher-dimensional spaces where high bits of information can be encoded. Once it has been encoded using the orthogonal elements of the basis, it can be transmitted and decoded back for its reading.

A recent demonstration was performed using four vector beams to encode two bits of information. The information was encoded using the identity matrix and the three Pauli operators to the polarization degree of freedom and decoded with high efficiency using a Mach–Zehnder interferometer and q-plates [14]. A similar technique for manipulating vector beams with Pauli operators was demonstrated in [17]. The vector beams were generated using dove prisms and a vortex phase plate inside a Sagnac interferometer, and decoded with the help of a spatial light modulator. Further, the two dove prisms were used to perform rotations on the spatial degree of freedom. Crucially, computer-controlled devices provide more flexibility in the generation of vector modes, compared to other devices, such as *q*-plates or metamaterials. In one of the pioneering experiments that took advantage of this flexibility, a hexadecimal encoding of a 64×64 pixels image was demonstrated, which used 16 vector beams generated on a single SLM [18], Here, the grey value of each pixel was converted to a hexadecimal number and associated to one of the 16 vector beams. The image was encoded and transmitted pixel by pixel by switching the phase pattern displayed on the SLM. A similar approach proposed a multi-ary encoding/decoding in a high-dimensional space consisting on multiple possible states of hybrid vector beams [17]. In this approach, high-dimensionality encoding was achieved by transforming the OAM modes of vector beams under the assistance of polarization. The encoded vector modes were decoded with very low mode crosstalk using a modified Mach–Zehnder interferometer and spatial light modulators.

10.1.3 Underwater optical communications

There is also an increasing interest for underwater optical communications for scientific, commercial, and military applications. Traditionally, underwater communications were with acoustic signals, which are limited in both bandwidth and speed. Alternatively, the use of radio frequency signals represents another approach, which is of low cost and offers relatively high communication speed but suffers from strong attenuation in water. Hence, underwater wireless optical communication

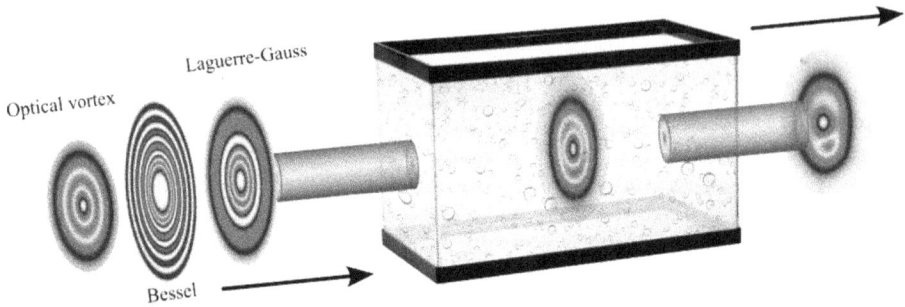

Figure 10.6. Underwater optical communications in the presence of dynamic water bubbles. Here, three different sets of optical modes are transmitted trough a water tank.

(UWOC) systems with optical waves in the blue–green wavelength regions provides a potential solution, as it can propagate through clean water with relatively low attenuation. In addition, it provides higher bandwidth resources, which translates in higher communication capacities, lower power consumption as well as higher levels of security. Along this line and in the context of OAM, the simultaneous transmission of multiple orthogonal spatial modes, each carrying an independent data stream, was demonstrated recently [19]. Here, the transmission of data at 40 Gbit s^{-1} was demonstrated with four OAM multiplexed beams in the green wavelength. Further, a detailed investigation of the degrading effects caused by scattering/turbidity, water current, and thermal gradient-induced turbulence revealed that distortions are associated to thermal gradients and losses to turbidity. Importantly, the authors also demonstrated that inter-channel crosstalk induced by thermal gradients could be reduced via multi-channel equalization.

In a more recent experiment, an OAM-multicasting (one-to-many) data information transfer was demonstrated in an underwater environment, to distribute information between multiple users. In the experiment, a water tank 2 m long was used to transmit a 4-fold optical wave endowed with OAM at 520 nm. The beam was designed to multicast four OAM modes, OAM-6, OAM-3, OAM + 3, OAM + 6, each carrying 1.5-Gbaud 8-ary quadrature amplitude modulation (8-QAM) with orthogonal frequency-division multiplexing (OFDM) [20]. Studies have been also carried out to analyse the performance of underwater optical communications of structured light modes in the presence of obstructions [21]. In the experiment the authors evaluated the performance of three different sets of structured light modes in the presence of dynamic bubbles and static obstructions, namely, optical vortices, diffraction-free and self-healing Bessel–Gauss modes, as schematically depicted in figure 10.6. The experiments demonstrated that in the presence of dynamic bubbles, most structured light beams perform almost similarly, while in the presence of static obstacles, Bessel–Gauss modes presented the best performance compared to the other modes.

10.2 Secure quantum communications with structured light

Structured light beams have also attracted significant attention towards the implementation of secure quantum communication links. Traditional quantum systems rely on the use of light's polarization to encode quantum bits (qubits) of information for the implementation of quantum protocols [22, 23]. In addition, polarization entanglement has enabled cryptographic protocols for the implementation of secure quantum communications through quantum key distribution (QKD) [24, 25]. Nonetheless, the increasing demand for faster and more robust quantum key distribution protocols, with higher capabilities to prevent third party attacks, the use of higher-dimensional QKD has become a topic of late. The main reason is that the security of QKD protocols scales up with the dimension (d) of the encryption basis. In addition, it also increases the transmission key rates, with each photon capable of carrying up to $\log_2 (d)$ bits of information [26]. As such, the use of structured modes embedded with OAM have gained considerable attention, with laboratory demonstrations reaching dimensions as high as $d = 7$ [27, 28]. For example, in [28] the authors reported a QKD system based on a seven-dimensional alphabet encoded in the OAM of light and the mutual unbiased basis of angular position modes.

Data encryption allows a sender to transmit information to a desired recipient in a secure way, provided an encryption algorithm and a secure communication channel are guaranteed. Here, the first and most fundamental step is the distribution of a cryptographic key in a secure way, to guarantee the encrypted information can only be deciphered by the intended recipient. In this way, traditional cryptographic methods rely on complex mathematical algorithms, with the security relying mainly on the computational power, which in the beginnings was though to be highly secure. Nonetheless, with the recent advances of computer technology, it is nowadays easier for hackers to break their security. A possible alternative is quantum mechanics-based protocols, which provide the means for implementing cryptographic tasks, which are almost possible using only classical communication systems. Here, the no-cloning theorem is of particular relevance, which guarantees that data encoded in a quantum state cannot be copied, since any attempt to read the encoded data will modify the quantum state and alert for intruders during key distribution [29, 30]. In other words, the fundamental laws of quantum mechanics, which include not only the non-cloning theorem but also Heisenberg's uncertainty principle, provides the means to detect the presence of an evesdropper in the communication channel [29, 30]. In this way, quantum communication is the encoding of information using quantum states (qbits) since they are more powerful than the classical bits of information. In the context of optics, the quantum states are transported by single photons, encoded for example in the polarization degree of freedom, which forms a two-dimensional space. Amongst the most common protocols for QKD, the BB84 has become one of the most popular [31], taking the name from their inventor, Bennet and Brassard in the year 1984, with an experimental demonstration in 1989 [32]. Improving the information capacity and transmissions distances of QKD protocols is amongst the current challenges in quantum communication [33–36].

As such, high-dimensional QKD protocols of dimensions $d > 2$ using structured light modes to encode the information have become a topic of late [12, 37, 38]. Some of the main reasons are that the information capacity per photon, the error threshold, and the robustness to noise are directly proportional to the dimension d [39–41]. Advances in the use of structured light modes for high-dimensional QKD, in particular using OAM modes, have been achieved in free space [28, 42], optical fibers [43], underwater [44–46] and also in hybrid polarization-OAM states [47–49]. Crucially, structured light modes can also bring additional benefits, such as their inherent properties, as in the case of Bessel–Gauss modes, with self-healing properties at both the classical and quantum levels [49–51]. For interested readers there is a very interesting tutorial about high-dimensional cryptography with structured light modes and its simulation with classical structured light modes in [52].

10.2.1 Quantum key distribution with polarization

To better explain the advantages of high-dimensional QKD, let us compare it with the 2-dimensional case using the polarization degree of freedom of light, where the sender (Alice) encodes her bits of information using the two orthogonal states of polarization, horizontal and vertical, represented by the kets $\{|H\rangle, |V\rangle\}$, respectively, conforming a 2-dimensional basis denoted by ψ. For the transmission, the bits values 0 and 1 are assigned to each polarization state, say $|H\rangle = 0$ and $|V\rangle = 1$, respectively. Hence, an encoded message can be deciphered by the receptor (Bob) using also the same polarization basis $\{|H\rangle, |V\rangle\}$. Nonetheless, the message can also be intercepted by an intruder, an eavesdropper (Eve), who can then send exactly the same message to the recipient without alerting any of the involved parties about their presence. Hence, to prevent hackers (the eavesdropper) getting access to the message, or at least knowing about their presence, an additional polarization basis is also used to encode the message. Say the basis ϕ defined by the diagonal anti-diagonal polarization states $\{|D\rangle, |A\rangle\}$, respectively, with the bit 1 assigned to the ket $|D\rangle$ and the bit 0 to $|A\rangle$. The elements of both polarization bases are related by

$$|A\rangle = \frac{1}{\sqrt{2}}(|H\rangle - |V\rangle) \text{ and } |D\rangle = \frac{1}{\sqrt{2}}(|H\rangle + |V\rangle) \qquad (10.1)$$

Now Alice has the option to encode the message in either of the two bases, ψ or ϕ, in a random way and without disclosing her choices, as illustrated in figure 10.7(a). In a similar way, Bob has the option to measure each bit using either of the two bases, which he also picks randomly. In this way, only in the event when Bob picks the same basis as Alice will he measure a correct bit with the maximum probability. On the contrary, when the measuring and encoding bases are different, the probability of measuring correctly the encoded bit will reduce to only 50%. An example of the encoding-measuring procedure is illustrated in figure 10.7(b). Hence, Bob can only deduce correctly the bit Alice sent when he picks the same basis in which it was encoded. Exactly the same happens for Eve and it is precisely this condition that reveals her presence.

Once the transmission is complete, both parties reveal in a public channel their choices for generation/measuring each transmitted bit. This allows them to discard the bit values in which both bases do not match, which forms the sifted key. If no intruders are present during the transmission, Alice and Both will be left with an average set of 50% of the total transmitted bits. Nonetheless, the presence of an intruder will introduce additional errors, which in the end will reveal their presence

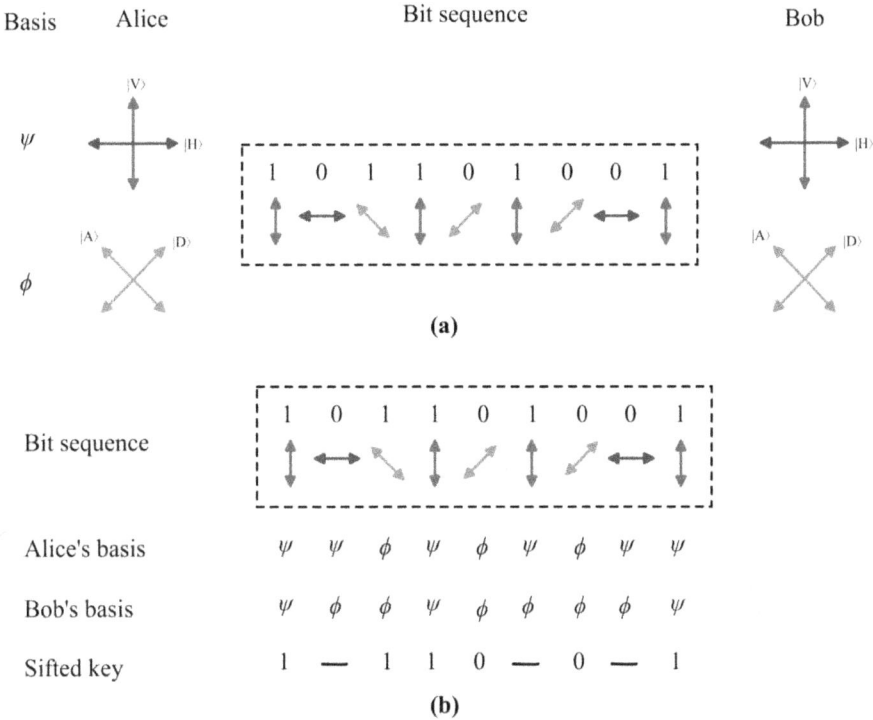

| Basis | Alice | Bit sequence | | Bob |

(a)

(b)

Figure 10.7. (a) Two unbiased polarization bases, ψ and ϕ, are used to encode the message. (b) Alice picks a basis to encode each bit in a random way and without disclosing her choice publicly. In the same way, Bob picks a basis in which to measure each bit. A sifted key is generated from the bits in which the generation and detection basis coincide with each other.

to Alice and Bob. More precisely, given that Eve intercepts and resends each and all of the photons, she will measure on average also 50% of the bits incorrectly. Hence, on average, Bob is left with only 1/4 correct bits and 1/4 wrong bits, out of the total sent bits. As a final step and to find out whether or not an eavesdropper has intercepted the message, Alice and Bob compare through a public channel a subset of their sifted 1/2 bits, from which the error can be estimated, allowing Alice and Bob to decide if whether or not to proceed with the transmission using the sifted key. It is worth emphasising that the subset, which was disclosed publicly, is discarded from the final key.

10.2.1.1 Important parameters of QKD

The first parameter of importance for QKD is the error rate Q, which quantifies the security of a quantum channel. More precisely, the error rate measures the probability that Bob has for detecting wrong states, provided he measures in the correct basis. A standard procedure to measure Q is by constructing the transfer matrix of the system given by

$$T = \begin{pmatrix} |\langle \psi_i^{\text{Bob}}| \hat{U} |\psi_j^{\text{Alice}}\rangle|^2 & |\langle \psi_i^{\text{Bob}}| \hat{U} |\phi_j^{\text{Alice}}\rangle|^2 \\ |\langle \phi_i^{\text{Bob}}| \hat{U} |\psi_j^{\text{Alice}}\rangle|^2 & |\langle \phi_i^{\text{Bob}}| \hat{U} |\phi_j^{\text{Alice}}\rangle|^2 \end{pmatrix}. \tag{10.2}$$

This matrix captures the procedure described above of Alice preparing a state in either of the two bases, ψ and ϕ, and Bob measuring also in any of these two bases, where $i, j = \{1, 2\}$ represents the two polarization states in each basis. The error rate for each basis can be computed through the expressions

$$Q_\psi = 1 - \frac{1}{2}\sum_{i=1}^{2}|\langle \psi_i^{\text{Bob}}| \hat{U} |\psi_i^{\text{Alice}}\rangle|^2, \tag{10.3}$$

$$Q_\phi = 1 - \frac{1}{2}\sum_{i=1}^{2}|\langle \phi_i^{\text{Bob}}| \hat{U} |\phi_i^{\text{Alice}}\rangle|^2. \tag{10.4}$$

Here, $\hat{U}$ represents the communication channel and the second term of each equation, also known as the fidelity F, represents the probability Bob measures the state $|\psi_i^{\text{Bob}}\rangle$ ($|\phi_i^{\text{Bob}}\rangle$), provided Alice sent the state $|\psi_i^{\text{Alice}}\rangle$ ($|\phi_i^{\text{Bob}}\rangle$). The error rate of the system is then measured as $Q = \text{mean}\{Q_\psi, Q_\phi\}$.

Another important parameter is the amount of information that Alice and Bob share, which can be computed as [26]

$$I_{AB}(Q) = 1 + Q \log_2(Q) + (1 - Q)\log_2(1 - Q). \tag{10.5}$$

By way of example, for an ideal communication channel, which does not perturb the transmitted states, $\hat{U}$ becomes the identity operator and therefore $F = 0$ and $Q = 0$ and the maximum amount of information that Alice and Bob can share for each photon sent is 1 bit. Nonetheless, the errors introduced to the transmission channel by the extraction of information by Eve is proportional to the amount of information she extracts. This can be quantified through the cloning fidelity $F_E(Q)$, the ability Eve has to copy the transmitted states, [26]

$$F_E(Q) = \frac{1}{2}(1 + 2\sqrt{Q(1 - Q)}). \tag{10.6}$$

From this expression it is evident that the maximum cloning fidelity can be reached for $Q = 1/2$. In other words, Eve can achieve a maximum cloning fidelity by introducing an error of 50% to the transmission channel.

Finally, a parameter of interest for the transmission of information is the secret key rate, which estimates the amount of secure information that that can be exchanged, which can be computed as [53]

$$R_\Delta(Q) = 1 + 2(1 - Q)\log_2(1 - Q) + 2Q \log_2(Q), \tag{10.7}$$

which must be positive to ensure a secure transmission of information, which is achieved for $Q < 11\%$. In the ideal case $Q = 0$, the system reaches its maximum rate of secret information transmission per photon.

10.2.2 Quantum key distribution in higher dimensions

Achieving QKD in higher dimensions has become a topic of relevance for improving the robustness of QKD protocols, achieving quantum channels with longer propagation distances, as well as for achieving higher key transmission rates, which are directly related to the losses in the communication channel [54]. Hence, the parameter introduced above for 2D QKD has been deduced for higher dimensions [26]. The mutual information shared between Alice and Bob has now the form,

$$I_{AB}(d, Q) = \log_2(d) + (1 - Q)\log_2(1 - Q) + (Q)\log_2\left(\frac{Q}{d - 1}\right). \tag{10.8}$$

Notice that the amount of secure information increases with the dimension d in a logarithmic way, the higher the dimension, the higher the amount of information that can be send in every single photon. The cloning fidelity associated to Eve takes now the form,

$$F_E(d, Q) = \frac{1}{d}(1 + (d - 2)Q + 2\sqrt{(d - 1)Q(1 - Q)}). \tag{10.9}$$

In this case, the maximum cloning fidelity will be $(d - 1)/d$. The mutual information shared between Alice and Eve can be computed from the expression,

$$I_{AE}(d, Q) = \log_2(d) + (F_E - Q)\log_2\left(\frac{F_E - Q}{1 - Q}\right) + (1 - F_E)\log_2\left(\frac{1 - F_E}{(d - 1)(1 - Q)}\right). \tag{10.10}$$

From this equation it can be seen that $I_{AE}(d, Q)$ increases with the dimension of the basis. Finally, the secret key rate for a two basis protocol can be quantified from [55]

$$R_\Delta(d, Q) = \log_2(d) + 2(1 - Q)\log_2(1 - Q) + 2Q\log_2\left(\frac{Q}{d - 1}\right). \tag{10.11}$$

In a nutshell, higher-dimensional states can carry more information and QKD protocols employing higher dimensions are also more robust to noise compared to the 2D case. Finally, the higher error rate threshold achieved ensures QKD protocols with higher security. The spatial degree of freedom of photons provides us with the means to implement QKD protocols in high dimensions since the tools to prepare and measure classical structured light modes have already been extended to the quantum regime [27, 28, 56, 57]. Along this line, structured light modes endowed with OAM have become popular as many tools for their generation and detection are now ubiquitous [58–62].

10.2.3 Quantum communications with vector beams

In recent times vector beams have gained popularity in quantum communications, for example, as carriers for polarization encoded qubits in alignment free QKD [63]. In general, polarization-entangled protocols for quantum communications require that both the sender and receiver share a common reference frame, limiting, for example, the realization of a long-haul communication with structured light modes.

In [63] the authors proposed a quantum system for the transmission of information encoded in the polarization state of single-photon invariants to arbitrary rotations. This scheme, which exploits the rotation-invariance of hybrid polarization-OAM degrees of freedom inherent to vector beams, removes the need to align the detectors for the recovery of information. A mayor advance in QKD, relying on the use of vector beams, was the experimental demonstration of high-dimensional quantum cryptography under more realistic conditions [48]. This experiment was performed in a 0.3 km free-space link over the city of Ottawa and under the presence of moderate atmospheric turbulence. In the experiment, a four-dimensional quantum state based on a high-dimensional BB84 protocol was demonstrated, with a quantum bit error rate of 11% and a secret key rate of 0.65 bits per sifted photon.

Finally, even though high-dimensional QKD protocols using structured light can in principle increase the transmissions distances as well as the information capacity, perturbations in the quantum channel, such as atmospheric turbulence in free space or defects in optical fibers, an entanglement decay can occur. Paradoxically, this entanglement decay can be corrected once the quantum link has been established. Crucially, the mathematical similarity between vector beams and quantum-entangled photons goes beyond a mathematical similarity, as it was evidenced recently in an experiment which showed that quantum entanglement decay of a pair of photons is identical to that of vector beams [64]. In the experiment, performed in identical conditions with entangled photons and vector beams, a comparison of the entanglement decay was performed, as illustrated in figure 10.8(a). In the quantum case, two photos entangled in OAM where produced through spontaneous parametric down conversion, one of which was sent through turbulence (simulated with an SLM) and the other was kept undisturbed. In the classic case with vector beams, all the beam is sent through the simulated atmospheric turbulence, nonetheless, only the spatial degree is modified, due to the fact that atmospheric turbulence is not birefringent. In both cases, the quantum decay was monitored as a function of the

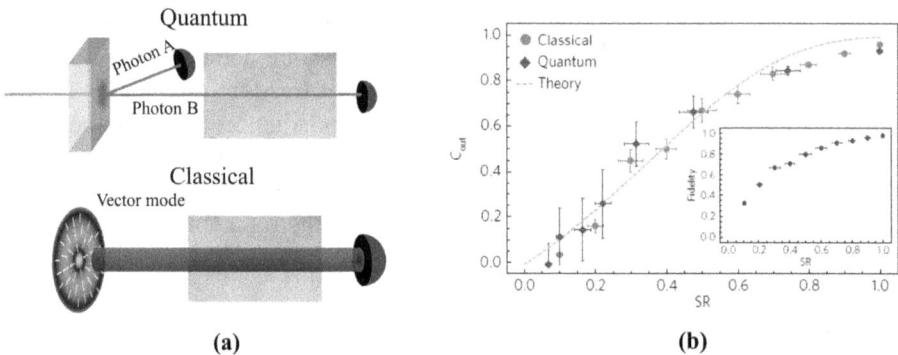

Figure 10.8. (a) Two entangled photos are generated with spontaneous parametric down conversion, one of which is set to atmospheric turbulence and the other is kept in the laboratory (top). In a similar way, a vector mode is sent to the same atmospheric turbulence but only the spatial degree of freedom is distorted by the medium. (a) Entanglement decay as a function of turbulence strength, measured through the SR parameter, the smaller, the stronger the turbulence.

atmospheric turbulence strength, as shown in figure 10.8(b). Here, the turbulence strength was measured through the Strehl Ratio (SR); the smaller the value, the stronger the turbulence. The experiment showed an identical entanglement decay for both cases, with the advantage that vector beams are constituted by many photons and therefore are easier to align. Beyond the fundamental insights this experiment brought to light, it settled the basis for practical applications; for example, it provided a platform to perform the characterization of a quantum channel using vector light modes instead of single photons. Further, it also paved the way to perform quantum error correction in short and long-haul optical communication links not only of free-space channels but also in optical fibers.

Finally, in recent time it was demonstrated that vector beams propagate through disturbing channels maintaining their vectorial inhomogeneity unaltered, even though their spatial amplitude, phase and polarization structure, seem to distort [15]. This demonstration solved a standing debate on the robustness of vectorial light in complex media, and at the same time it demonstrated that the distortion is only a matter of perspective. Furthermore, a basis-independent measure of concurrence revealed that indeed it remains constant when propagating through unitary complex media, where one of the two degrees of freedom, say polarization, remains unaffected. This realization allowed in recent time a novel technique for multi-bit optical communication using the vectorness of vectorial light as the information carrier [65]. The key to this technique is that the concurrence C is invariant to beam distortions and therefore the full range of C values (from 0 to 1) can split into a discrete number of optical channels. For demonstrating this technique, the authors used 50 channels.

References

[1] Kao K and Hockham G 1966 Dielectric-fiber surface waveguides for optical frequencies *Proc. Inst. Electr. Eng.* **113** 1151–8
[2] Richardson D J 2010 Filling the light pipe *Science* **30** 327–8
[3] Bozinivic N, Ren Y Y, Tur Y, Kristensn M, Huang P, Willner H and Ramachandran S 2013 Terabit-scale orbital angular momentum mode division multiplexing in fibers *Science* **340** 1545–8
[4] Wang J *et al* 2012 Terabit free-space data transmission employing orbital angular momentum multiplexing *Nature Photon.* **6** 488–96
[5] Willner A E *et al* 2015 Optical communications using orbital angular momentum beams *Adv. Opt. Photon.* **7** 66–106
[6] Richardson D J, Fini J M and Nelson L E 2013 Space-division multiplexing in optical fibers *Nat. Photon.* **7** 354–62
[7] Liu J *et al* 2018 Direct fiber vector eigenmode multiplexing transmission seeded by integrated optical vortex emitters *Light: Sci. Appl.* **7** 17148
[8] Willner A E 2018 Vector-mode multiplexing brings an additional approach for capacity growth in optical fibers *Light: Sci. Appl.* **7** 18002
[9] Trichili A, Rosales-Guzmán C, Dudley A, Ndagano B, Ben Salem A, Zghal M and Forbes A 2016 Optical communication beyond orbital angular momentum *Sci. Rep.* **6** 27674

[10] Xie G *et al* 2016 Experimental demonstration of a 200-gbit/s free-space optical link by multiplexing Laguerre–Gaussian beams with different radial indices *Opt. Lett.* **41** 3447–50

[11] Gibson G, Courtial J, Padgett M J, Vasnetsov M, Pas'ko V, Barnett S M and Franke-Arnold S 2004 Free-space information transfer using light beams carrying orbital angular momentum *Opt. Express* **12** 5448–56

[12] Ndagano B, Nape I, Cox M A, Rosales-Guzmán C and Forbes A 2018 Creation and detection of vector vortex modes for classical and quantum communication *J. Light. Technol.* **36** 292–301

[13] Milione G *et al* 2015 4 × 20 gbit/s mode division multiplexing over free space using vector modes and a q-plate mode (de) multiplexer *Opt. Lett.* **40** 1980–3

[14] Milione G, Nguyen T A, Leach J, Nolan D A and Alfano R R 2015 Using the nonseparability of vector beams to encode information for optical communication *Opt. Lett.* **40** 4887–90

[15] Nape I *et al* 2022 Revealing the invariance of vectorial structured light in complex media *Nat. Photon.* **16** 538–46

[16] Barreiro J T, Wei T C and Kwiat P G 2008 Beating the channel capacity limit for linear photonic superdense coding *Nat. Phys.* **4** 282–6

[17] Li P, Wang B and Zhang X 2016 High-dimensional encoding based on classical non-separability *Opt. Express* **24** 15143–59

[18] Zhao Y and Wang J 2015 High-base vector beam encoding/decoding for visible-light communications *Opt. Lett.* **40** 4843–6

[19] Ren Y *et al* 2016 Orbital angular momentum-based space division multiplexing for high-capacity underwater optical communications *Sci. Rep.* **6** 33306

[20] Zhao Y, Xu J, Wang A, Lv W, Zhu L, Li S and Wang J 2017 Demonstration of data-carrying orbital angular momentum-based underwater wireless optical multicasting link *Opt. Express* **25** 28743–51

[21] Zhao Y, Wang A, Zhu L, Lv W, Xu J, Li S and Wang J 2017 Performance evaluation of underwater optical communications using spatial modes subjected to bubbles and obstructions *Opt. Lett.* **42** 4699–702

[22] Ursin R *et al* 2007 Entanglement-based quantum communication over 144 km *Nat. Phys.* **3** 481–6

[23] Hübel H, Vanner M R, Lederer T, Blauensteiner B, Lorünser T, Poppe A and Zeilinger A 2007 High-fidelity transmission of polarization encoded qubits from an entangled source over 100 km of fiber *Opt. Express* **15** 7853–62

[24] Jennewein T, Simon C, Weihs G, Weinfurter H and Zeilinger A 2000 Quantum cryptography with entangled photons *Phys. Rev. Lett.* **84** 4729–32

[25] Poppe A *et al* 2004 Practical quantum key distribution with polarization entangled photons *Opt. Express* **12** 3865–71

[26] Cerf N J, Bourennane M, Karlsson A and Gisin N 2002 Security of quantum key distribution using d-level systems *Phys. Rev. Lett.* **88** 127902

[27] Mafu M *et al* 2013 Higher-dimensional orbital-angular-momentum-based quantum key distribution with mutually unbiased bases *Phys. Rev. A* **88** 032305

[28] Mirhosseini M, Magaña-Loaiza O S, O'Sullivan M N, Rodenburg B, Malik M, Lavery M P J, Padgett M J, Gauthier D J and Boyd R W 2015 High-dimensional quantum cryptography with twisted light *New J. Phys.* **17** 033033

[29] Wootters W K and Zurek W H 1982 A single quantum cannot be cloned *Nature* **299** 802–3

[30] Shor P W and Preskill J 2000 Simple proof of security of the bb84 quantum key distribution protocol *Phys. Rev. Lett.* **85** 441

[31] Bennett C H and Brassard G 1984 Quantum cryptography: public key distribution and coin tossing *Proc. IEEE Int. Conf. on Computer System Signal Processing (Bangalore, India)* pp 175–9

[32] Bennett C H and Brassard G 1989 Experimental quantum cryptography: the dawn of a new era for quantum cryptography: the experimental prototype is working *SIGACT News* **20** 78–80

[33] Stucki D, Walenta N, Vannel F, Thew R T, Gisin N, Zbinden H, Gray S, Towery C R and Ten S 2009 High rate, long-distance quantum key distribution over 250 km of ultra low loss fibers *New J. Phys.* **11** 075003

[34] Schmitt-Manderbach T *et al* 2007 Experimental demonstration of free-space decoy-state quantum key distribution over 144 km *Phys. Rev. Lett.* **98** 010504

[35] Gobby C, Yuan Z and Shields A 2004 Quantum key distribution over 122 km of standard telecom fiber *Appl. Phys. Lett.* **84** 3762–4

[36] Liao S K *et al* 2017 Satellite-to-ground quantum key distribution *Nature* **549** 43

[37] Erhard M, Fickler R, Krenn M and Zeilinger A 2018 Twisted photons: new quantum perspectives in high dimensions *Light: Sci. Appl.* **7** 17146

[38] Forbes A and Nape I 2019 Quantum mechanics with patterns of light: progress in high dimensional and multidimensional entanglement with structured light *AVS Quantum Sci.* **1** 011701

[39] Bechmann-Pasquinucci H and Tittel W 2000 Quantum cryptography using larger alphabets *Phys. Rev. A* **61** 062308

[40] Cerf N J, Bourennane M, Karlsson A and Gisin N 2002 Security of quantum key distribution using d-level systems *Phys. Rev. Lett.* **88** 127902

[41] Ali-Khan I, Broadbent C J and Howell J C 2007 Large-alphabet quantum key distribution using energy-time entangled bipartite states *Phys. Rev. Lett.* **98** 060503

[42] Ndagano B and Forbes A 2018 Characterization and mitigation of information loss in a six-state quantum-key-distribution protocol with spatial modes of light through turbulence *Phys. Rev. A* **98** 062330

[43] Cozzolino D *et al* 2019 Orbital angular momentum states enabling fiber-based high-dimensional quantum communication *Phys. Rev. Appl.* **11** 064058

[44] Hufnagel F, Sit A, Bouchard F, Zhang Y, England D, Heshami K, Sussman B J and Karimi E 2020 Investigation of underwater quantum channels in a 30 meter flume tank using structured photons *New J. Phys.* **22** 093074

[45] Bouchard F *et al* 2018 Quantum cryptography with twisted photons through an outdoor underwater channel *Opt. Express* **26** 22563–73

[46] Chen Y *et al* 2020 Underwater transmission of high-dimensional twisted photons over 55 meters *PhotoniX* **1** 5

[47] Wang F X, Chen W, Yin Z Q, Wang S, Guo G C and Han Z F 2019 Characterizing high-quality high-dimensional quantum key distribution by state mapping between different degrees of freedom *Phys. Rev. Appl.* **11** 024070

[48] Sit A *et al* 2017 High-dimensional intracity quantum cryptography with structured photons *Optica* **4** 1006

[49] Nape I, Otte E, Vallés A, Rosales-Guzmán C, Cardano F, Denz C and Forbes A 2018 Self-healing high-dimensional quantum key distribution using hybrid spin-orbit bessel states *Opt. Express* **26** 26946–60

[50] McLaren M, Mhlanga T, Padgett M J, Roux F S and Forbes A 2014 Self-healing of quantum entanglement after an obstruction *Nat. Commun.* **5** 3248 EP

[51] Otte E, Nape I, Rosales-Guzmán C, Vallés A, Denz C and Forbes A 2018 Recovery of nonseparability in self-healing vector bessel beams *Phys. Rev.* A **98** 053818

[52] Otte E, Nape I, Rosales-Guzmán C, Denz C, Forbes A and Ndagano B 2020 High-dimensional cryptography with spatial modes of light: tutorial *J. Opt. Soc. Am.* B **37** A309–23

[53] Ferenczi A and Lütkenhaus N 2012 Symmetries in quantum key distribution and the connection between optimal attacks and optimal cloning *Phys. Rev.* A **85** 052310

[54] Takeoka M, Guha S and Wilde M M 2014 Fundamental rate-loss tradeoff for optical quantum key distribution *Nat. Commun.* **5** 5235

[55] Sheridan L and Scarani 2010 Security proof for quantum key distribution using qudit systems *Phys. Rev.* A **82** 030301

[56] Gröblacher S, Jennewein T, Vaziri A, Weihs G and Zeilinger A 2006 Experimental quantum cryptography with qutrits *New J. Phys.* **8** 75

[57] Vallone G, D'Ambrosio V, Sponselli A, Slussarenko S, Marrucci L, Sciarrino F and Villoresi P 2014 Free-space quantum key distribution by rotation-invariant twisted photons *Phys. Rev. Lett.* **113** 060503

[58] Forbes A, Dudley A and McLaren M 2016 Creation and detection of optical modes with spatial light modulators *Adv. Opt. Photon.* **8** 200–27

[59] Berkhout G C G, Lavery M P J, Courtial J, Beijersbergen M W and Padgett M J 2010 Efficient Sorting of orbital angular momentum states of light *Phys. Rev. Lett.* **105** 153601

[60] Lavery M P J, Robertson D J, Sponselli A, Courtial J, Steinhoff N K, Tyler G A, Wilner A E and Padgett M J 2013 Efficient measurement of an optical orbital-angular-momentum spectrum comprising more than 50 states *New J. Phys.* **15** 013024

[61] Walsh G F 2016 Pancharatnam-Berry optical element sorter of full angular momentum eigenstate *Opt. Express* **24** 6689

[62] Ruffato G, Girardi M, Massari M, Mafakheri E, Sephton B, Capaldo P, Forbes A and Romanato F 2018 A compact diffractive sorter for high-resolution demultiplexing of orbital angular momentum beams *Sci. Rep.* **8** 10248

[63] D'Ambrosio V, Nagali E, Walborn S P, Aolita L, Slussarenko S, Marrucci L and Sciarrino F 2012 Complete experimental toolbox for alignment-free quantum communication *Nat. Commun.* **3** 961

[64] Ndagano B *et al* 2017 Characterizing quantum channels with non-separable states of classical light *Nat. Phys.* **13** 397–402

[65] Singh K, Nape I, Buono W T, Dudley A and Forbes A 2023 A robust basis for multi-bit optical communication with vectorial light *Laser Photon. Rev.* **17** 2200844

IOP Publishing

Optical Vortices

Fundamentals and applications

Yuanjie Yang, Yu-Xuan Ren and Carmelo Rosales-Guzmán

Chapter 11

Optical trapping with optical vortices

11.1 Principle of optical tweezers

In 1986, Ashkin invented the optical tweezers, and subsequently demonstrated the trapping and manipulation of tobacco mosaic viruses and *Escherichia coli* bacteria using the single-beam optical tweezers [1]. A single cell had also been trapped in the single optical tweezers using an infrared laser beam [2]. Single beam optical gradient traps became an indispensable tool to grasp and manipulate microparticles. This is formed by focusing the collimated beam with a high numerical aperture objective lens. A slight tilt of the collimated beam at the back focal plane of the objective may result in the translation of the trapping position. By splitting and recombining the beam with a polarizing beam splitter, and introducing a slight tilt angle between the two beams at the back focal plane of the objective, dual-trap optical tweezers can be created. In some applications, more traps with reconfigurable positions and flexible patterns are desired. This can, however, be achieved by rapid scanning of the beam among traps with an acousto-optic deflector using time-sharing. A more elaborate method is to use the spatial light modulator to create a static optical potential that has the designed geometry.

Later on, the optical tweezers have been used to study a series of single molecule proteins, nucleic acids, and enzymes, and provide important and fine tools in the single molecule arsenal [3]. There are a series of review articles on high-resolution optical tweezers and single molecule biophysics [3, 4, 5]. Although the single-beam optical tweezers have great contributions in biology studies, especially single molecule biophysics [3], the complex manipulation of the biological cells are in high demand. Although advanced techniques are developing to increase the manipulation throughput, e.g., the photonic nanojet mediated backaction force [6, 7], traditional holographic optical tweezers (HOT) with structured light are still popular to manipulate multiple cells into a designed pattern, both of the same or different species, to form building blocks similar to primary tissue [8].

The HOT combines optical trapping and holography to manipulate and control the microparticles in parallel or with complex motion trajectories. Historically, HOT started from the design of the diffractive optical elements and relied on the craftsmanship of the micro-/nano-fabrication [9]. With advanced liquid crystal technology, the laser beam can be tailored to an arbitrary shape with spatial phase control [10, 11]. Such a tailored beam produces optical trapping with ab complex energy landscape and temporally-variable manipulation. In general, we categorized all the optical trapping experiments with spatial beam control as HOT. Thus, HOT includes array optical trapping, higher-order spatial modes, and interference patterns. Some applications require the manipulation of numbers of particles simultaneously and individual control of each particle. The array of optical tweezers also allows the organization of particles into complex structures in 3D and intelligent sorting of the microscopic particles [9].

11.2 Optical trapping with nondiffracting beams

The higher-order beam diffracts against propagation, which means the beam size will change along the propagation axis while keeps the shape. There exist beams with both size and shape invariant during propagation. Another feature for such a non-diffracting beam is self-healing, which implies that the beam can self-recover the shape after partial blocking of the beam. This feature can be understood by the geometric optics under catastrophe theory, specifically, the breaking of the caustic structure of an Airy beam will result in the ultimate self-healing of the beam pattern [12].

In this section, Bessel and Airy beams are discussed as examples for trapping with nondiffracting beams. A Bessel beam keeps a Bessel distribution in the transverse plane while preserves the shape during propagation. A self-reconstructing Bessel beam can manipulate the microparticles in multiple planes along a long distance [13]. The beam itself first distorts and then self-reconstructs against propagation [13].

The particle jet is propagating in the opposite direction to the hollow Bessel-like beam. On the left-hand side of figure 11.1, the particles will be guided in the hollow beam if the particle jet is aligned with the Bessel-like beam. Once the particle jet is misaligned with the beam, the particle jet will further find the trajectory to follow the beam path. The concept has also been experimentally corroborated, as evidenced in the right-hand side of figure 11.4. The particles are deposited on a sticky gel-pack microscope slide for better visualization. The laser beam is transversally misaligned by 75 µm, with respect to the jet axis. The top spot was deposited on the slide while the beam was in the 'off' state. The lower spot in presence of the beam demonstrates the shift of particle path by radiation pressure imposed by the beam.

An Airy beam was first experimentally produced by spectrum shaping of the wavefront of a collimated beam using an SLM loading the cubic phase pattern [15, 16]. The optical Airy beam has been demonstrated the excellence in optical particle clearing and the light-sheet fluorescence microscopy [17]. Baumgartl *et al* demonstrated the 'snowblowing' experiment using the nondiffracting Airy beam [18]. The Airy beam was exposed on a suspension of colloidal particles for 2 min. The field of view (FOV) was colored in four sections to highlight the blowing effect. The Airy beam cleared the section by conveying the particles along 3D parabolic trajectories

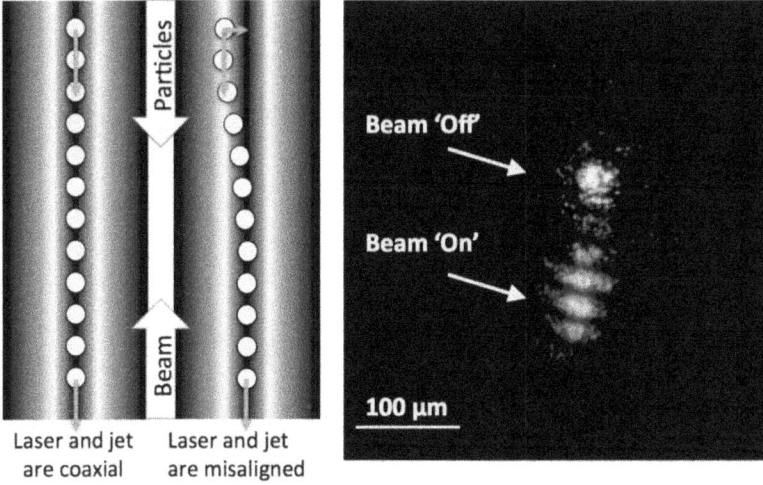

Figure 11.1. The concept of a beam diverging experiment with hollow Bessel-like beam [14]. The left panel illustrates the beam is either aligned or misaligned with the beam axis, while the right panel shows the positions of particles deposited on a sticky gel-pack microscope slide, in the absence (top) and presence (bottom) of the Bessel-like beam. Adapted with permission from [14]. © 2013 Optical Society of America.

to the diagonal section. Such particle clearing is reversible, and the particle will be directed back to the green section when the Airy beam orientation is reversed.

11.3 Optical trapping with vortex beams

Beams of light with helical wavefronts carry OAM [19–21], which can be transferred to the trapped particle resulting in the orbital rotation of the particle [22]. A helical mode $\psi_l(\mathbf{r})$ distinguished by polar angle-dependent phase factor reads

$$\psi_l(\mathbf{r}) = u(r, z)e^{-\mathrm{i}kz}e^{-\mathrm{i}l\theta}, \tag{11.1}$$

where $\mathbf{k} = k\mathbf{z}$, is the wave vector, $u(r, z)$ is the beam radial profile and l is an integer winding number termed topological charge, which can be defined as [23, 24]

$$l = \frac{1}{2\pi} \oint_c \mathrm{d}\mathbf{r} \cdot \nabla \chi = \frac{1}{2\pi} \oint_c \mathrm{d}\chi, \tag{11.2}$$

where $\chi = l\theta$ is the phase of the optical field.

The helical beam can be produced via conversion of the Gaussian beam with mode converters. However, the SLM converts the helical modes with reconfigurable topological charge. The high-resolution SLM makes it possible to produce helical modes with large topological charge. The reported topological charge can go up to $l = 90$ for a DMD and $l = 200$ for a liquid crystal SLM [24, 25].

Figure 11.2(a) shows a typical HOT setup to study the trapping behavior of microparticles in the vortex potential. The inset displays the vortex phase mask with topological charge $l = 40$. Figure 11.2(b) shows the intensity profile captured by a CCD camera for the vortex beam. The central spot is the diffraction-limited focus of a

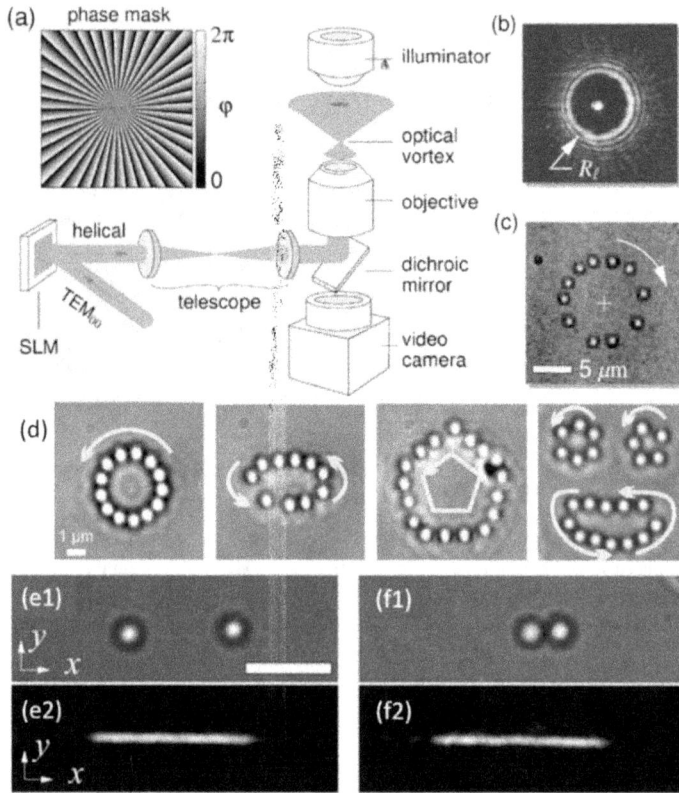

Figure 11.2. Optical phase gradient mediated force. (a) The experimental setup, where the inset displays the phase mask for a vortex beam with topological charge $l = 40$. (b) The intensity profile of the experimentally produced vortex beam. (c) Time-lapse image of a single micro-sized polymer sphere orbiting around the vortex beam. (d) Various binary structures with transverse scattering forces moving the microparticles along the bright lines [28]. (e, f) Line traps with repelling (e) and attraction (f) potential. (e1, f1) show the microscopic image of two polymer sphere in the line trap, the profiles of which are shown in (e2, f2). Reprinted with permission from [19], Copyright (2003) by the American Physical Society. Reprinted with permission from [31], Copyright (2008) by the American Physical Society. Reprinted from [28], Copyright (2008), with permission from Elsevier.

separate coaxial TEM_{00} mode that coincides with the optical axis. Figure 11.2(c) shows the time-lapse image of a single colloidal sphere orbiting around the optical vortex. With full amplitude and phase modulation [26], the vortex phase structure can be embedded into many transverse intensity patterns including pedal-like structure [27]. Figure 11.2(d) shows the movement of microspheres along various curves due to the transverse scattering force produced by the vortex phase [28]. Figure 11.2(e) is a holographic line trap carrying a phase gradient in the x-direction with a repelling potential. Figure 11.2(f) shows the line trap with attracting potential.

Higher-order spatial mode may trap and confine particles with low (hollow sphere) or complex (metal) indices with optimized efficiency [29]. Mathieu beams, described in elliptic coordinates, were demonstrated its excellence in particle organization and assembly [30]. In this section, we show that the interference of the LG beam with a plane wave can produce complex energy landscapes for optical manipulation.

A single-ringed ($p = 0$) Laguerre–Gaussian (LG) beams are described by the azimuthal index l, which corresponds to the number of complete cycles (2π) of phase around the mode circumference [32]. An LG mode with $l = 3$ can be thought of as consisting of phase fronts with a triple start helix. Interference of such a beam with a plane wave converts the azimuthal phase variation into an azimuthal intensity variation, resulting in a pattern with l spiral arms [32]. These spiral arms are determined by the mismatch in both the Gouy phase and the respective wavefront curvatures, causing the pattern to vary azimuthally as the interference propagates. Objects can be trapped in the spiral arms of the pattern due to optical gradient force. The variation of the optical path length induces the rotation of the pattern and thus the trapped object thanks to the helical nature of the phase fronts. The rotation of microparticles constitutes an all-optical microstirrer and has the potential for applications in optically-driven micromachines and motors. This technique does not depend on intrinsic properties of the particle and offers important applications in optical micromachines.

By interfering LG beams with opposite topological charges (index l and $-l$, opposite beam helicity), a spatial pattern of $2l$ spots arranged on the circumference of a ring can be produced. This is fulfilled by passing the LG beam through a Mach–Zehnder interferometer with a dove prism in one arm to invert the handedness of the helical wavefront. These interference patterns could be applied to create the extended 3D crystalline structure [33]. MacDonald *et al* have stacked as many as 16 spheres (5 μm in diameter) and moved the chain as a whole structure. The stacks of spheres can be inclined by tilting the tweezing beam to an angle of 5°. The trapped structure can be controllably and continuously rotated by a frequency difference between the two beams [33]. The rotation of the pattern around its axis of symmetry can be induced using the angular Doppler effect [34], which is a simple but effective way to introduce the relatively small frequency difference required (from less than 1 Hz to 1 kHz) between two light beams for the controlled rotation of the interference pattern. A frequency difference results in the rotation of the whole interference pattern of spots at a frequency of Ω/l, where Ω is the rotation rate of a half-wave plate (HWP) in one arm [33].

Moreover, the LG beam with OAM can be utilized to rotate the biological cells. Figure 11.3 shows the video snapshots for the biconcave red blood cells rotation in the OAM beam with topological charge switched from (i) $l = 15$ to (ii) $l = -15$. The reported rotation frequency was ~12 rpm at ~15 mW of trapping beam power [35]. Figure 11.3(ii) shows the reversed rotation of the same RBC when the topological charge is $l = -15$. This allows the change of the rotation direction of the micro-motor by altering the helicity of the trapping beam. The control over the rotation can facilitate the bi-directional operation for micro-machine components like micromotors [35].

i

ii

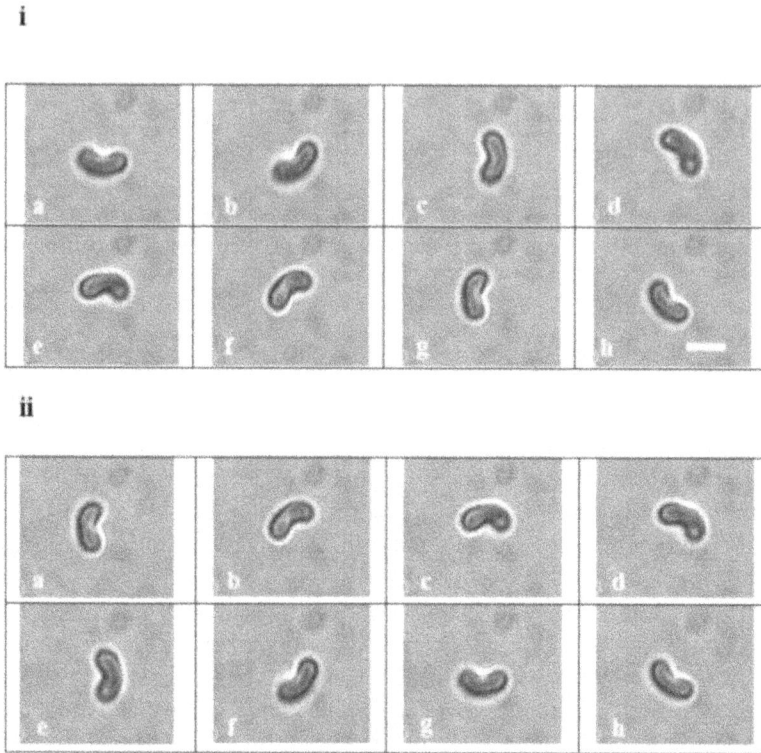

Figure 11.3. Optical rotation of a trapped RBC through OAM transfer with magnitude of topological charge of 15. The snapshots of the video shows reversal rotation as the topological charge of the LG beam switched from (i) $l = 15$ to (ii) $l = -15$. The cell rotates by approximately 45° with respect to the previous frame with a time separation of ∼625 ms. Scale bar: ∼ 5 µm. Adapted with permissions from [35]. © 2011 Optical Society of America.

11.4 Holographic optical tweezers

The ability to independently manipulate multiple particles in multiple traps was first demonstrated by use of the galvo-scanning mirror or acousto-optic beam deflectors (AOD). The holographic optical tweezers (HOT) were first demonstrated with the diffractive optical element (DOE) to tailor the beam into desired trapping potential [9]. The DOE pattern can be etched on the silica substrate, such that the DOE provides a phase modulation to the beam.

The fundamental problem in HOT is to design a 2D phase distribution $\Phi^{in}(r)$ with which the incident beam evolves to a desired pattern. Figure 11.4(a) depicts the relationship of the electric field between the input and the focal planes [9]. The monochromatic plane wave modulated by a phase $\Phi^{in}(r)$ at the input reads, $E^{in}(r) = A^{in}(r) \exp [\Phi^{in}(r)]$, where amplitude and imposed phase are real-valued functions. The electric field at the focal plane reads $E^f(\rho) = A^f(\rho) \exp [\Phi^f(\rho)]$. These fields constitute a Fourier transform pair,

$$E^f(\rho) = \frac{k}{2\pi f} \int d^2 r E^{\text{in}}(r) e^{-i k r \cdot \rho / f}, \qquad (11.3)$$

where f is the focal length of the lens and $k = 2\pi/\lambda$ is the wave number. Since there is no analytical solution for the phase distribution, usually an iterative algorithm, e.g., adaptive-addictive algorithm, is applied to search and optimize for the phase [9]. The iterative algorithms can also be used to design 3D structured patterns [36].

A typical binary version of the DOE pattern for hexagonal tweezers array is shown in figure 11.4(b). The trapped microspheres in the array are shown in figure 11.4(c). With modern SLM, the phase pattern can be made with more steps in the form of quasi continuous pattern. Figure 11.4(d) shows the phase pattern for a 20×20 array tweezers. The tweezers array distribution is shown in figure 11.4(e) with trapped microsphere image in inset [25]. The SLM tailors the wavefront to the

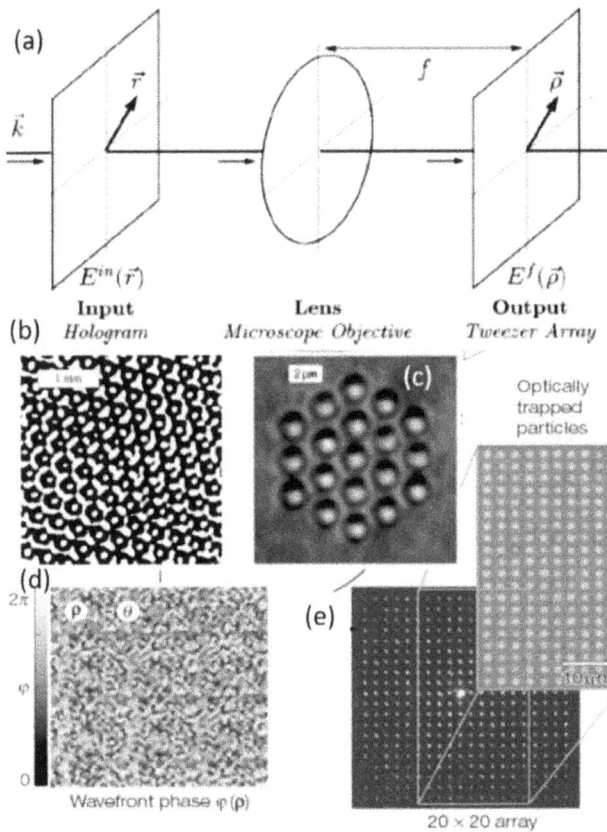

Figure 11.4. Scheme for array optical tweezers. (a) The geometry to calculate the phase modulation pattern for a desired arrangement of optical traps [9]. (b) Diffractive optical element (black regions shift the phase by π radians) etched on the fused silica substrate for hexagonal array tweezers. (c) 1 μm diameter silica spheres trapped in hexagonal tweezers. (d) Holographic phase pattern to produce 20×20 array tweezers. (e) Intensity distribution for 20×20 tweezers and the trapped microparticles (inset) [25]. Reproduced with permission from American Institute of Physics [9], and reprinted from [25], Copyright (2002), with permission from Elsevier.

desired trapping potential, and compensates for the wavefront distortion and attenuation [37, 38]. Use of the motion statistics of the optically trapped multiple particles can also probe the aberration of the optical system [39]. In real biological applications, the propagation of the beam inside the sample will experience strong scattering-induced distortion. A more exciting feature of the SLM is that the high refresh rate enables dynamic and interactive control of holographic optical tweezers [25].

In addition to array tweezers that form discrete optical potentials, the optical tweezers can be extended into 3D with the holographic technique. Some 2D or 3D continuous optical potentials, e.g., 3D spiral and ring-like structures, provide a strong axial intensity gradient to trap objects [40]. Figures 11.5(a) and (b) show the 3D profile of a spiral both in theory and experiment. An optical solenoid characterizes a 3D intensity pattern where the intensity spirals around the central axis with wavefront carrying an independent helical pitch [41]. Solenoid beams trap the microscopic objects in 3D while their phase-gradients drive the particle along the spiral. More sophisticated optical potentials become possible by applying and additional inclination of the ring-like structure. A pair of interlocking bright rings in the form of a Hopf link has been produced. The interpenetrating rings act as 3D optical traps and can organize colloidal silica spheres (figure 11.5(e)). Interestingly, the trapped spheres could freely pass each other along the knotted rings [42]. It is worth to mention that with the advanced SLM, the HOT can be made interactive; for instance, the microparticles can be controlled by multi-touch screen by moving, adding or removing the traps [36, 43].

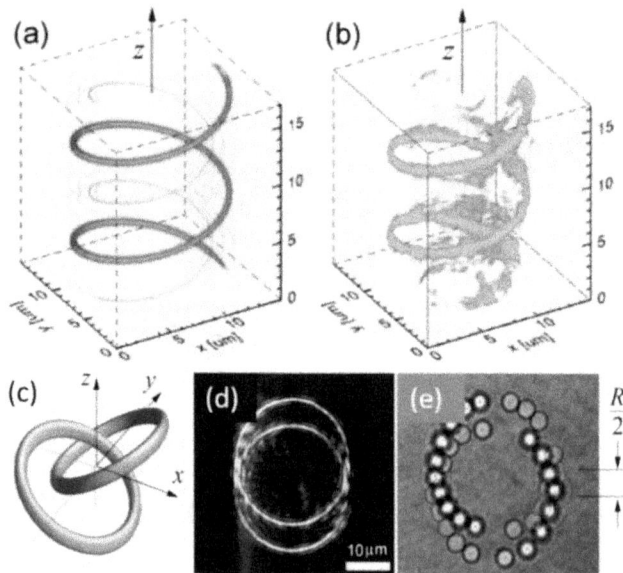

Figure 11.5. Optical manipulation with 3D curvilinear structure. (a) Calculated 3D intensity distribution of a solenoid beam propagating in the z-direction. (b) Volumetric rendering of the measured intensity [41]. (c) Schematic representation of the three-dimensional interwound rings. (d) Experimental intensity distribution of two tilted ring traps simultaneously with opposite inclination. (e) Colloidal silica spheres trapped in 3D within the focused Hopf link. Adapted with permission from [42]. © 2010 Optical Society of America and [43]. © 2011 Optical Society of America.

Optical tweezers bring the colloidal chemistry research into single particle level. Dynamic interaction between two microparticles can be studied by placing two particles using dual-trap optical tweezers and recording the dynamical trajectories after switching off the tweezers [44]. Array optical tweezers not only assemble and construct colloidal structures, but also can be used to study the many-body colloidal interaction [45], such as force and mobility measurement among multiple particles [46], and the manipulation of particles with complex geometry, like the bioconcave red blood cells [47].

11.5 Optical manipulation with metallic nanoparticles

The light may induce a dipole in the metallic nanoparticle and allows the trapping of metallic particle, which offers exciting applications in trapping combined with surface polariton and evanescent waves [48]. Recent advancement in radially polarized beam for the trapping of metallic nanoparticles provides opportunities to study the nonlinear light-matter interaction [49, 50]. The refraction and the dominant gradient force are the bases for stable optical trapping of dielectric particles, while, the laser beam can induce relatively large scattering and absorption forces on metal particles close to their localized surface plasmon resonances.

In the early days, it was generally accepted that the stable optical trapping of metal nanoparticles cannot be obtained readily [51]. However, in 1994 Svoboda and Block demonstrated that metal Rayleigh particles can be trapped stably in three dimensions [52]. In their experiment, the near-infrared laser beam was used to trap the gold nanoparticle, i.e., the optical trapping was achieved off-resonance. Interestingly, in 2008, Dienerowitz et al experimentally demonstrated the metal nanoparticles can be trapped by laser beam close to their plasmonic resonance [53]. They showed that a vortex beam (Laguerre–Gaussian beam with annular profile) can confine the metal nanoparticles in the dark region of the beam centre. Since the vortex beam carries OAM, they observed the rotation of particles as well.

Later, Lehmuskero et al optically trapped gold nanoparticles and set them into orbital rotation with an orbiting frequency of 86 Hz using an LG vortex beam carrying OAM [55]. In their experiment, to achieve a stable trap, a q-plate (figure 11.6(b)) was adopted to produce a vortex beam with uniform intensity distribution and a circular polarized laser beam was used to induce circular symmetric optical gradient forces. Recently, the dynamics of electrodynamically coupled metal nanoparticles in optical ring vortex trap was studied [54]. Figure 11.6 shows the schematic of the focused Bessel–Gauss optical vortex beam and ring trap over a gold nanoplate. They used a retroreflection geometry with a gold nanoplate mirror to generate a constant-radius optical vortex. Note the Au nanoplate overfills the panel. The curved arrows in (c) and (d) indicate the direction of rotation of the nanoparticles in the optical vortex ring trap. They demonstrated that, comparing with a glass coverslip, the retroreflection geometry can significantly increase the spatial confinement and optical drive force, and a superior trap can be created accordingly.

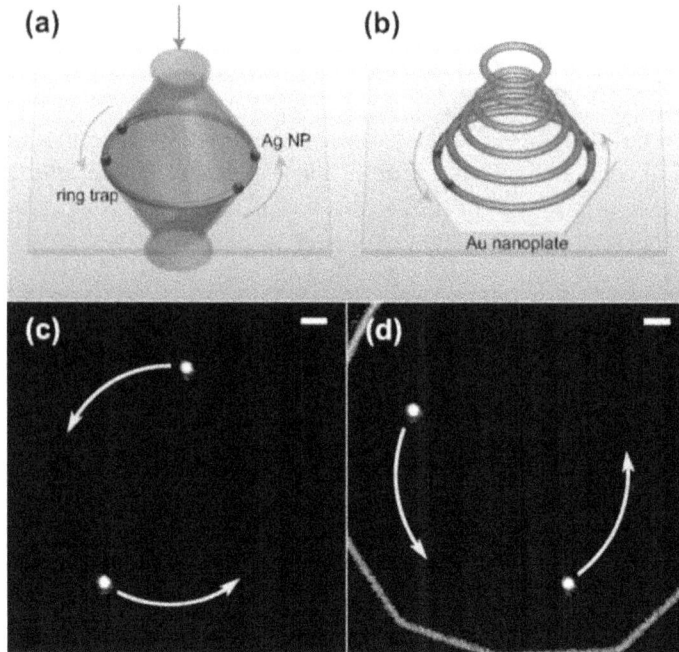

Figure 11.6. Schematic of the focused optical beam and ring trap over (a) glass coverslip and (b) an Au nanoplate. A dark-field microscopic image of two Ag nanoparticles trapped (c) over glass and (d) over an Au nanoplate mirror. The white scale bars in (c) and (d) is 1 μm. Reprinted figure with permission from [54], Copyright (2017) by the American Physical Society.

11.6 Optical trapping with structured light

The interference of LG beams with a plane wave suggests the trapping with structured light. In general, many higher-order spatial modes can be used for optical trapping. In this section, we demonstrate the Ince–Gaussian (IG) beam, as a transition mode between the HG and LG beams, for optical trapping. Experimentally, the higher-order spatial modes can also be produced extracavity with good quality using advanced diffractive optics [56], liquid crystal SLM [11], and digital micromirror devices (DMD) [57], but also can be created intra-cavity by inserting the SLMs inside the cavity to make digital lasers [58]. Figure 11.7 shows the experimentally produced (a–c) even IG $IG^e_{7,5}$, and (d–f) odd $IG^o_{8,6}$ with ellipticities (a) and (d) $\varepsilon = 0.8$, (b) and (e) $\varepsilon = 10$, (c) and (f) $\varepsilon = 1000$ using a DMD device [59].

IG mode is a more general class and can be utilized to assemble and organize particles on microscale level into highly ordered assemblies. Woerdemann *et al* utilized the IG beam on a microscopic scale and enabled advanced manipulation of microparticles [60]. Figures 11.7(g)–(j) demonstrate mode pattern (top) and microparticles trapped in IG mode (bottom) of order (g) $IG^o_{5,5}$, (h) $IG^o_{2,2}$, (i) $IG^o_{4,2}$, (j) $IG^e_{14,14}$.

Although the biological cells have a refractive index above that of the surrounding environment (positive polarizability), the refractive index contrast is very weak.

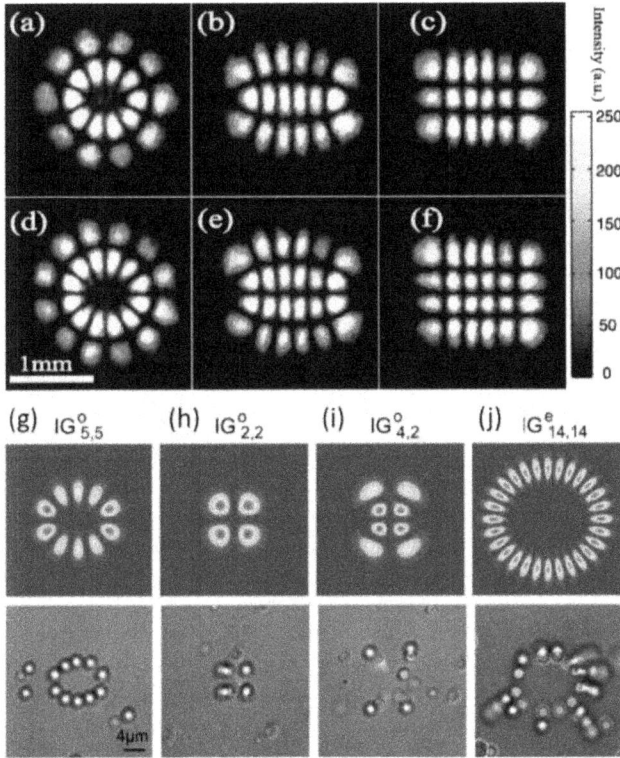

Figure 11.7. Optical trapping with Ince–Gaussian beam. (a–c) demonstrate profiles for even IG $IG^e_{7, 5}$, and (d–f) show the patterns for $IG^o_{8, 6}$. The ellipticities are (a) and (d) $\varepsilon = 0.8$, (b) and (e) $\varepsilon = 10$, (c) and (f) $\varepsilon = 1000$. Beam profiles and respective trapping of microsphere in the beam for (g) $IG^o_{5, 5}$, (h) $IG^o_{2, 2}$, (i) $IG^o_{4, 2}$, (j) $IG^e_{14, 14}$. Reproduced with permission from APS [59, 60].

Hence, the trapping of biological cells is less stable as compared with the standard polymer sphere. However, the cells may assume different geometry, i.e., rod shape or irregular shape, holographic optical tweezers could provide multiple traps to hold different parts of the biological cells with weak refractive index contrast. For instance, the dual-trap holographic tweezers could hold two ends of a bacteria and align the bacteria horizontally (schematics in figure 11.8(a)), in contrast, the bacterium is aligned vertically in a single-beam optical trap (figure 11.8(b)). Figure 11.8(c) shows the T-cell trapped in the single-beam optical trap, which clearly suggests that the cell experiences fierce rotation while the stage is moving. Figure 11.8(d) shows the position distribution of a cell (black) and a polymer sphere (red). The T-cell is more scattered around the tweezers in contrast to the polymer sphere. Holographic dual-trap tweezers were able to hold the T-cell more stably than a single optical trap [61]. One benefit of using the multiple optical tweezers is to align the orientation of the cell for specific applications, e.g., super-resolution nanoscopy towards a certain direction [62]. This is another advantage to combine the HOT with structured light and the interdisciplinary methods, e.g., super-resolution microscopy and Raman spectroscopy.

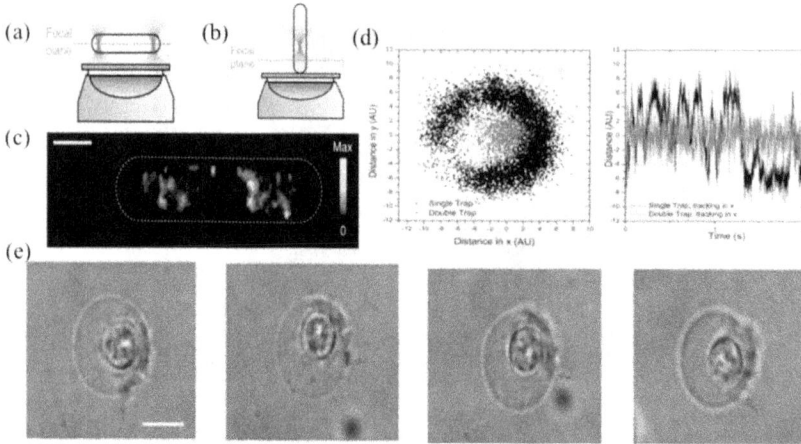

Figure 11.8. Strategies for stable optical trapping of rod-shaped bacteria. Schematics for (a) holographic dual-trap optical tweezers and (b) conventional single-trap optical tweezers. (c) The T-cell under single-beam optical tweezers experiences rotation in the presence of stage motion. (d) The locations of a single cell (black) and a standard polymer sphere (red) in a single optical tweezers, and the positional traces (right) [61]. (e) The combination of *d*STORM and optical trapping allows isotropic super-resolution of 2D localization microscopy for each orientation of the rod-shaped bacterium [62]. (f) The schematics of tug-of-war tweezers for the study of bacteria disassembly. Adapted with permission from [61]. © 2014 Optical Society of America, Reproduced from [62]. CC BY 4.0.

Figure 11.8(e) shows the super-resolution image of the rod-shaped bacteria hold in dual-trap optical tweezers [62]. Finally, with the structured tug-of-war tweezers, the rod-shaped bacteria can be disassembled to combat the biofilm formation [63].

Moreover, optical tweezers and shaped structured light provide indispensable and noncontact means of optical manipulation. For *in vitro* applications, optical manipulation now enables the assembly of different biophotonic components and devices for further biomedical researches, for example biophotonic waveguides, biosensors and biodetectors, bio-microlens and biomagnifiers, and bio-micromotors.

11.7 Mind-controlled optical rotation

The conventional optical manipulation adopts human hands to control the software and design the control sequences on the optical tweezer setup. The recent progress in brain-computer interfaces (BCIs) allows the direct recording and dissection of the brainwaves to control optical intensity distributions. Prior efforts include the use of brainwaves to control the metasurface to alter the scattering pattern off the metasurface according to the user's brainwaves via Bluetooth [64]. It is highly unexplored on the microscopic level. Peng and colleagues demonstrated the use of brainwaves to control both the magnitude and the rotation direction of the microscopic particles [65].

Figure 11.9(a) shows the schematics of the mind-controlled optical rotation [65]. The human brainwaves are collected by the TGAM, and the attention value is extracted to encode the topological charge of the optical beam. Figure 11.9(b) showcases the assignment of different topological charge into the attention value.

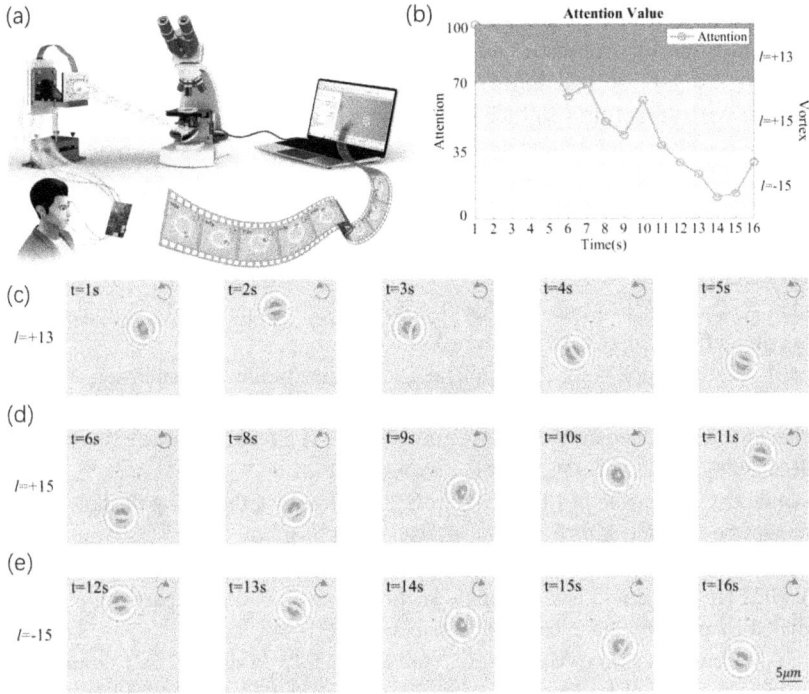

Figure 11.9. Mind-controlled optical rotation [65]. (a) Schematic setup of the mind-controlled optical rotation. (b) The attention value recorded by the brain-computer interface. The orbital rotation of dielectric microparticles three levels of attention values for OAM of (c) $l = +13$, (d) $l = +15$, and (e) $l = -15$. Reprinted with permission from [65], Copyright (2024), American Chemical Society.

The magnitude of the topological charge is directly used to synthesize a computer-generated hologram. The hologram is further sent to the SLM to shape the beam into optical vortices. The vortex beams are then used by the microscope to form a rotational trap. Figures 11.9(c)–(e) demonstrate the microscopic images from the video series suggesting that the microparticles can rotate according to the strength of brainwaves. Notably, both the rotation magnitude (c and d) and the rotation direction (d and e) can be controlled. From this demonstration, we foresee that the brainwaves may further offer even finer control of the orbitation rotation, and more complex manipulation of microscopic and even nanoscopic particles.

11.8 Conclusion

The optical vortex has changed the fashion of optical manipulation, allowing the rotational control of microparticles in liquid and vacuum environment for biomedical and fundamental researches. The optical vortex beam, however, will continue to offer rotational degrees of freedom to gyrate the micro-/nanoparticles, and will be useful in nanomotors in biomedical research.

Through the demonstration of the structured light and the optical vortex beam in the optical trapping community, one would expect that the optical vortex with the spiral phase structure may further offer complex degrees of freedom to manipulate particles with synthetic materials, particles of complex spatial structures, and even biological particles with the soft optical hand. We are confident that the optical trapping and manipulation with structured light will further expand in the coming decades.

References

[1] Ashkin A and Dziedzic J M 1987 Optical trapping and manipulation of viruses and bacteria *Science* **235** 1517–20

[2] Ashkin A, Dziedzic J M and Yamane T 1987 Optical trapping and manipulation of single cells using infrared laser beams *Nature* **330** 769–71

[3] Fazal F M and Block S M 2011 Optical tweezers study life under tension *Nat. Photon.* **5** 318–21

[4] Bustamante C J, Chemla Y R, Liu S and Wang M D 2021 Optical tweezers in single-molecule biophysics *Nat. Rev. Methods Primers* **1** 1–25

[5] Bustamante C, Macosko G J L and Wuite S 2000 Grabbing the cat by the tail: manipulating molecules one by one *Nat. Rev. Mol. Cell Biol.* **1** 130–6

[6] Ren Y-X, Yip G G K, Zhou L-M, Qiu C-W, Shi J, Zhou Y, Mao H, Tsia K K and Wong K K Y 2022 Hysteresis and balance of backaction force on dielectric particles photothermally mediated by photonic nanojet *Nanophotonics* **11** 4231–44

[7] Ren Y-X, Zeng X, Zhou L-M, Kong C, Mao H, Qiu C-W, Tsia K K and Wong K K Y 2020 Photonic nanojet mediated backaction of dielectric microparticles *ACS Photonics* **7** 1483–90

[8] Zhang Z, Kimkes T E P and Heinemann M 2019 Manipulating rod-shaped bacteria with optical tweezers *Sci. Rep.* **9** 19086

[9] Dufresne E R, Spalding G C, Dearing M T, Sheets S A and Grier D G 2001 Computer-generated holographic optical tweezer arrays *Rev. Sci. Instrum.* **72** 1810–6

[10] Reicherter M, Haist T, Wagemann E U and Tiziani H J 1999 Optical particle trapping with computer-generated holograms written on a liquid-crystal display *Opt. Lett.* **24** 608–10

[11] Grier D G 2003 A revolution in optical manipulation *Nature* **424** 810–6

[12] Vaveliuk P, Martínez-Matos Ó, Ren Y-X and Lu R-D 2017 Dual behavior of caustic optical beams facing obstacles *Phys. Rev. A* **95** 063838

[13] Garcés-Chávez V, McGloin D, Melville H, Sibbett W and Dholakia K 2002 Simultaneous micromanipulation in multiple planes using a self-reconstructing light beam *Nature* **419** 145–7

[14] Eckerskorn N, Li L, Kirian R A, Küpper J, DePonte D P, Krolikowski W, Lee W M, Chapman H N and Rode A V 2013 Hollow Bessel-like beam as an optical guide for a stream of microscopic particles *Opt. Express.* **21** 30492–9

[15] Siviloglou G A and Christodoulides D N 2007 Accelerating finite energy Airy beams *Opt. Lett.* **32** 979–81

[16] Siviloglou G A, Broky J, Dogariu A and Christodoulides D N 2007 Observation of accelerating airy beams *Phys. Rev. Lett.* **99** 213901

[17] Vettenburg T, Dalgarno H I C, Nylk J, Coll-Llado C, Ferrier D E K, Cizmar T, Gunn-Moore F J and Dholakia K 2014 Light-sheet microscopy using an airy beam *Nat. Methods* **11** 541–4

[18] Baumgartl J, Mazilu M and Dholakia K 2008 Optically mediated particle clearing using airy wavepackets *Nat. Photon.* **2** 675–8

[19] Curtis J E and Grier D G 2003 Structure of optical vortices *Phys. Rev. Lett.* **90** 133901

[20] Allen L, Beijersbergen M W, Spreeuw R J C and Woerdman J P 1992 Orbital angular momentum of light and the transformation of Laguerre–Gaussian laser modes *Phys. Rev. A* **45** 8185–9

[21] Allen L, Padgett M J and Babiker M 1999 IV the orbital angular momentum of light *Prog. Opt.* **39** 291–372

[22] Friese M E J, Enger J, Rubinsztein-Dunlop H and Heckenberg N R 1996 Optical angular-momentum transfer to trapped absorbing particles *Phys. Rev. A* **54** 1593

[23] Dennis M R, O'Holleran K and Padgett M J 2009 Singular optics: optical vortices and polarization singularities *Progress in Optics* ed E Wolf (Amsterdam: Elsevier) pp 293–363

[24] Chen Y, Fang Z-X, Ren Y-X, Gong L and Lu R-D 2015 Generation and characterization of a perfect vortex beam with a large topological charge through a digital micromirror device *Appl. Opt.* **54** 8030–5

[25] Curtis J E, Koss B A and Grier D G 2002 Dynamic holographic optical tweezers *Opt. Commun.* **207** 169–75

[26] Jesacher A, Maurer C, Schwaighofer A, Bernet S and Ritsch-Marte a M 2008 Full phase and amplitude control of holographic optical tweezers with high efficiency *Opt. Express* **16** 4479–86

[27] Curtis J E and Grier D G 2003 Modulated optical vortices *Opt. Lett.* **28** 872–4

[28] Jesacher A, Maurer C, Fürhapter S, Schwaighofer A, Bernet S and Ritsch-Marte M 2008 Optical tweezers of programmable shape with transverse scattering forces *Opt. Commun.* **281** 2207–12

[29] O'Neil A T and Padgett M J 2001 Axial and lateral trapping efficiency of Laguerre–Gaussian modes in inverted optical tweezers *Opt. Commun.* **193** 45–50

[30] Alpmann C, Bowman R, Woerdemann M, Padgett M and Denz C 2010 Mathieu beams as versatile light moulds for 3D micro particle assemblies *Opt. Express* **18** 26084–91

[31] Roichman Y, Sun B, Roichman Y, Amato-Grill J and Grier D G 2008 Optical forces arising from phase gradients *Phys. Rev. Lett.* **100** 013602

[32] Paterson L, MacDonald M P, Arlt J, Sibbett W, Bryant P E and Dholakia K 2001 Controlled rotation of optically trapped microscopic particles *Science* **292** 912–4

[33] MacDonald M P, Paterson L, Volke-Sepulveda K, Arlt J, Sibbett W and Dholakia K 2002 Creation and manipulation of three-dimensional optically trapped structrues *Science* **296** 1101–3

[34] Garetz B A 1981 Angular Doppler effect *J. Opt. Soc. Am.* **71** 609–11

[35] Dasgupta R, Ahlawat S, Verma R S and Gupta P K 2011 Optical orientation and rotation of trapped red blood cells with Laguerre–Gaussian mode *Opt. Express* **19** 7680–8

[36] Sinclair G, Leach J, Jordan P, Gibson G, Yao E, Laczik Z J, Padgett M J and Courtial J 2004 Interactive application in holographic optical tweezers of a multi-plane Gerchberg-Saxton algorithm for three-dimensional light shaping *Opt. Express* **12** 1665–70

[37] Wulff K D, Cole D G, Clark R L, DiLeonardo R, Leach J, Cooper J, Gibson G and Padgett a M J 2006 Aberration correction in holographic optical tweezers *Opt. Express* **14** 4169–74

[38] Preciado M A, Dholakia K and Mazilu M 2014 Generation of attenuation-compensating airy beams *Opt. Lett.* **39** 4950–3

[39] Roichman Y, Waldron A, Gardel E and Grier D G 2006 Optical traps with geometric aberrations *Appl. Opt.* **45** 3425–9

[40] Woerdemann M, Alpmann C, Esseling M and Denz C 2013 Advanced optical trapping by complex beam shaping *Laser Photonics Rev.* **7** 839–54

[41] Lee S-H, Roichman Y and Grier D G 2010 Optical solenoid beams *Opt. Express* **18** 6988–93

[42] Shanblatt E R and Grier D G 2011 Extended and knotted optical traps in three dimensions *Opt. Express* **19** 5833–8

[43] Grieve J A, Ulcinas A, Subramanian S, Gibson G M, Padgett M J, Carberry D M and Miles M J 2009 Hands-on with optical tweezers: a multitouch interface for holographic optical trapping *Opt. Express* **17** 3595–602

[44] Crocker J C and Grier D G 1994 Microscopic measurement of the pair interaction potential of charge-stabilized colloid *Phys. Rev. Lett.* **73** 352–5

[45] Larsen A E and Grier D G 1997 Like-charge attractions in metastable colloidal crystallites *Nature* **385** 230–3

[46] Merrill J W, Sainis S K, Blawzdziewicz J and Dufresne E R 2010 Many-body force and mobility measurements in colloidal systems *Soft Matter* **6** 2187–92

[47] Ren Y-X, Lamstein J, Zhang C, Conti C, Christodoulides D N and Chen Z 2023 Biophotonic rogue waves in red blood cell suspensions *Photon. Res* **11** 1838–46

[48] Lehmuskero A, Johansson P, Rubinsztein-Dunlop H, Tong L and Käll M 2015 Laser trapping of colloidal metal nanoparticles *ACS Nano* **9** 3453–69

[49] Rui G and Zhan Q 2014 Trapping of resonant metallic nanoparticles with engineered vectorial optical field *Nanophotonics* **3** 351–6

[50] Jiang Y, Narushima T and Okamoto H 2010 Nonlinear optical effects in trapping nanoparticles with femtosecond pulses *Nat. Phys.* **6** 1005–9

[51] Ashkin A 1980 Applications of laser radiation pressure *Science* **210** 1081–8

[52] Svoboda K and Block S M 1994 Optical trapping of metallic Rayleigh particles *Opt. Lett.* **19** 930–2

[53] Dienerowitz M, Mazilu M, Reece P J, Krauss T F and Dholakia K 2008 Optical vortex trap for resonant confinement of metal nanoparticles *Opt. Express* **16** 4991–9

[54] Figliozzi P, Sule N, Yan Z, Bao Y, Burov S, Gray S K, Rice S A, Vaikuntanathan S and Scherer N F 2017 Driven optical matter: dynamics of electrodynamically coupled nanoparticles in an optical ring vortex *Phys. Rev.* E **95** 022604

[55] Lehmuskero A, Li Y, Johansson P and Käll M 2014 Plasmonic particles set into fast orbital motion by an optical vortex beam *Opt. Express* **22** 4349–56

[56] Kennedy S A, Szabo M J, Teslow H, Porterfield J Z and Abraham E R I 2002 Creation of Laguerre–Gaussian laser modes using diffractive optics *Phys. Rev.* A **66** 043801

[57] Ren Y-X, Lu R-D and Gong L 2015 Tailoring light with a digital micromirror device *Ann. Phys.* **527** 447–70

[58] Tsai K-F and Chu S-C 2018 Generating laser output with arbitrary lateral shape by using multi-point beam superposition method in digital lasers *Laser Phys.* **28** 075801

[59] Ren Y-X, Fang Z-X, Gong L, Huang K, Chen Y and Lu R-D 2015 Dynamic generation of Ince–Gaussian modes with a digital micromirror device *J. Appl. Phys.* **117** 133106

[60] Woerdemann M, Alpmann C and Denz C 2011 Optical assembly of microparticles into highly ordered structures using Ince–Gaussian beams *Appl. Phys. Lett.* **98** 111101

[61] McAlinden N, Glass D G, Millington O R and Wright A J 2014 Accurate position tracking of optically trapped live cells *Biomed. Opt. Express* **5** 1026–37

[62] Diekmann R, Wolfson D L, Spahn C, Heilemann M, Schüttpelz M and Huser T 2016 Nanoscopy of bacterial cells immobilized by holographic optical tweezers *Nat. Commun.* **7** 13711

[63] Bezryadina A S, Preece D C, Chen J C and Chen Z 2016 Optical disassembly of cellular clusters by tunable 'tug-of-war' tweezers *Light: Sci. Appl.* **5** e16158–8

[64] Zhu R *et al* 2022 Remotely mind-controlled metasurface via brainwaves *eLight* **2** 10

[65] Peng L, Yao J, Bai Y, Sun Y, Zeng J, Ren Y-X, Xie J, Hu Z, Zhang Q and Yang Y 2024 Mind-controlled optical manipulation *ACS Photonics* **11** 1213–20

IOP Publishing

Optical Vortices

Fundamentals and applications

Yuanjie Yang, Yu-Xuan Ren and Carmelo Rosales-Guzmán

Chapter 12

Biomedical imaging with optical vortex

In this chapter, light beam sculpturing has benefited many of the applications listed. We introduce representative biomedical applications using the optical vortex and nondiffracting beams, for instance, in multiphoton microscopes (MPM), and optical manipulation. Specifically, the optical vortex beam can improve the resolution in volumetric two-photon microscopy, and the Airy beam can improve the axial resolving ability in volumetric microscopy. We will show some representative examples of how the sculptured optical vortex beams are used in biomedical applications.

12.1 Grant challenges in biomedical imaging

The structured light has been applied to address critical challenges in biomedical application, including optical trapping and optical imaging. In this chapter, we demonstrate how the challenges in nonlinear microscopy can be addressed through structured light. Nonlinear microscopy includes two-photon and three-photon microscopy, second harmonic generation, and coherent anti-Stokes Raman scattering (CARS) microscopy, etc [1]. The nondiffracting Bessel and Airy beams have been applied in two-photon microscopy. In 2018, three-photon microscopy (3PM) with Bessel beam excitation was developed for neuronal applications [2–4]. The CARS microscopy also benefits from the nondiffracting beam to collect volumetric chemical projection images [5]. In this chapter, grand challenges in nonlinear microscopy will be demonstrated by using the nondiffracting structured beams.

With the structured needle-like beam, the scanning time for volumetric imaging can be greatly reduced with greatly improved imaging speeds. The structured light can also improve the axial resolution in volumetric imaging. Moreover, the optical coherence tomography can also benefit from the structured light. We demonstrate the use of structured light and optical vortex to address the grand challenges in volumetric nonlinear microscopy, e.g., laser-scanning two-photon microscopy (TPM) [6, 7]. Traditional laser confocal microscopy can only image in a very

doi:10.1088/978-0-7503-5844-6ch12

limited depth, i.e., $\sim 200\,\mu m$. TPM helps with the understanding of deep tissue activities in neuron science [8–11], owing to the deep penetration and high signal-to-background ratio. Multiphoton microscopy can image structure much deeper due to the use of longer wavelength and multiphoton excitation [12]. However, the use of longer wavelength results in larger excitation point-spread-function, and reduced spatial resolution.

In the following sections, we introduce representative biomedical applications using the nondiffracting beams, for instance, how the volumetric imaging under TPM performs faster by using the nondiffracting beams. Examples include the multiphoton microscope (MPM) [13], light-sheet fluorescence microscope (LSFM) [14, 15], and optical manipulation [16]. The optical vortex beam, with phase singularity and null intensity would also benefit the biomedical imaging and micromanipulation.

12.2 Volumetric microscopy with structured light

Biological structures have complex structure in space, and require raster scanning to map the spatial morphology. Traditional wide-field fluorescence microscopy can map the volumetric information by simply scanning the biological tissue layer by layer, however, it can only image structures within limited depth, especially when the strong scattering exists in most biological tissues. Two-photon microscopy can improve the depth imaging ability through nonlinear excitation, but relies on point-by-point scanning to map the image [12]. For volumetric imaging, the number of voxels increases dramatically as the image goes from two dimensions (2D) to three dimensions (3D). In various 3D imaging applications, e.g., the neurons, the sample structure is sparse and 2D mapping of the volumetric information is enough to understand the temporal events, e.g., the neuronal activity. It is possible to use the structured nondiffracting beam to perform the projected image of the whole volume with much reduced scanning time.

Both the Bessel [17, 18] and Airy beam [19, 20] can extend the depth of field (eDOF) through projection of the volume onto a single 2D image. This approach not only provides a fast speed volumetrically [17], but also suggests larger penetration depth in tissues owing to the self-healing nature of the nondiffracting beam [20, 21]. The nondiffracting Bessel beam preserves the shape over a distance much longer than the Rayleigh range of the carrying Gaussian beam, and has been applied to perform multiphoton volumetric imaging for sparse sample with unprecedented scanning speed.

The nondiffracting Bessel beam improves the volumetric detection speed. The point-spread functions (PSFs) of the 3PM (figure 12.1(a)) and 2PM (figure 12.1(b)) show needle-like structure, and have been in good agreement with experiment (figures 12.1(c) and (d)). The transverse profiles after integration of the Bessel beam along the z-axis suggests fluorescence side lobes (figures 12.1(e), 3PM; 12.1(g), 2PM), which are also corroborated by experiment (figure 12.1(f, h)).

The nondiffracting Airy beam extends an axial distance much longer than the traditional Gaussian beam. The nondiffracting Airy beam increases the acquisition

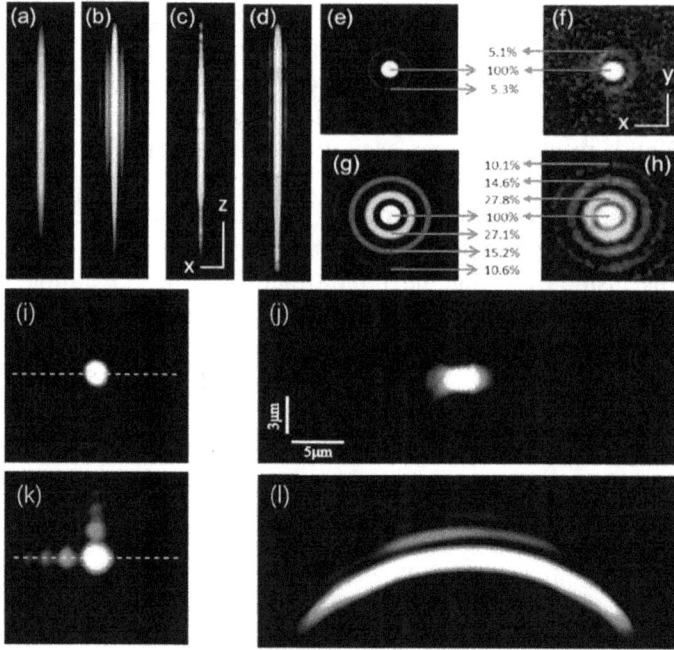

Figure 12.1. (a and b) Theoretical PSFs of 3PM and 2PM in the x–z-plane. (c and d) The experimental PSFs. Scale bars, z, 100 μm, x, 2 μm. (e and g) Transverse cross sections from integration over the z-axis of the Bessel beam (e, 3PM; g, 2PM), (f and h) experimental results. Scale bar, 2 μm. Images of 1 μm fluorescent sphere under (i and j) Gaussian and (k and l) Airy TPM. The (i) lateral and (j) axial PSFs under Gaussian mode. The lateral (c) and axial (d) PSFs under Airy mode. Adapted with permission from [2]. © 2018 Optical Society of America under the terms of the OSA Open Access Publishing Agreement, and adapted with permission from [20]. © 2019 Optical Society of America.

speed by projecting the axial information onto a 2D image [20, 22, 23]. The spatial spectrum modulation with a CPM relies on a pupil function,

$$P(u, v) = p(u, v)e^{ic(u^3 + v^3)}, \qquad (12.1)$$

where $p(u, v) = 1$, for $u^2 + v^2 < R^2$, and $p(u, v) = 0$, elsewhere, with R the aperture radius. The cubic phase term $c(u^3 + v^3)$ sculptures the beam into an Airy beam, with the self-bending regulated by the regulation factor c. Such a cubic phase could be imprinted on a liquid crystal SLM or a DMD [24]. Alternatively, the commercially available cubic phase mask (CPM, Holo/Or, PE-206-I-Y-A, 9.2 mm) can endure a higher damage threshold, suitable for femtosecond laser applications [25].

The traditional Gaussian beam maps the volumetric specimen layer by layer; in contrast, the TPM with a nondiffracting Airy beam captures a volumetric image in an individual frame with much improved speed. Moreover, the nondiffracting beam can also acquire a sample structure behind a strongly scattered medium. Figures 12.1(i)–(k) show the transverse two-photon images of the polymer sphere under Gaussian and Airy beam excitation, in sequence. The sideview profiles of the Gaussian suggest limited regions of high intensity within the Rayleigh range

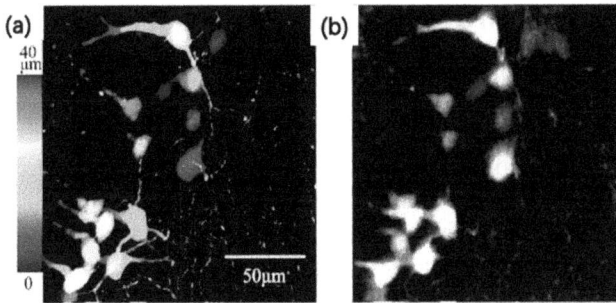

Figure 12.2. (a), Projection of image stack of a mouse brain slice with depth coded by color. Step size: 0.5 μm. (b), Single volumetric frame of mouse brain slice captured with Airy TPM. Adapted with permission from [20]. © 2019 Optical Society of America.

(figure 12.1(j)), while the Airy beam PSF suggests a bent trajectory (figure 12.1(l)). The Airy beam extends the axial range around six times longer than a conventional Gaussian beam, while preserving a comparable lateral size.

A 40 μm thick brain specimen from Thy1-YFPH mice is scanned layer-by-layer (figure 12.2(a)) under Gaussian excitation with color encoding the axial dimension. In contradistinction, all the neurons can be simultaneously mapped in a single projected volumetric frame under the Airy mode (figure 12.2(b)). The axial range of the Airy beam can cover all the structures in the whole slice. The non-diffraction nature of the Airy beam offers higher signal-to-noise ratio (SNR) than traditional confocal microscopy, especially in turbid media [26]. The resolution is well preserved since the small structure in the slice can still be preserved in the image under Airy mode.

Conventional TPM scans a volumetric sample pixel-by-pixel at a pretty low volumetric speed. The nondiffracting beam greatly improves the scanning speed of TPM. The nondiffracting Bessel beam with elongated focus significantly extends the axial imaging range, and thus increases the acquisition speed by mapping the axially elongated structure onto a two-dimensional (2D) image [2–4, 22, 23]. The hybridization of the Bessel module and a two-photon mesoscope allows the imaging of volumetric neural activity on the mesoscale with a synaptic resolution [27]. Visualizing neural circuits over large tissue volume at subcellular resolution and high speed is challenging due to the light scattering and slow scanning [18]. Such a hybridized instrument allows longer penetration depth while maintaining decent lateral resolution.

The elongated Airy beam can also project the volumetric structure onto a single frame without axial scan without significant degradation of the lateral resolution. Although Airy beam microscopy captures all the axial images within the effective focal length, the bending feature was once thought to be detrimental to resolve a perfect 3D image. However, the bending degree of freedom carries the information of the axial location, and this benefit will be further explored in the next sections. All in all, these advantages make Airy TPM a potential tool for real-time recording of the deep biological activities in a large volume.

12.3 Reolustion improvement in nonlinear microscope

To improve the resolution of multiphoton microscopy, the excitation beam has to be structured and the focus of the beam would become smaller. One approach is to suppress the size of PSF of the excitation beam. As a demonstration, the hollow Gaussian beam (HGB) can improve the resolution [28]. As the beam order increases, the beam size at the focus of a high-NA lens decreases. Meanwhile, the effectively excited two-photon fluorescence becomes much narrower in space. Additionally, the HGB enjoys an improved signal-to-background ratio (SBR) at depth due to the insufficient dosage to excite multiphoton signals with the outer beam structure.

The HGB can be simply and cost-effectively created through OAM annihilation using vortex phase plates with opposite topological charge. The electric field of the HGB can be described as

$$E_n(r, 0) = E_0 \left(\frac{r^2}{\omega_0^2} \right)^{|n|} \exp\left(-\frac{r^2}{\omega_0^2} \right),$$

(12.2)

where n is the magnitude of topological charge of the vortex phase plate and thus the order of HGB, E_0 is a constant electric field, and ω_0 is the beam waist.

Traditional optical vortex beams can be focused into a ring-shaped pattern owing to the existence of their singular phase structure. In contradistinction, the ring-shaped charge-free HGB focuses on a narrower spot with respect to a conventional Gaussian beam, and the focus size decreases with the beam order. This feature can be utilized to improve the spatial resolution and SNR through tuning mode order [28]. The vortex phase plates (VPP1, VPP-m1064, VIAVI Solutions) convert the Gaussian beam to an optical vortex from a custom-made mode-locked fiber laser (180 fs, 30 MHz, and centred at 1065 nm) with an efficiency of ~95%. Another vortex phase plate (VPP2) with opposite phase winding with respect to VPP1 annihilates the topological charge and generates the HGB beam. The SNR improves through the suppressing of the out-of-focus background signal.

The performance of TPM enhanced with HGB was evaluated by imaging fluorescence nanospheres (F8800, Life Technologies Ltd). Figure 12.3 shows the experimentally mapped PSF with (a) a Gaussian beam and HGB with orders of (b) $l = 1$, (c) $l = 4$, (d) $l = 8$, respectively. Clearly, the size of the profiles shrinks as the mode order increases suggesting the improvement of spatial resolution owing to the HGB. The FWHM of the fluorescent profile for the Gaussian beam and HGB of orders 1, 4 and 8 are ~679, 553, 500, and 432 nm, respectively. Corresponding lateral-resolution enhancements are ~18.48%, 26.24% and 36.35% with respect to the Gaussian beam.

HGB concentrates the energy more onto the major lobe, and maximizes the efficiency of the generated two-photon signal and thus SNR. With the HGB beam, the lateral resolution can be greatly improved. The side lobe of the focused HGB possesses tiny power and has negligible contribution to two-photon excitation and image quality.

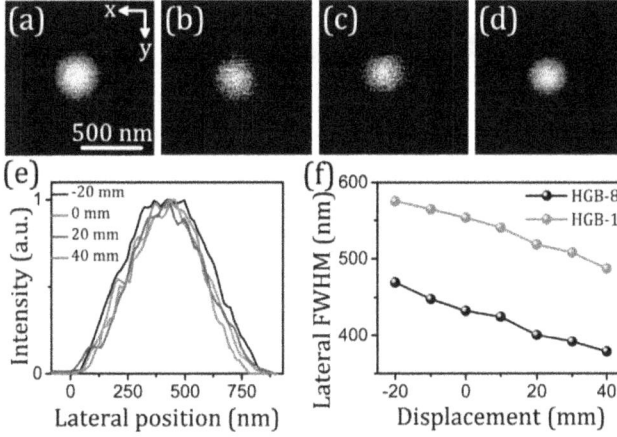

Figure 12.3. Experimentally measured lateral PSFs. Illumination utilizes (a) Gaussian, (b) HGB-1, (c) HGB-4, and (d) HGB-8 beams, respectively. (e) Lateral beam profiles suggest improved resolution for higher mode order. (f) Lateral resolution improvement by shifting the lens. Adapted with permission from [29]. © 2022 Optica Publishing Group.

Another possibility is to explore the concept of stimulated excitation depletion (STED) microscopy. The resolution is enhanced by using multiple orders of Bessel beams to perform laser mode switching (LMS). The equivalent PSF in the two-photon microscopy will be a thin needle-like beam, produced by subtraction of the fluorescence excited by the 1st order Bessel beam from the fluorescence excited by the 0th order Bessel beam. As a result, the lateral resolution can be improved by 28.6% over the axial depth of 56 μm. The side lobe of the Bessel beam contributes to the background since the third-order Bessel beam well matches the side lobes of zeroth-order Bessel beam in a lateral pattern. It is possible to extinct the background through the subtraction.

The TPM resolution is estimated through the excitation PSF with $r = 0.61\lambda/\mathrm{NA}$. It is limited by the wavelength, i.e., on the order of $\sim 1\,\mu\mathrm{m}$. In cylindrical coordinates, the Bessel beam has an electric field of [29],

$$E_l(r,\ \varphi,\ z) = E_0 \exp{(ik_z z)} J_l(k_r r) \exp{(il\varphi)}, \qquad (12.3)$$

where r and φ are the radial and azimuthal coordinates, J_l is the lth-order Bessel function, and k_z and k_r are the longitudinal and radial wave vectors, respectively. The azimuthal phase $\exp{(il\varphi)}$ contributes to a singularity for topological charge l greater than 1. The focal field of a Bessel beam under high-NA objective is described by [30],

$$\begin{aligned}
S_{\mathrm{high}}(r,\ \varphi) \propto {}& [1 + \cos{(\alpha)}]^2\, J_0^2\,(kr\sin{(\alpha)}) + [1 - \cos{(\alpha)}]^2\, J_2^2\,(kr\sin{(\alpha)}) \\
& + [2\sin^2{(\alpha)} J_0(kr\sin{(\alpha)}) J_2(kr\sin{(\alpha)}) \cos{(2\varphi)}] \qquad (12.4)\\
& + 4 J_1^2\,(kr\sin{(\alpha)}) \cos^2{(\varphi)},
\end{aligned}$$

where $k = 2\pi/\lambda$ is the wave vector, λ is the wavelength, and α is the half cone angle. Equation (12.4) implies that the transverse profile keeps invariant along the axial position.

The subtraction between the two images achieved by the zeroth- and high-order Bessel beams suppresses the image background. The subtracted PSF for the final image is [31]

$$\text{PSF}_{\text{sub}} = \text{PSF}_{\text{0th-Bessel}} - g \times \text{PSF}_{l\text{th-Bessel}}, \tag{12.5}$$

where g is a regulation factor. Equation (12.5) is the typical equation for image processing [32–34]. The resolution enhancement over the extended DOF can be achieved through the subtraction of the zeroth- and first-order Bessel beams [34]. The mechanism is to suppress the size of the dark spot of the first-order Bessel beams.

The side lobe from the Bessel beam could be inhibited by a higher-order Bessel beam through subtraction between the two images [34]. In contrast to the conventional 2D switching laser modes (SLAM), the volumetric SLAM (vSLAM) features with extended depth and improved transverse resolution [34]. The side lobes mainly contribute to the image background. Because 2PE (3PE) fluorescence takes the cubic (quadratic) dependence on the excitation power, the PSF with a Bessel beam under 3PE has much narrower side lobes than it does under 2PE. The lateral profile of the third-order Bessel beam matches well with the side lobes of the zeroth-order Bessel beam, which is ideal to conduct the image subtraction for cancelling the image background introduced by the side lobes of the zeroth-order Bessel beam [35]. On the other hand, the first-order Bessel beam also enhances the lateral resolution over the extended DOF with zeroth-order Bessel beams. As a consequence, the equivalent beam in v-SLAM is a much thinner needle, produced from the subtraction from the straw-like first-order Bessel beam. In contrast to the zeroth-order Bessel beam, the transverse resolution for the v-SLAM increases by 28.6% and preserves the axial depth over 56 μm.

Figures 12.4(a)–(c) show the transverse PSF of a two-photon microscope under NAs of 0.5 and 0.7. Non-diffracting vortex beams extend the DOF, for instance, both the zeroth-order and the third-order Bessel beams have axially elongated profiles and maintain invariant in lateral size. The subtraction of the straw-like third-order beam from the needle-like beam results in a thinner needle-like profile, thus the lateral resolution is enhanced over the entire DOF. The side lobes are still distinguishable for the transverse profiles of a fluorescent 1 μm bead (figures 12.4(a1–b1) and (a2–b2)). The peak intensity ratio between the first side lobe and the major lobe of the zeroth-order Bessel beam is 3.9% (6.3%) for an effective NA of 0.5 (0.7). Considering the lateral profiles of the third-order (figured 12.4(b1–b2)) and the zeroth-order (figures 12.4(a1–a2)) Bessel beams, the sizes of the third-order Bessel beams are comparable to the side lobes of zeroth-order Bessel beams under the same effective NA. The subtraction coefficients are chosen to be 0.15 (0.2) under the effective NA of 0.5 (0.7), respectively. The brightness of figures 12.4(a1–c2) is enhanced by ×7 for clear visualization of the side lobes. In contrast to the zeroth-order Bessel beams (figures 12.4 (a1–a2)), the side lobes are eliminated in the subtracted images (figures 12.4(c1–c2)).

The 1 μm bead was also scanned along the z-axis for obtaining the axial beam profiles for the zeroth- and third-order Bessel beams. The axial intensity distributions of the zeroth-order, third-order and subtracted Bessel beams under the effective

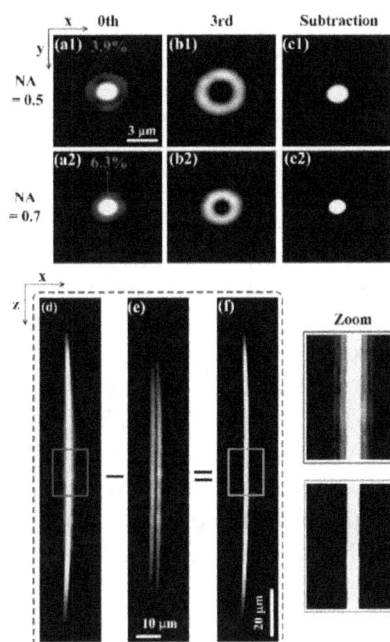

Figure 12.4. (a) Intensity distributions of the zeroth-order, third-order and subtracted Bessel beams in x–y-plane under the effective objective NA of 0.5 (a1–c1) and 0.7 (a2–c2). Axial intensity distributions of the zeroth-order (d), third-order (e) and subtracted (f) Bessel beams in x–z-plane for an effective NA of 0.7. The zoom-in images correspond to the 0th-order and subtracted Bessel beams. Copyright © 2021, IEEE, Reprinted with permission from [35].

NA of 0.7 are shown in figures 12.4(d)–(f). The same axial length of the zeroth- and third-order Bessel beams (figures 12.4(d) and (e)) ensures the complete subtraction over the extend DOF. The zoomed-in images show the background inhibition caused by the concentric side lobes, and only the central lobe is maintained over the entire extended DOF. The comparison between 2PE and 3PE highlights the advantage of rapid volumetric imaging with an enhanced SBR in deep brain tissue *in vivo* using nondiffracting vortex beams.

12.4 Axial resolution in volumetric microscopy

The Bessel beam or SAB suggest needle-like focus, and perform volumetric imaging of sparse samples by projecting volumetric information on a single 2D image [2, 4, 23]. The drawback of these modalities lies in the deteriorated axial resolution. The self-accelerating Airy beam bents in space and offers potential ability to resolve the depth of sample structures. The axial depth associated with the bending PSF allows the dissection of the axial location.

A dual-scan volumetric two-photon microscopy (TPM) with a pair of Airy beams of opposite self-acceleration has been demonstrated to resolve the depth. The cubic phase would modulate a fundamental Gaussian beam into an Airy beam [36]. The cubic phase in reciprocal space (k_x, k_y) is determined by [37]

$$\phi(k_x, k_y) = \exp\left[\frac{i}{3}(2\pi b)^3(k_x^3 + k_y^3)\right],\qquad(12.6)$$

where k_x and k_y, are the coordinates in the reciprocal space. The adjustment of the pixel size Δ greatly influences the lateral beam size and the axial length of the Airy beam. By reversing the sign of the scaling parameter b, the cubic phase on the SLM screen switches to the phase for MABs (insets of figures 12.5(c) and (d)). To map the axial location, a pair of Airy beams with opposite self-bending direction were implemented to scan the volumetric specimen under two-photon microscopy [19].

Mirrored Airy beams (MABs) with opposite bending directions are utilized to capture two-photon images and reconstruct the volume with axial resolution. Specifically, the fluorescent specimen was sequentially scanned by the MABs, and the deflection magnitude reflects the depth. The fluorescence profiles of the MABs are characterized by a single fluorescent nanosphere and the depth-resolving capability is confirmed by imaging multiple nanospheres at varying depths. The axial field range is beyond 32 µm.

Figure 12.5(a) illustrates the principle of depth-resolved volumetric microscopy with mirrored SABs. The CPM converts the fundamental Gaussian beam to an Airy beam. A paired Airy beam with the CPM rotated by 180° (figure 12.5(b)) excites the fluorescence image with opposite bending directions. The paired Airy beams bend towards opposite directions as corroborated by the PSF mapped with a single fluorescent microsphere (figures 12.5(c) and (d)). The relative lateral positions of the PSFs are depth-dependent. Since the CPM for producing an Airy beam is anti-symmetric, i.e., $\phi = cx^3$, the mirrored CPM produces Airy beams with opposite self-acceleration (figures 12.5(a) and (b)).

The real volumetric image can be resolved through optimization. Specifically, when an optical instrument assumes a shift-invariant response, the collection of

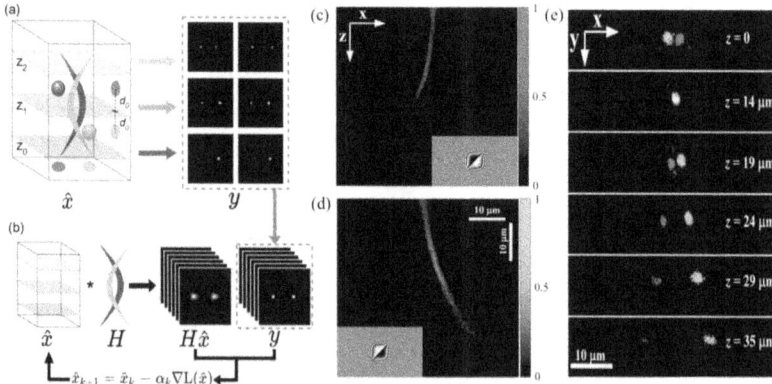

Figure 12.5. Schematics of depth-resolved TPM with self-bending Airy beams. (a) Depth-resolving concept. (b) Images created by the convolution of the retrieved volume and PSFs. The loss function are optimized by a gradient descent algorithm [38]. Two-photon profile for a fluorescence microsphere with (c) left (magenta), and (d) right (green) Airy beams. Insets, CPMs. (e) Synthetic image of the microsphere at respective depth [19]. Adapted with permission from [19]. © 2019 Optical Society of America, and adapted with permission from [38]. © 2020 Optical Society of America.

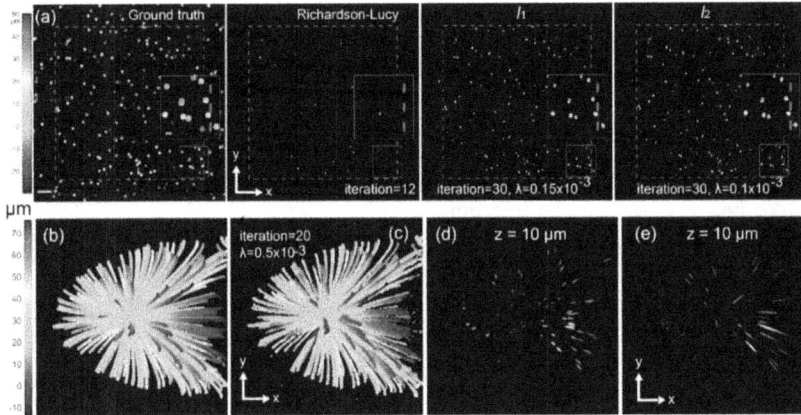

Figure 12.6. Maximum intensity projection (MIP) for image sequence with average (peak) SNR of 3.32 dB (23.88 dB) at six axial locations. (a) Depth color-coded MIPs towards the z-direction for the ground truth, reconstructions with Richardson–Lucy deconvolution, ℓ_1- and ℓ_2-GD. Magenta box marks an area of 512×512 pixels. Scale bars, 5 μm. (b–e) MIP of a simulated microtubule network along the z-direction with color-coded depth for (b) ground truth and (c) optimization. An image section at $z = 10$ μm of (d) the ground truth and (e) the optimization. Adapted with permission from [40]. © 2020 Optical Society of America.

signal Y at a 3D position $p \in \mathbb{R}^3$ can be evaluated as the convolution between the PSF H and the sample X,

$$Y(p) = \iiint_{\mathbb{R}^3} H(p - r)X(r)\, dr. \tag{12.7}$$

Here, $y \in \mathbb{R}^m$ and $x \in \mathbb{R}^n$ are vectors acquired by raster scan at respective planes. The matrix multiplication Hx performs the discrete convolution of equation (12.7) with matrix $H \in \mathbb{R}^{m \times n}$. The resultant image overlaid with noise approximates $y = \text{Pois}(Hx) + \mathcal{N}(0, \sigma^2)$, where $\text{Pois}(\lambda)$ is a variable with Poisson distribution of parameter λ and $\mathcal{N}(0, \sigma^2)$ is a normal variable with zero mean and variance σ^2. PSFs were modelled using the Airy beam and normalization through $\sum H = 1$.

Figure 12.6 demonstrates the outcomes with different algorithms to process the mirrored Airy beam images, including Richardson–Lucy deconvolution and ℓ_1- and ℓ_2-gradient decent (GD). For both ℓ_1- and ℓ_2-GD, the lateral and axial positions are well consistent with the ground truth. Richardson–Lucy deconvolution suggests a shorter effective axial extension than ℓ_1- and ℓ_2-GD. Furthermore, ℓ_1 regularization has higher contrast and better suppression of artifacts. The depth-resolving range of the volumetric TPM can be further improved by optimizing the CPM and the excitation laser power. Therefore, combined with the advanced algorithms, the self-bending beams improve the axial resolution in volumetric imaging, and may address more challenges in volumetric neuronal imaging.

12.5 Structured light to resolve the axial information

In this section, we will introduce the tomographic-encoded multiphoton (TEMP) microscopy using the structured Bessel droplet beam. In contrast to the paired Airy beam, more generally, the Bessel beam can also resolve the axial information by periodic structuring. More specifically, the Bessel droplets excite the specimen periodically along the depth. Recall that the electric field of the fundamental Bessel beam reads [29] $E(r, z) = E_0 \exp(ik_z z)J_0(k_r r)$, where J_0 is the zeroth-order Bessel function of the first kind, and k_z and k_r are the longitudinal and transverse wave vectors, respectively. The interference of two Bessel beams results in the droplet-like beam with the electric field

$$E_d(r, z) = E_1 \exp(ik_{z1}z)J_0(k_{r1}r) + E_2 \exp(ik_{z2}z)J_0(k_{r2}). \qquad (12.8)$$

Assume $E_1 = E_2$, the intensity distribution of the Bessel is

$$I(0, z) = |E_d^i|^2 = 2I_0[1 + \cos((k_{z1} - k_{z2})z]], \qquad (12.9)$$

where $I_0 = 2E_1^2 = 2E_2^2$, denoting a constant intensity distribution along the axial direction. Equation (12.9) indicates that the Bessel droplet periodically repeats the intensity maxima along z-axis with a space interval of $\Delta z = 2\pi/\Delta k_z$ [39] (figure 12.7 (a)), owing to the modulation of the cosine function, where $\Delta k_z = k_{z1} - k_{z2}$. Here, Δk_z is defined as the beating frequency of the two Bessel beams.

Figure 12.7(a) shows the PSF of the Bessel droplet under single-photon, two-photon, and three-photon fluorescence in sequence. Clear periodic side lobes appear in linear single-photon excitation, while nonlinear 2P and 3P excitation suggests inhibition of side lobes owing to the nonlinear excitation. Moreover, the beating frequency Δk_z can be regulated through the radii r_1 and r_2 of the two rings (figure 12.7(b)). A series of Bessel droplets with variable Δk_z are used to sequentially scan the volumetric sample, and acquire a set of 2D projected images. Thanks to the modulation in the cosine function, by converting the images in the spatial frequency domain to the spatial domain through an inverse fast Fourier transform (iFFT), the volumetric image can be reconstructed in the real space.

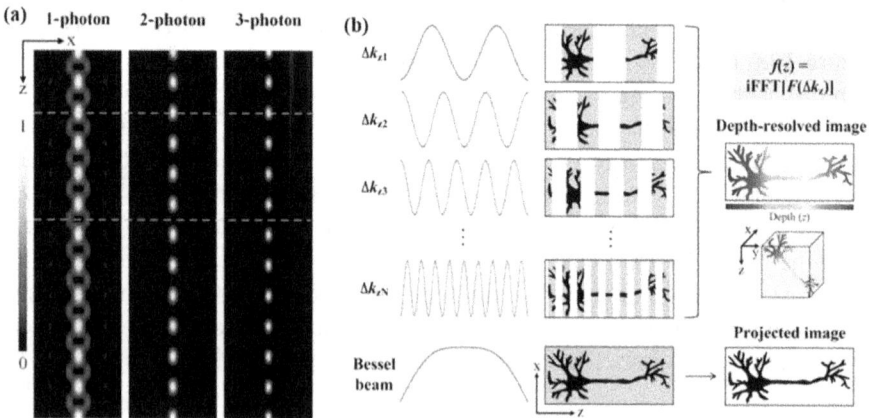

Figure 12.7. Concept of TEMP. (a) Longitudinal fluorescence emission under single-, two-, and three- photon excitation by droplet beam. (b) Principle of depth-resolved nonlinear TEMP and the image reconstruction procedure. Adapted with permission from [40], Copyright (2022) American Chemical Society.

Mathematically, for a 3D sample $f(x, y, z)$ scanned by the Bessel droplet, a series of 2D projection images $F(x, y, \Delta k_z)$ can be obtained for various Δk_z. Thus, the volumetric sample can be constructed in the spatial frequency domain as [21]

$$F(x, y, \Delta k_z) = \int_{z_{\min}}^{z_{\max}} I_0(z)f(x, y, z)[\cos(\Delta k_z z) + 1]\mathrm{d}z. \qquad (12.10)$$

Equation (12.10) shows the z-dependent cosine function in Δk_z space. The 3D image $f(x, y, z')$ can be reconstructed through iFFT,

$$\begin{aligned}
g(x, y, z') &= \frac{1}{2\pi} \int_{\Delta k_{z,\min}}^{\Delta k_{z,\max}} F(x, y, \Delta k_z) \exp(i\Delta k_z z')\mathrm{d}\Delta k_z \\
&= \left[f(x, y, z')I_0(z') + f(x, y, -z')I_0(-z') + \delta(z') \cdot 2 \int_{z_{\min}}^{z_{\max}} f(x, y, z)I_0(z)\mathrm{d}z \right] \\
&\quad * \frac{1}{2}\mathrm{sinc}\left[\frac{(\Delta k_{z,\max} - \Delta k_{z,\min})z'}{2} \right],
\end{aligned} \qquad (12.11)$$

where Δk_z ranges from $\Delta k_{z,\min}$ to $\Delta k_{z,\max}$, and '*' denotes convolution. The three terms in the square bracket express the retrieved image, mirror image, and direct current (DC) term. The reconstructed 3D image $f(x, y, z')$ is achieved by normalizing to the illumination function $I_0(z')$.

The Bessel droplet not only benefits from the absence of side lobes and smaller beam focus for better image resolution, and the nonlinear excitation leads to a qualitatively improved performance. First, the TEMP microscopy has better purity of the Bessel droplet, allowing more accurate reconstruction of the 3D image. Secondly, multiple illuminations from a series of Bessel droplets enhance the image SBR, leading to a superior ability to visualize weak fluorescence neurons. Thirdly, the TEMP microscopy reconstructs from a smaller number of raw images than that in 1 P excitation for a similar image SBR. This implies that the axial scanning speed can be further improved in TEMP microscopy.

A 50-μm-thick mouse brain tissue was prepared for imaging. Figure 12.8(a) shows the images scanned with a Bessel droplet beam, Gaussian beam, and Bessel beam in sequence. The neurons extend over a thickness of 26 μm. The neurons over a larger axial range can be mapped by droplet and Bessel beams, while the traditional Gaussian beam only image neurons within a narrow axial range. The images collected with the Bessel droplet and Gaussian beams are well consistent. The neurons in the volumes are well resolved along the axial direction through Bessel droplet with a comparable axial resolution to the Gaussian excitation. Figures 12.8 (b) and (c) show the intensity of the neurons in two regions. Four clear peaks are distinguishable in the Bessel droplet image (figure 12.8(b)), while Peaks 2 and 3 are missing in the Gaussian image. Similar performance was corroborated in figure 12.8 (c), and Peak 2 is only resolved in the Bessel droplet image. Figure 12.8(d) shows the zoomed-in images in the region marked by dashed box in figure 12.8(a).

In contrast to recently reported axial scan technologies, e.g., laser pulses multiplexing [41], multiple laser paths for different layer excitations [42, 43], multiple detectors [44], or special designed optical structures [45], the TEMP microscopy only

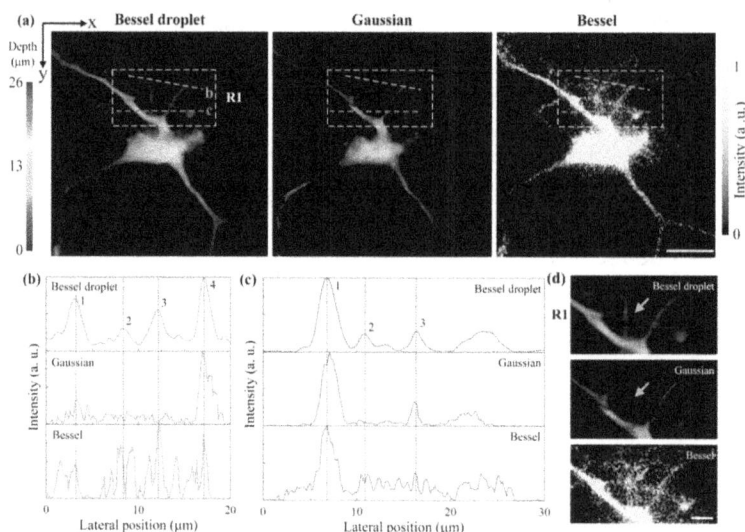

Figure 12.8. Neuron imaging in mouse brain slice. (a) Volumetric stacks scanned by the Bessel droplet, projection of the Gaussian stacks, and the image by the zeroth-order Bessel beam. (b–d) Comparisons of the imaging performance in region R1. Scale bar: (a) 15 μm and (d) 5 μm. Post-objective power: ~70 mW (Bessel droplet and Bessel), ~13 mW (Gaussian). Adapted with permission from [40], Copyright (2022) American Chemical Society.

inserts an SLM or DMD as a reflecting mirror in the excitation beam. The tomographic-encoded MPM using the Bessel droplets, termed TEMP microscopy, allows fast axial scanning, less photodamage and photobleaching, and high resolution and high contrast. The axial scanning speed is limited by the refresh rate of the spatial light modulator (SLM), however, this can be improved with the state-of-the-art DMD with a speed up to ~20 kHz [24]. The TEMP microscopy holds great potential for fast volumetric multiphoton imaging, especially for highly scattering samples.

12.6 Improve the signal-to-background ratio

The image quality can be evaluated through signal-to-background ratio (SBR). The TPM greatly improves the SBR as compared with wide-field fluorescence microscopy. However, the side lobes of the nondiffracting beams would contribute to the background in TPM. For instance, the side lobes of the Bessel beam are detrimental to multiphoton volumetric imaging. The combination of the Bessel beam and the three-photon excitation was confirmed to increase the SBR, owing to the higher-order nonlinearity in the 3PM than that in the 2PM [2, 4], but requires much higher power density. Recent advances suggest that the effect of the side lobes can be effectively eliminated by using a pair of Bessel beams with different order, resulting in much better SBR.

The lateral intensity of a zeroth-order Bessel beam suggests a bright spot in the center surrounded by a series of concentric rings (side lobes) [46]. The concept of the stimulated emission depletion microscopy can be extended to volumetric imaging using Bessel beams. The performance of the Bessel-beam-scanned image mainly depends on the excitation of the central lobe owing to its dominant intensity, and the full width of the half-maximum (FWHM) determines the lateral resolution of the 2PM [47]. The side lobes are thought to be redundant in imaging due to the contribution to image background [17]. The implementation of image subtraction requires two images scanned by beam of different order. The basic aim to enhance the contrast for the volumetric image is to maintain the major lobe of zeroth-order Bessel beam while removing the artefact from the side lobes. Therefore, a beam with lateral profile matching the side lobes of the zeroth-order Bessel beam would be optimum for the subtraction.

Proper choice of the Bessel beam would effectively remove the artefact through image subtraction. The image scanned by the third-order Bessel beam can be subtracted from the image with the zeroth-order Bessel beam. The positions of the side lobes for the zeroth-order Bessel beam well match the rings of the third-order Bessel beam, leading to an efficient background removal. The PSF subtraction also improves the SNR because the lateral rings of the third-order Bessel beam matched well with the side lobes of the zeroth-order Bessel beam [35]. The excellent cancellation is attributed to the fact that the lateral profile of the third-order Bessel beam is well matched with the side lobes of the zeroth-order Bessel beam, leading to an efficient image subtraction for cleaning the background. This background subtraction technique provides a simple and versatile tool for increasing the SBR in the volumetric 2PM, especially for tightly focused modalities.

12.7 Conclusion

The structured light has advances in biomedical imaging, to improve the spatial resolution, increase the SBR, and offer volumetric imaging ability. In particular, the spatial vortex phase in the nondiffracting beam can effectively suppress the PSF in nonlinear microscope. We foresee that the nondiffracting beam will further contribute to the biomedical applications, in optical imaging and microscopy.

References

[1] He H, Zhou M, Qiao T, Lai H M, Ran Q, Ren Y-X, Ko H, Zheng C, Tsia K K and Wong K K Y 2022 890-nm-excited SHG and fluorescence imaging enabled by an all-fiber mode-locked laser *Opt. Lett.* **47** 2710–3
[2] Chen B *et al* 2018 Rapid volumetric imaging with Bessel-beam three-photon microscopy *Biomed. Opt. Express* **9** 1992–2000
[3] Lu R, Tanimoto M, Koyama M and Ji N 2018 50 Hz volumetric functional imaging with continuously adjustable depth of focus *Biomed. Opt. Express* **9** 1964–76
[4] Rodríguez C, Liang Y, Lu R and Ji N 2018 Three-photon fluorescence microscopy with an axially elongated Bessel focus *Opt. Lett.* **43** 1914–7
[5] Chen X, Zhang C, Lin P, Huang K-C, Liang J, Tian J and Cheng J-X 2017 Volumetric chemical imaging by stimulated Raman projection microscopy and tomography *Nat. Commun.* **8** 15117

[6] Denk W, Strickler J H and Webb W W 1990 Two-photon laser scanning fluorescence microscopy *Science* **248** 73

[7] Helmchen F and Denk W 2005 Deep tissue two-photon microscopy *Nat. Methods* **2** 932–40

[8] Oheim M, Beaurepaire E, Chaigneau E, Mertz J and Charpak S 2001 Two-photon microscopy in brain tissue: parameters influencing the imaging depth *J. Neurosci. Methods* **111** 29–37

[9] Svoboda K and Yasuda R 2006 Principles of two-photon excitation microscopy and its applications to neuroscience *Neuron* **50** 823–39

[10] Wang B G, König K and Halbhuber K J 2010 Two-photon microscopy of deep intravital tissues and its merits in clinical research *J. Microsc.* **238** 1–20

[11] Kobat D, Horton N G and Xu C 2011 *In vivo* two-photon microscopy to 1.6-mm depth in mouse cortex *J. Biomed. Opt.* **16** 106014

[12] Xu C and Webb W W 2002 Multiphoton excitation of molecular fluorophores and nonlinear laser microscopy *Topics in Fluorescence Spectroscopy* (Berlin: Springer) pp 471–540

[13] Zipfel W R, Williams R M and Webb W W 2003 Nonlinear magic: multiphoton microscopy in the biosciences *Nat. Biotechnol.* **21** 1369

[14] Hosny N A *et al* 2020 Planar Airy beam light-sheet for two-photon microscopy *Biomed. Opt. Express* **11** 3927–35

[15] Vettenburg T, Dalgarno H I C, Nylk J, Coll-Llado C, Ferrier D E K, Cizmar T, Gunn-Moore F J and Dholakia K 2014 Light-sheet microscopy using an Airy beam *Nat. Methods* **11** 541–4

[16] Baumgartl J, Mazilu M and Dholakia K 2008 Optically mediated particle clearing using Airy wavepackets *Nat. Photonics* **2** 675–8

[17] Lu R, Sun W, Liang Y, Kerlin A, Bierfeld J, Seelig J D, Wilson D E, Scholl B, Mohar B and Tanimoto M 2017 Video-rate volumetric functional imaging of the brain at synaptic resolution *Nat. Neurosci.* **20** 620–8

[18] Lu R, Liang Y, Meng G, Zhou P, Svoboda K, Paninski L and Ji N 2020 Rapid mesoscale volumetric imaging of neural activity with synaptic resolution *Nat. Methods* **17** 291–4

[19] He H, Kong C, Tan X-J, Chan K Y, Ren Y-X, Tsia K K and Wong K K 2019 Depth-resolved volumetric two-photon microscopy based on dual Airy beam scanning *Opt. Lett.* **44** 5238–41

[20] Tan X-J, Kong C, Ren Y-X, Lai C S W, Tsia K K and Wong K K Y 2019 Volumetric two-photon microscopy with a non-diffracting Airy beam *Opt. Lett.* **44** 391–4

[21] Gong L, Lin S and Huang Z 2021 Stimulated Raman scattering tomography enables label-free volumetric deep tissue imaging *Laser Photon. Rev.* **15** 2100069

[22] Dufour P, Piché M, De Koninck Y and McCarthy N 2006 Two-photon excitation fluorescence microscopy with a high depth of field using an axicon *Appl. Opt.* **45** 9246–52

[23] Thériault G, De Koninck Y and McCarthy N 2013 Extended depth of field microscopy for rapid volumetric two-photon imaging *Opt. Express* **21** 10095–104

[24] Ren Y-X, Lu R-D and Gong L 2015 Tailoring light with a digital micromirror device *Ann. Phys.* **527** 447–70

[25] Wei X, Kong C, Samanta G K, Tsia K K and Wong K K Y 2016 Self-healing highly-chirped fiber laser at 1.0 μm *Opt. Express* **24** 27577–86

[26] Nagar H and Roichman Y 2019 Deep-penetration fluorescence imaging through dense yeast cells suspensions using Airy beams *Opt. Lett.* **44** 1896–9

[27] Lu R *et al* 2017 Video-rate volumetric functional imaging of the brain at synaptic resolution *Nat. Neurosci.* **20** 620–8

[28] Ul Alam S, Kumar Soni N, Srinivasa Rao A, He H, Ren Y-X and Wong K K Y 2022 Two-photon microscopy with enhanced resolution and signal-to-background ratio using hollow Gaussian beam excitation *Opt. Lett.* **47** 2048–51

[29] Arlt J and Dholakia K 2000 Generation of high-order Bessel beams by use of an axicon *Opt. Commun.* **177** 297–301

[30] Botcherby E, Juškaitis R and Wilson T 2006 Scanning two photon fluorescence microscopy with extended depth of field *Opt. Commun.* **268** 253–60

[31] Dehez H, Piché M and De Koninck Y 2013 Resolution and contrast enhancement in laser scanning microscopy using dark beam imaging *Opt. Express* **21** 15912–25

[32] Segawa S, Kozawa Y and Sato S 2014 Resolution enhancement of confocal microscopy by subtraction method with vector beams *Opt. Lett.* **39** 3118–21

[33] Yoshida M, Kozawa Y and Sato S 2019 Subtraction imaging by the combination of higher-order vector beams for enhanced spatial resolution *Opt. Lett.* **44** 883–6

[34] He H, Kong C, Chan K Y, So W L, Fok H K, Ren Y-X, Lai C S W, Tsia K K and Wong K K Y 2020 Resolution enhancement in an extended depth of field for volumetric two-photon microscopy *Opt. Lett.* **45** 3054–7

[35] He H, Ren Y X, Chan R K Y, So W L, Fok H K, Lai C S W, Tsia K and Wong K K Y 2021 Background-free volumetric two-photon microscopy by side-lobes-cancelled Bessel beam *IEEE J. Sel. Top. Quantum Electron.* **27** 6801307

[36] Siviloglou G, Broky J, Dogariu A and Christodoulides D 2007 Observation of accelerating Airy beams *Phys. Rev. Lett.* **99** 213901

[37] Latychevskaia T, Schachtler D and Fink H-W 2016 Creating Airy beams employing a transmissive spatial light modulator *Appl. Optics* **55** 6095–101

[38] Chan R K Y, He H, Ren Y-X, Lai C S W, Lam E Y and Wong K K Y 2020 Axially resolved volumetric two-photon microscopy with an extended field of view using depth localization under mirrored Airy beams *Opt. Express* **28** 39563–73

[39] Ruffner D B and Grier D G 2012 Optical conveyors: a class of active tractor beams *Phys. Rev. Lett.* **109** 163903

[40] He H, Dong X, Ren Y-X, Lai C S W, Tsia K K and Wong K K Y 2022 Tomographic-encoded multiphoton microscopy *ACS Photonics* **9** 3291–301

[41] Beaulieu D R, Davison I G, Kılıç K, Bifano T G and Mertz J 2020 Simultaneous multiplane imaging with reverberation two-photon microscopy *Nat. Methods* **17** 283–6

[42] Weisenburger S, Tejera F, Demas J, Chen B, Manley J, Sparks F T, Traub F M, Daigle T, Zeng H and Losonczy A 2019 Volumetric Ca^{2+} imaging in the mouse brain using hybrid multiplexed sculpted light microscopy *Cell* **177** 1050–66

[43] Cheng A, Gonçalves J T, Golshani P, Arisaka K and Portera-Cailliau C 2011 Simultaneous two-photon calcium imaging at different depths with spatiotemporal multiplexing *Nat. Methods* **8** 139–42

[44] Badon A, Bensussen S, Gritton H J, Awal M R, Gabel C V, Han X and Mertz J 2019 Video-rate large-scale imaging with multi-*Z* confocal microscopy *Optica* **6** 389–95

[45] Chakraborty T, Chen B, Daetwyler S, Chang B-J, Vanderpoorten O, Sapoznik E, Kaminski C F, Knowles T P, Dean K M and Fiolka R 2020 Converting lateral scanning into axial focusing to speed up three-dimensional microscopy *Light-Sci. Appl* **9** 1–12

[46] McGloin D and Dholakia K 2005 Bessel beams: diffraction in a new light *Contemp. Phys.* **46** 15–28

[47] Thibon L, Lorenzo L E, Piché M and De Koninck Y 2017 Resolution enhancement in confocal microscopy using Bessel–Gauss beams *Opt. Express* **25** 2162–77

IOP Publishing

Optical Vortices
Fundamentals and applications
Yuanjie Yang, Yu-Xuan Ren and Carmelo Rosales-Guzmán

Chapter 13

Metrology with structured light

In this chapter we will introduce another application of structured light fields. The first in the field of laser remote sensing, where light beams with a nonhomogeneous transverse phase distribution have been used to measure directly the velocity component along the transverse plane. We will begin by describing the general concept and theoretical background about the use of structured light beams to directly measure the transverse velocity component, using what has been termed the transverse Doppler shift. Afterwards we will focus on light beams endowed with orbital angular momentum, which feature an azimuthally-varying phase, that allow us to determine the velocity of rotation along the transverse plane. We will continue our description by introducing a technique that allows us to determine directly both velocity components, longitudinal and transverse, in a single shot using vector beams. To finalise this chapter, we will describe a technique that allows us to measure directly the vorticity in fluids, which also relies on the use of light beams endowed with OAM.

13.1 Laser remote sensing

13.1.1 The longitudinal Doppler shift

To begin with, let us recall that common laser remote sensing systems, which allow us to monitor the location and velocity of moving targets, are based on the longitudinal Doppler effect, the perceived change in frequency of waves caused by the relative motion between a transmitting source and a detector, first observed by Christian Doppler in 1842 [1]. Here, a Gaussian beam with an almost constant transverse phase profile impinging on a moving target is reflected or scattered back by the target with a time-varying phase given by

$$\Psi(\mathbf{r}, t) = 2kz(t), \tag{13.1}$$

where $k = |\mathbf{k}| = 2\pi/\lambda$ is the wavenumber and λ is the wavelength. Further, $z(t)$ is the time-dependent relative displacement along the propagation direction z, between

doi:10.1088/978-0-7503-5844-6ch13
13-1

the target and the target emitter. As such, a target moving with constant velocity $\mathbf{v}$ illuminated by a Gaussian light beam with its propagation direction making an angle θ with respect to the velocity vector of the target, induces on the reflected signal an optical frequency shift

$$\Delta f = \frac{2|\mathbf{v}|\cos\theta}{\lambda}. \tag{13.2}$$

In this way, the velocity of the target can be determined remotely from the frequency shift as

$$|\mathbf{v}| = \frac{\lambda\Delta f}{2\cos\theta}. \tag{13.3}$$

Notice that for a target moving in the direction perpendicular to the propagation direction of light $\theta = 90°$, and therefore it generates no frequency shift. Hence, in the classical longitudinal Doppler effect, the transverse velocity component is undetectable only the longitudinal velocity component can be determined, i.e., the velocity component along the line-of-sight. Nonetheless, the transverse velocity component can be induced, for example, by measuring the longitudinal Doppler shift for a large set of directions [2]. As such, there have been proposed several velocity retrieval techniques to estimate not only two-dimensional but also three-dimensional vector fields based on the longitudinal Doppler principle. Nonetheless, many of these techniques involve averaging over a large set of data, causing poor spatial and temporal resolution, and in general does not provide local information about the velocity field. In addition, their implementation can become cumbersome as many of these techniques involve fast mechanical realignment of the direction of propagation of the laser beam.

Importantly, the theory of relativity shows that the Doppler effect is sensitive to transverse velocities as well - *the relativistic transverse Doppler effect* [3]. Unfortunately, the observed frequency shifts are relatively small ($\approx v^2/c^2$) and therefore impractical in laser radar system applications.

13.1.2 The transverse Doppler shift

A major breakthrough in laser remote sensing techniques was made by Belmonte and Torres in 2011, who pioneered a novel technique based on the use of structured light, capable of measuring the transverse velocity components in a direct way [4]. The key idea for its implementation relies on the use of structured light waves, light beams with different phase values along the transverse plane, for example, an azimuthally-varying phase, characteristic of OAM-carrying modes. In this way, the phase of the backscattered or reflected light from a target moving along the transverse plane contains information in the form of a Doppler shift about its transverse position and velocity at each instant. Note that the light beam has to be of a size comparable to the transverse trajectory of the moving target. Further, previous knowledge of the target's trajectory allows one to properly engineer the transverse phase of the illuminating light beam, allowing one to simplify the measurement of the transverse velocity, as will be detailed below.

If a moving target passes with a velocity $\mathbf{v}[\mathbf{r}(t)]$ through a region illuminated by a paraxial light beam $\mathbf{E}(\mathbf{r}, t)$ with a structured phase $\Psi[\mathbf{r}(t)]$ of the form

$$\mathbf{E}(\mathbf{r}, t) = E_0 \exp\{ikz + i\Psi[\mathbf{r}(t)] - 2i\pi ft\}, \tag{13.4}$$

it generates a burst of optical echoes. Here E_0 is the complex amplitude of the beam and $\mathbf{r}_\perp$ is the transverse position across the beam wavefront. If the backscattered light is coherently detected, for example by interference with a reference field, a time-varying phase of the form

$$\Theta = 2kzt + \Psi[\mathbf{r}_\perp(t)], \tag{13.5}$$

will be produced. It is easy to see that the time variation of the total phase is

$$\frac{\partial\Theta}{\partial t} = 2k\mathbf{v}_z + \nabla_\perp\Psi[\mathbf{r}(t)] \cdot \mathbf{v}_\perp[\mathbf{r}(\mathbf{t})]. \tag{13.6}$$

The first term of equation ((13.6)) is precisely the usual longitudinal Doppler shift, whereas the second is a new Doppler frequency shift, associated with the transverse velocity $\mathbf{v}_z$ of the target, which has the specific form

$$\Delta f_\perp = \frac{1}{2\pi}\nabla_\perp\Psi[\mathbf{r}(t)] \cdot \mathbf{v}_\perp[\mathbf{r}(t)], \tag{13.7}$$

where $\nabla_\perp\Psi[\mathbf{r}(t)]$ is the transversal gradient of the phase. Note that this new transverse Doppler component is independent of the optical frequency [5]. Two specific examples of targets moving along the transverse plane will be presented next: a particle moving with a constant linear velocity $\mathbf{v}_\perp$ and a particle rotating with an angular velocity Ω.

In the first case, a target moving with a constant velocity $\mathbf{v}_\perp$ along the $\hat{x}$ direction, perpendicular to the beam's direction, an appropriate structured phase to illuminate it with is a linear phase of the form $\Psi = \gamma x$, where γ is a constant term that defines the periodicity of the linear phase. As a result, the light scattered back from the moving target will be endowed with a continuously varying phase that contains information about the position and transverse velocity of the target. A coherent detection will therefore evidence a frequency shift such as the one described by equation (13.7). Computation of $\nabla_\perp\Psi = \gamma\hat{x}$ shows immediately that an appropriate choice of the structured phase simplifies the mathematics. Hence, the frequency shift is

$$\Delta f = \frac{|\gamma\mathbf{v}_\perp|}{2\pi}. \tag{13.8}$$

In the second case, for an object rotating with a constant angular velocity Ω, the suggested structured phase is that with an azimuthal dependence of the form $\exp(i\ell\phi)$, characteristic of OAM modes. The transverse Doppler shift for this case

$\nabla_\perp\Psi$ can be computed more easily in cylindrical coordinates, where the angular velocity is given by $\mathbf{v} = \Omega\rho\hat{\phi}$ and the transverse gradient $\nabla_\perp$ is defined as

$$\nabla_\perp\Psi = \frac{\partial(\ell\phi)}{\partial\rho}\hat{\rho} + \frac{1}{\rho}\frac{\partial(\ell\phi)}{\partial\phi}\hat{\phi} = \frac{\ell}{\rho}\hat{\phi}. \tag{13.9}$$

Hence, the backscattered light will be shifted in frequency according to

$$\Delta f_\perp = \frac{|\ell\omega|}{2\pi}. \tag{13.10}$$

Notice that in neither of the two cases the transverse Doppler shift depends on the wavelength of the illuminating beam. In addition, previous knowledge about the target's trajectory allows a proper tailoring of the structured light beam to simplify the mathematics involved as illustrated in figure 13.1 for the two cases presented above.

An alternative description can be given in terms of the Poynting vector, for this we consider the Doppler frequency shift generated by a moving surface

$$\Delta f = \frac{1}{\lambda}(\hat{d_1} - \hat{d_2}) \cdot (v), \tag{13.11}$$

where $\hat{d_1}$, $\hat{d_2}$ is the unit vector of the scattered light and $\mathbf{v}$ is the velocity of the moving surface. For a paraxial incident beam whose vector potential $\mathbf{A}$ is of the form $\mathbf{A} = \hat{x}\mathbf{u}(x, y, z)\exp(ikz)$, the Poynting vector $\mathbf{S}$ will be given by

$$\mathbf{S} = \frac{|E|^2}{2\eta}\left(\hat{z} + \frac{1}{k}\nabla_\perp\Psi\right), \tag{13.12}$$

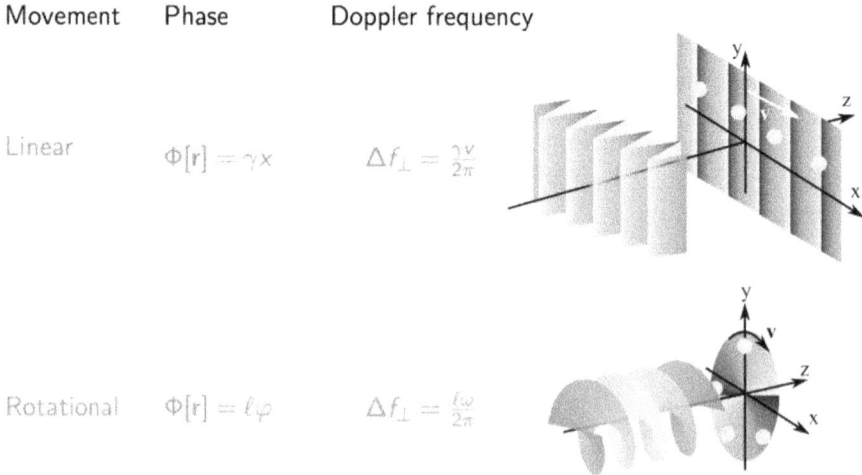

Movement	Phase	Doppler frequency	
Linear	$\Phi[r] = \gamma x$	$\Delta f_\perp = \frac{\gamma v}{2\pi}$	
Rotational	$\Phi[r] = \ell\varphi$	$\Delta f_\perp = \frac{\ell\omega}{2\pi}$	

Figure 13.1. Transverse Doppler shift for two specific types of motion: (top) linear and (bottom) rotational. In both cases, the structured phase is adapted to the type of motion to simplify the computation of the transverse Doppler frequency.

where η is the vacuum impedance. Since for a paraxial beam the longitudinal component of S is much larger than the transverse component, one can write

$$\hat{d} = \hat{z} + \frac{1}{k}\nabla_\perp \Psi. \tag{13.13}$$

Combination of equations (13.11) and (13.13) for $\hat{d}_2 = -\hat{z}$ yields a Doppler shift

$$\Delta f = \frac{1}{\lambda}\frac{\nabla_\perp \Psi}{k} \cdot \mathbf{v}, \tag{13.14}$$

which after some simplifications yields again equation (13.7).

To experimentally measure the rotational Doppler described above one can illuminate the moving target with an engineered structured light beam. To exemplify this, if a target is rotating at constant angular velocity, an azimuthally-varying phase such as the one present in Laguerre–Gaussian vortex beams adapt perfectly to their motion (figure 13.1(a)). The desired information is then extracted from the interference of the backscattered light and a reference beam, as illustrated in figure 13.2(a). Afterwards, the interference signal is measured, as illustrated in figure 13.2(b), and its Fourier transform is computed to extract the transverse frequency, as illustrated in figure 13.2(c), and from this the transverse velocity component of the moving target can be determined.

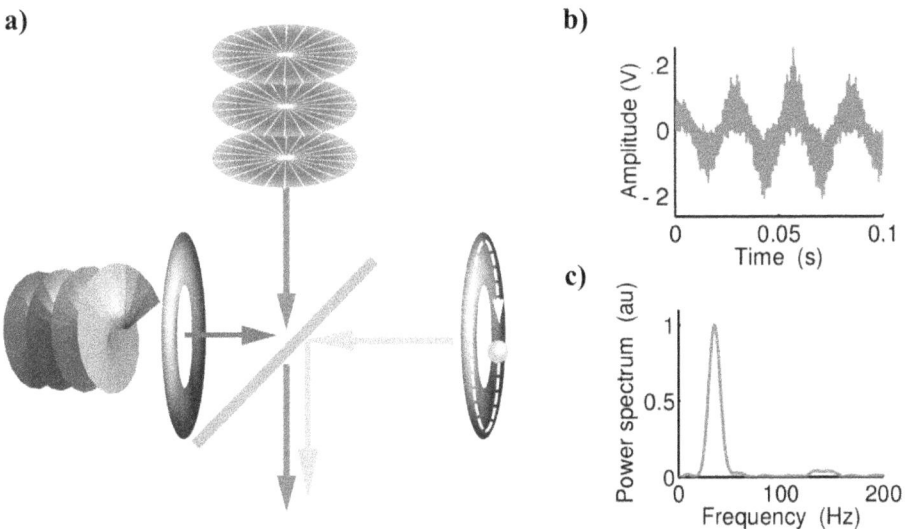

Figure 13.2. (a) A rotating target is illuminated with an optical field embedded with orbital angular momentum, the light scattered in the backwards direction is interfered with a reference beam, which for simplicity can be a Gaussian beam. (b) Interference signal of the backscattered light and the reference beam. (c) Transverse frequency shift, obtained by Fourier transforming the recovered signal.

13.2 Other techniques

13.2.1 Determining all velocity components in a full 3D helical motion

The realization that structured light could be use to directly determine the transverse velocity component of moving targets paved the path for a wide variety of experiments, the first to demonstrate this technique experimentally [6–9]. In one of these pioneering experiments, a direction-sensitive technique based on frequency shifts induced by dynamic structured light beam illumination was proposed to measure the direction of rotation of the target [9]. In the proof-of-concept experiment, applied to a rotating target, the illuminating beam with a helical phase was dynamically rotated clockwise and anticlockwise and from the downshifted or upshifted frequency the target's direction of rotation was determined. Further, in [10] it was demonstrated that the rotational Doppler shift obtained from the backscattering of a rotating target illuminated with white light is, unlike the linear Doppler effect, achromatic (wavelength-independent). In other words, all spectral components are shifted by the same frequency. The use of high-order Bessel beams has also represented a step forward in the rotation Doppler shift, as they enable us to determine the rotation of targets in the presence of obstacles due to their self-healing property [11] (figure 13.3).

More complicated experiments have also been proposed to determine all velocity components in a full 3D helical motion [8]. In this two-step technique the target is first illuminated with a Gaussian mode to measure its longitudinal velocity. Afterwards, the illumination beam is switched to an OAM-carrying beam to measure its velocity of rotation. This technique is limited to how fast the two beams can be shifted, which in the case of a spatial light modulator is a maximum switching rate of 60 Hz.

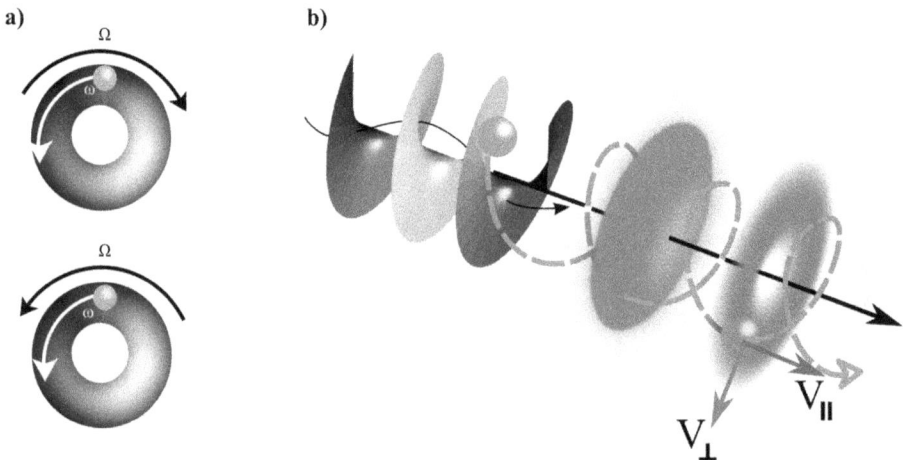

Figure 13.3. (a) A direction-sensitive technique based on frequency shifts induced by dynamic structured light beam illumination allows us to measure the direction of rotation of the target. (b) A two-step technique allows us to determine all velocity components in a full 3D helical motion. For this, the target is first illuminated with a Gaussian beam to measure its longitudinal velocity. Afterwards, the illumination beam is switched to an OAM-carrying beam to measure its velocity of rotation.

13.2.2 Single-shot determining all velocity components in a full 3D helical motion

An advanced version of this technique based on the use of a single interrogating beam and with the capability to determine simultaneously the longitudinal and angular speed of a cooperative target was proposed recently [12]. The key idea of this technique consists of striking the target with a properly engineered vector beam, classically entangled in its polarization and spatial degrees of freedom, as illustrated in figure 13.4(a). More precisely, the illuminating beam is formed by two orthogonal spatial modes encoded on two orthogonal polarization states. Hence, by assuming that the light scattered back from the target preserves its polarization state, which is true for cases larger than the wavelength of the illuminating beam, the information stored in each polarization state can be unambiguously separated upon detection. Figure 13.4(b) illustrates this technique schematically, where two frequency peaks are observed, each associated to each Doppler shift (longitudinal and rotational).

As a final comment for this section, a technique to perform real-time sensing of the position and velocity of opaque objects moving in two dimensions was recently proposed [13]. This technique also exploits the nonseparability of complex vector modes, classically entangled in their spatial and polarization degree of freedom. More precisely, a moving target traversing across the previously described light beam features an instantaneous position that is a function of the spatial-polarization nonseparability of the beam. Hence, the moving target disturbs the spatial degree of freedom of the beam but not the global polarization state. Crucially, a correlation between the resulting spatial modulation with the global polarization state allows us to obtain information of the position and velocity of the target.

13.2.3 Determination of the vorticity of fluids

Immediate applications of the rotational Doppler shift were proposed. For example, in [14] this technique was used to determine the angular velocity of a spinning micron-sized particle trapped inside optical tweezers. In the context of OAM

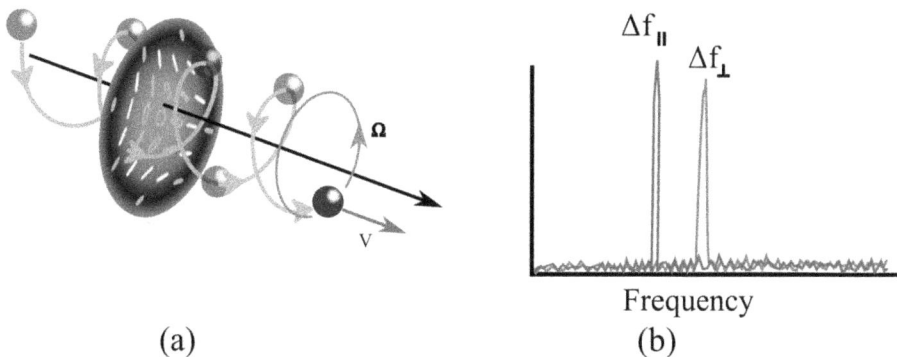

(a)   (b)

Figure 13.4. (a) A target describing a helical motion, with longitudinal and rotational motion illuminated with a vector beam, classically entangled in its spatial and polarization degree of freedom. (b) Frequency spectrum associated to each motion, longitudinal and rotational.

characterization techniques, an OAM complex spectrum analyser was proposed and demonstrated experimentally for the simultaneous measurements of the power and phase distributions of OAM modes, a technique completely based on the rotational Doppler effect [15]. In the context of fluid dynamics, the rotational Doppler shift was used to measure directly the vorticity ω of fluids [16]. Vorticity is a measure of the amount of angular rotation of a material point about a particular position in a flow field, which may be regarded as a measure of the local angular velocity of the fluid, the tendency of a fluid to rotate. A proper determination of vorticity is of paramount importance in a wide variety of research fields spanning areas such as biology, complex motions in the oceanic and atmospheric boundary layers, and microfluidics, amongst others. Nonetheless, the precise measurement of flow vorticity represents a real challenge since it is commonly computed as the curl of the velocity vector $\mathbf{U}$, that is $\omega = \nabla \times \mathbf{U}$. Hence, uncertainty errors on the velocity directly affects the accuracy of the measurement of vorticity. In addition, determining the full velocity vector in a fluid is not an easy task. Crucially, the technique proposed in [16] is capable of directly measuring the local vorticity of fluids, as it does not rely on measuring the velocity vector. To this end, the fluid is illuminated with a light beam endowed with OAM and relies on measuring the transverse Doppler shift, see figure 13.5(a).

More precisely, the backscattered light generates a spectrum of frequency shifts that contains information about the velocity in the fluid at every point along the ring of light (see figure 13.5 (b)). Hence, the measured frequency spectrum in this case is a mixture of different frequencies whose Fourier transform yields a histogram of

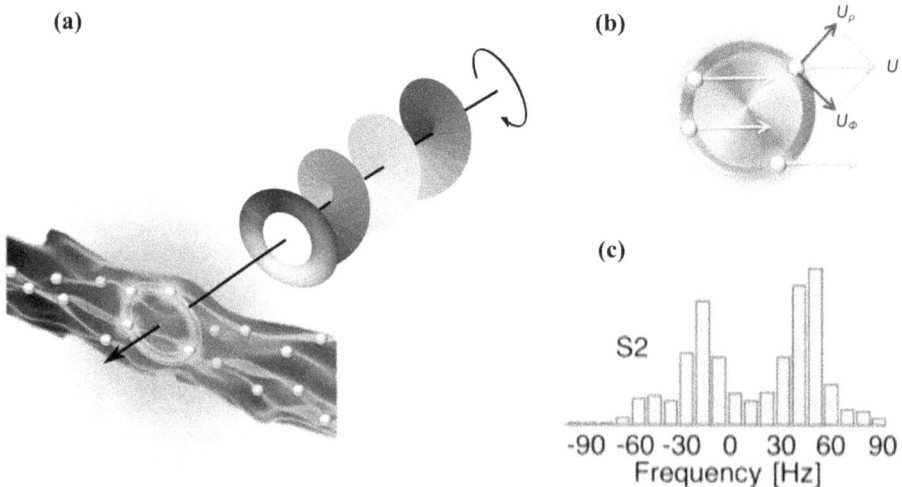

Figure 13.5. (a) A fluid is illuminated with a light beam endowed with orbital angular momentum. (b) The section of the illuminated fluid scatters light in the backward direction, which contains information about its vorticity. (c) The measured frequency spectrum is formed by a mixture of different frequencies whose Fourier transform yields a histogram of Doppler frequency components.

Doppler frequency components, as illustrated in figure 13.5(c). Crucially, the centroid of this histogram is directly related to the vorticity of the fluid as

$$\omega = \frac{2}{\ell} \int_0^{2\pi} f_\perp(\rho_0, \phi) \mathrm{d}\phi \tag{13.15}$$

where ℓ is the topological charge of the OAM beam. Importantly, the line integral in the previous equation is precisely the frequency centroid $\langle f_\perp(\rho_0, \phi)\rangle$, that is,

$$\langle f_\perp \rangle = \frac{2}{2\pi} \int_0^{2\pi} f_\perp(\rho_0, \phi) \mathrm{d}\phi. \tag{13.16}$$

Therefore, the vorticity is given by

$$\omega = \frac{4\pi}{\ell} \langle f_\perp \rangle. \tag{13.17}$$

References

[1] Measures R M 1992 *Laser Remote Sensing: Fundamentals and Applications* (Malabar, FL: Krieger Publishing Company)
[2] Durst F, Howe B M and Richter G 1982 Laser-Doppler measurement of crosswind velocity *Appl. Opt.* **21** 2596–607
[3] Sommerfeld A 1954 *Lectures on Theoretical Physics: Optics* (New York, NY: Academic)
[4] Belmonte A and Torres J P 2011 Optical Doppler shift with structured light *Opt. Lett.* **36** 4437–9
[5] Lavery M P J, Barnett S M, Speirits F C and Padgett M J 2014 Observation of the rotational Doppler shift of a white-light, orbital-angular-momentum-carrying beam backscattered from a rotating body *Optica* **1** 1–4
[6] Rosales-Guzmán C, Hermosa N, Belmonte A and Torres J P 2013 Experimental detection of transverse particle movement with structured light *Sci. Rep.* **36** 2815
[7] Lavery M P J, Robertson D J, Sponselli A, Courtial J, Steinhoff N K, Tyler G A, Wilner A E and Padgett M J 2013 Efficient measurement of an optical orbital-angular-momentum spectrum comprising more than 50 states *New J. Phys.* **15** 013024
[8] Rosales-Guzmán C, Hermosa N, Belmonte A and Torres J P 2014 Measuring the translational and rotational velocities of particles in helical motion using structured light *Opt. Express* **22** 16504–9
[9] Rosales-Guzmán C, Hermosa N, Belmonte A and Torres J P 2014 Direction-sensitive transverse velocity measurement by phase-modulated structured light beams *Opt. Lett.* **18** 5415–8
[10] Lavery M P J, Barnett S M, Speirits F C and Padgett M J 2014 Observation of the rotational doppler shift of a white-light, orbital-angular-momentum-carrying beam backscattered from a rotating body *Optica* **1** 1–4
[11] Fu S, Wang T, Zhang Z, Zhai Y and Gao C 2017 Non-diffractive Bessel-Gauss beams for the detection of rotating object free of obstructions *Opt. Express* **25** 20098–108
[12] Hu X B, Zhao B, Zhu Z H, Gao W and Rosales-Guzmán C 2019 In situ detection of a cooperative target's longitudinal and angular speed using structured light *Opt. Lett.* **44** 3070–3

[13] Berg-Johansen S, Töppel F, Stiller B, Banzer P, Ornigotti M, Giacobino E, Leuchs G, Aiello A and Marquardt C 2015 Classically entangled optical beams for high-speed kinematic sensing *Optica* **2** 864–8

[14] Phillips D B, Lee M P, Speirits F C, Barnett S M, Simpson S H, Lavery M P J, Padgett M J and Gibson G M 2014 Rotational Doppler velocimetry to probe the angular velocity of spinning microparticles *Phys. Rev.* A **90** 011801

[15] Zhou H-L, Fu D-Z, Dong J-J, Zhang P, Chen D-X, Cai X-L, Li F-L and Zhang X-L 2017 Orbital angular momentum complex spectrum analyzer for vortex light based on the rotational Doppler effect *Light: Sci. Appl.* **6** e16251

[16] Belmonte A, Rosales-Guzmán C and Torres J P 2015 Measurement of flow vorticity with helical beams of light *Optica* **2** 1002–5

Chapter 14

Past, present, and future

In this chapter, we briefly summarize the major contributions of this book. The research on optical vortices is an old topic, but still expanding, especially on the applications in various disciplines. This book is by no means complete, and we hope to excite more interesting applications with optical vortices.

14.1 Overview of the past researches

Light beams exhibit lots of features, including amplitude, phase, frequency, polarization, and coherence. Light photons with circular polarization carry spin angular momentum, while photons with orbital angular momentum are associated with the spatial phase profile of the light beam. The optical vortex beam characterizes with the spatially spiral phase. The inclusion of such a spiral phase results in OAM on each photon. In this book, we have introduced the nature of optical vortices, the characterization of optical vortex and the interdisciplinary applications of optical vortex.

Over the past few decades, people have understood the spatial phase structure of optical field and characterized the feature using holographic techniques. Moreover, the spiral phase of the optical field has excited many important applications, including super-resolution microscopy, optical trapping and manipulation, and optical communication. In general, vortices appear in various systems, including water, wind, and fluids. An optical vortex has a ring-shaped intensity profile with a helical phase structure. Due to the nature of the helical phase, the intensity profile of the optical vortex has a central null. The optical vortex can be created through wavefront shaping, which precisely controls both the amplitude and phase of the output beam. Various research suggests that the transversal appearance of the vortex beam looks like an annular ring with radius determined by the topological charge.

We have extensively discussed the generation of conventional and novel optical vortex beams, and how to measure the OAM. The major components needed to generate an optical vortex are the liquid crystal SLM and DMD; however, the emerging technique of metasurfaces can create multi-color vortices. With proper

doi:10.1088/978-0-7503-5844-6ch14
14-1

design of the metasurface structure, the optical vortex can also be excited in the form of surface plasmonics, as the plasmonic vortex. The vortex beam also suggests coherence properties. The diversity of optical vortex generation and various forms of optical vortices offers great opportunities on the application side, including the optical quantum communication, optical trapping and manipulation, optical super-resolution microscopy and optical metrology.

14.2 Present researches on the optical vortex

In recent decades, generation of optical vortices with novel features has gained more interest. For instance, metasurfaces can create multi-color vortices, partially coherent vortices are also interesting in the optics community, and vector vortices suggest interesting polarization features. The optical vortex can also be generated in the form of spatio-temporal optical pulses.

Different generation mechanisms for optical vortex are still emerging, including the plasmonic metasurface, and partially coherent optical vortices. In contrast to the liquid crystal SLM, the DMD performs better in terms of the fill factor, refresh rate, polarization insensitivity, and broadband response. Emerging devices require advanced algorithms to design the modulation pattern, and state-of-the-art manufacturing facilities to create the modulation devices. For example, the geometric phase in the metasurface design allows the creation of advanced patterns on the metastructure and the final device relies on the laser lithography or ion beam etching.

Recent progress suggests that the optical vortex can be combined with other higher-order features in the optical beam, for instance, the vector vortex beam displays vector polarization in the spatial profile, where the vortex symmetric Airy beam (VSAB) overlays the helical phase structure with the spatially symmetric Airy beam. Moreover, the helical phase structure may interact with the higher-order spatial beam. The off-axis VSAB creates a hollow channel in the major lobe of SAB during propagation, and the self-acceleration feature of the Airy beam guides the vortex propagation. Multiple vortices in the major lobes of the SAB suggest protected propagation of vortices as evidenced by the non-interacting vortices embedded in the SAB.

The generation of the vortex beam depends on the modulation of light with a vortex phase. However, the amplitude and phase of the Bessel vortices strongly depend on the topological charge. The radius of the transverse intensity is highly desirable in many applications to be independent on the topological charge. By playing with the Bessel beam order and the vortex beam, one can select combination of Bessel vortex beams for high-resolution nonlinear microscopy. Moreover, narrow dark hollow beams with a large topological charge is strongly needed to increase the rotation speed of microparticles. People developed the perfect vortex with controllable OAM but invariant ring size. The perfect vortex characterized by its dark hollow and stable radius independent on its topological charge.

Although the vortex beam has been studied for several decades, recent decades have seen that this research area is still developing, and the proper use of vortex beams may excite many important applications, for instance to study the topological features in artificial photonic materials, and optical metrology.

14.3 Future perspectives

It is still unclear how the optical vortex will shape the future of optical applications. From the existing examples, we foresee that there is still room to unveil the fundamental aspects of optical vortices. On the application side, we would expect optical vortices will further strengthen their applications in optical metrology, optical quantum communication, biomedical imaging, and optical trapping and manipulation.

The field of optical vortices is growing and will continue to grow. For instance, the optical vortex beam in the ultraviolet regime, or even in the extremely ultraviolet beam, is potentially useful in semiconductor metrology. Although we try our best to include most recent advances on optical vortex, this book is by no means complete.

This book is intended for readers at the undergraduate level, and will also be good reference for graduate students who wish to learn about the optical vortex, the generation of optical vortices, and the application of optical vortices.